Readings for Environmental Literacy

Edited by

Michael L. McKinney and Parri Shariff

University of Tennessee, Knoxville

West Publishing Company

Minneapolis/St. Paul New York Los Angeles San Francisco

Cover Image: Wind turbines, Mojave, California. ©1996 Dean DeChambeau.

WEST'S COMMITMENT TO THE ENVIRONMENT

In 1906, West Publishing Company began recycling materials left over from the production of books. This began a tradition of efficient and responsible use of resources. Today, 100% of our legal bound volumes are printed on acid-free, recycled paper consisting of 50% new paper pulp and 50% paper that has undergone a de-inking process. We also use vegetable-based inks to print all of our books. West recycles nearly 27,700,000 pounds of scrap paper annually—the equivalent of 229,300 trees. Since the 1960s, West has devised ways to capture and recycle waste inks, solvents, oils, and vapors created in the printing process. We also recycle plastics of all kinds, wood, glass, corrugated cardboard, and batteries, and have eliminated the use of polystyrene book packaging. We at West are proud of the longevity and the scope of our commitment to the environment.

West pocket parts and advance sheets are printed on recyclable paper and can be collected and recycled with newspapers. Staples do not have to be removed. Bound volumes can be recycled after removing the cover.

Production, Prepress, Printing and Binding by West Publishing Company.

610 Opperman Drive
P.O. Box 64526
St. Paul, MN 55164–0526

Printed in the United States of America
03 02 01 00 99 98 97 96 8 7 6 5 4 3 2 1

ISBN 0–314–07569–0

Contents

Preface

Our concern for the environment shouldn't change, but our focus constantly shifts with each dramatic report of a new problem. Our solutions change as we understand our responsibilities and we develop new, or give up old, technology. Government administrations change and so do their policies. *Reading for Environmental Literacy* is designed to keep students and instructors abreast on what is happening to our environment today, and who is involved.

This collection of fifty recent articles from popular magazines and science journals presents new success stories, new concerns, and new approaches for sustainable use of our resources. The editors have carefully selected articles that present the different, and sometimes unpopular, viewpoints of what is often a controversial field. Each article begins with a summary of the ideas presented in the selection and ends with a set of questions to help identify the key points of the discussion.

Readings for Environmental Literacy is intended to supplement material a student might encounter when taking a course in environmental science. There is no guarantee that the authors of these articles are always right: science, and opinion, don't work that way. What is important is to understand that the problems—and sustainable solutions—are out there, and that you use the information to make your own decisions.

Acknowledgment

The editors and West Publishing wish to express their sincere thanks to the many magazines and journals that allowed us to reprint their articles.

Section One

Overview

1

Putting zoo-bred animals back into the wild is a very slow process. Animals such as the golden lion tamarin, the red wolf and the black-footed ferrets find it difficult to survive in the harsh environment of a natural habitat after being in safe captivity. Programs are facing problems such as the animals' reduced reactions to predators and their riendliness with humans. But because of these efforts, these animals are breeding in the wild, and their population is slowly increasing. This second generation now has a greater potential for survival because they are less vulnerable to predators, more territorial, and much more wary of humans.

BACK TO NATURE

Peter Radetsky

You can put a zoo-bred animal back into the wild. But that doesn't mean you're putting back anything like a wild animal.

A small face pokes from between the branches of an oak tree. It has a thick golden mane and looks oddly familiar, like the cowardly lion in *The Wizard of Oz*. Soon it's joined by another maned head, then two, three, four others. The animals are about the size of squirrels, with silky fur and dangling 12-inch tails. They scurry nimbly from branch to branch, 30 feet above the ground. "Look at the monkeys," shouts a grade-schooler in a Washington Redskins cap, dribbling ice cream onto his T-shirt. Seemingly oblivious of the human crowd, the little creatures dart about the treetops, calling to one another with whistles and trills.

These golden lion tamarins are a long way from their ancestral rain forests in Brazil. But they are going back. They were bred here at the National Zoo in Washington, D.C., as part of a plan to rescue the endangered species and return it to its tropical habitat. In the meantime, to help them make the transition from the zoo cages where they were born to the wild that awaits them, they are living on a couple of wooded acres just down the path from the bison exhibit.

This tamarin equivalent of a halfway house was conceived by Ben Beck, a primatologist, and Devra Kleiman, the zoo's assistant director of research, to address a vexing question in conservation biology: Once you've succeeded in breeding an endangered animal in captivity, how do you ensure that the animal will be fit to live in the wild? How much of its natural behavior will survive? The same worry underlies plans to return red wolves to Tennessee's Great Smoky Mountains and to restore black-footed ferrets to Wyoming's plains. For these efforts, the case of the golden lion tamarin is a prelude and an object lesson.

By the 1970s golden lion tamarins had almost vanished from Brazil, victims of their own appeal (the handsome animals were exported as pets) and of rampant deforestation. Hoping to reverse the trend, the Brazilian government set aside a 12,500-acre reserve in the rain forests north of Rio de Janeiro for some 100 wild survivors. In the United States, meanwhile, the National Zoo had become involved in breeding the animals in captivity. At the time, there were fewer than 80 golden lion tamarins in zoos around the world, and they were not doing well—deaths were exceeding births. Kleiman determined that one of their problems was diet. Zoos were feeding their tamarins traditional monkey fare: fruits and vegetables. But in the wild these animals also ate insects and birds' eggs. When, on Kleiman's recommendation, they were fed protein in the form of insect grubs, their health improved. Another problem, Kleiman found, was that because monkeys are usually social animals, zoos invariably caged tamarins in large groups. But if a group contained more than one adult female, Kleiman determined, only the dominant female gave birth. When tamarins were separated into breeding pairs, their birthrate shot up. By 1983 the worldwide zoo population of golden lion tamarins had risen to 370.

The following year, to bolster the wild population, Kleiman and Beck sent 13 tamarins back to the Brazilian reserve. They were in for a nasty surprise: the animals were almost helpless. "We thought that their primary problem would be finding natural foods," says Beck. In fact it was more a case of not knowing what to do with food when they found it. "Imagine a monkey that doesn't know how to peel a banana," Beck says with a sigh.

But the tamarins' main problem—"which we failed totally to anticipate," says Beck—was their disorientation. They were lost in the rain forest. Raised in cages, they simply didn't know how to plot a route through uncharted territory. Furthermore, they were used to climbing on sturdy lumber frames. Natural vegetation, which bent and swayed under their weight, literally floored them—they frequently fell. They preferred to crawl along the ground, a dangerous proposition for a small monkey. One fell prey to a snake, another to a feral dog. Seven others died or disappeared the first year—killed, perhaps, by such other predators as lizards and ocelots, or overcome by starvation, disease, and mishaps. "We had to go back, regroup, and revise our techniques," says Beck.

A first experiment in tamarin jungle-training met with mixed results. At the National Zoo several cages were stuffed with natural, springy vegetation, forcing monkeys to negotiate their way over the plants to find their food. In 1985 Beck's team released 11 tamarins in Brazil, 7 trained and 4 not. "For the first couple of weeks the trained group clearly had an advantage," says Beck. They moved more easily in the trees and found more food. "But the untrained group rapidly caught up," he adds. "In fact, the survival rate was the same for both groups—very low." Only two of the animals survived beyond a year.

Still, it was apparent that the tamarins could learn, and might learn better given more appropriate conditions. "That's how the outdoor exhibit here at the zoo originated," Beck says. Since 1986 tamarin families have lived in the trees in the warm-weather months from May through October. (Insects and fruits, including bananas in their skins, are provided for them in the trees, not on the ground.) Those that take to arboreal living are then transported to the rain forest to tackle the real world. Not only do these animals do better, Beck has found, but their offspring, who receive the benefits of their elders' experience, fare better still. As of this past January, Beck and his team had released 134 tamarins to the forest. Of those, 43 have survived. These 43 pioneers have given birth to 97 babies, of which 70 are still alive, for a grand total of 113 tamarins added to the existing wild population. Some 400 golden lion tamarins now roam the coastal forest of Brazil.

Curiously, says Beck, one thing that tamarins don't forget in captivity is how to recognize their most common predators. "Recognition of aerial predators is inborn," he says. "Anything that flies over, any dark shadow, and they give an alarm call. Then they either drop to the ground or run right to the center of a tree to avoid them. This behavior is genetically hardwired—they don't have to learn a thing about it." Tamarins are somewhat less wary of earthbound predators, perhaps because they can usually count on making a quick escape through the trees. Typically, when a tamarin encounters a threat such as a large snake, it approaches the animal nois-ily, as though alerting its family group to flee. (That first batch of disoriented, ground-hugging tamarins may have come to grief not because they failed to recognize their predators but because they failed to even detect them in time to make their clumsy getaway.)

Red wolves born in captivity don't have to learn to recognize wild predators, either, but for quite different reasons. Nothing in the wild preys on a wolf. Nor do they need to learn how to find food.

They'll readily hunt wild rodents, raccoons, rabbits, deer, even wild pigs. The problem with wolves is that they'll also readily hunt livestock, chickens, and turkeys. If their reintroduction is to be successful, red wolves must adapt to the wilderness, not to the all-too-inviting civilization surrounding it. The key is instilling in the animals a sense of territory and fear of humans.

Red wolves have ample reason to fear humans. By the late 1970s these lanky canids (which once ranged halfway across the country, from Texas to North Carolina) had been driven from their ranges and slaughtered to near extinction. In 1987, however, a federal program to restore the species got underway with the release of some captive-born animals in North Carolina, at Alligator River National Wildlife Refuge. Then in November 1991 a family of four red wolves was introduced as an experiment to Tennessee's Great Smoky Mountains National Park; for the first time in almost a century the howls of wolves were once more heard in the southern Appalachian mountains.

It was not a completely successful debut. Some of the animals weren't sufficiently savvy or cautious—in short, they were not wild enough. A large male took three turkeys from a farm just outside the park, says Chris Lucash, the U.S. Fish and Wildlife biologist who runs the Smokey Mountain project. "It wasn't that big a deal. It's not uncommon for a wild predator to take something like a domestic turkey. The problem was that the farmer came out and chased the wolf, but the wolf didn't leave. He sat down in the woods just outside the garden and ate the bird. When the guy went to work, the wolf came back and took another turkey. By the time we got there he had taken still another one, despite the presence of people and a barking dog—that's not a good wild animal. He had spent too much of his life in captivity. He was too comfortable with people. We recaptured him that day, and I didn't turn him loose again. He was a liability."

As planned, Lucash ended his experiment after ten months and recaptured the rest of the family. He also rethought his strategy, having decided it was necessary to instill in the wolves a sense of territory and a wariness of humans. Now before red wolves taste freedom, they're given an acclimation period in their own lupine halfway house. It consists of a pen 50 feet by 50 feet, with an 11-foot-high fence, set among the maple, oak, and pine trees of the Smoky Mountains. The wolves pass their first year there, amid the scents and seasonal shifts of the forest, mating and giving birth to pups.

Human contact is kept to a minimum. A caretaker who provides food and cleans the pen lives discreetly in a tent some 200 yards away. No visitors are allowed. Hikers or fisherman blundering into the area encounter a sign threatening a $20,000 fine for harassment of an endangered species. During most of 1992 two such pens on the western side of the park, separated by eight miles, provided transitional homes for two families of wolves, each consisting of an adult pair and their four pups.

The idea is to condition the animals to consider this plot their own. Food is plentiful in the Great Smokies, as is space—over 500,000 acres of it. If each group of wolves establishes a defined territory and sticks to it, not venturing into the ranches and farms bordering the park, one of the great hurdles facing the reintroduction will have been surmounted. Lucash envisions as many as eight family groups, possibly 50 animals, roaming the park, coexisting with one another, with farmers, and with the numerous visitors to the area.

Last October he took a step toward that goal by pulling open the gate to one of the pens. "I put a pig carcass outside the pen so they wouldn't just bolt out of the gate and run all over the place," says Lucash. "A week and a half later I gave them another carcass, and three weeks later another one. That gave them time to hone their hunting skills and not starve."

Just as Lucash and his colleagues hoped, the wolves began to consider the pen the heart of their territory. "They just hung out for the first week at the release area," says Lucash. "Then they started going out into the meadow and regrouping at the pen. They left a lot of scent marks there, defining their territory." And they began to hunt. "We saw them standing in a creek with a freshly killed 70-pound wild pig. We were pleasantly surprised by that."

In December Lucash opened the gate of the second pen. That meant 12 red wolves were now loose in country where just two years before there had been none. The strain began to show. "Some of them have been acting a little stupid," says Lucash. "We've seen them on the road. They're not afraid enough of vehicles to satisfy me. I'll probably have to do some negative reinforcement by shooting at them with exploding blank shells."

In January one of the pups released in December strayed out of the park into farm country; he was soon followed by his three siblings. Lucash and his crew returned the animals to the park, only to watch them stray again. The first pup wandered away three times. "They weren't necessarily misbehaving," says Lucash. "It's just that legally we can't let them out of the park. They cross this imaginary line and they're outlaws." Rather than perpetually retrieving them, Lucash decided to return the wandering pups to the pen.

That left eight wolves on the loose, and one of those has apparently made forays outside the park as well, judging from complaints Lucash has received about wolves taking chickens. "I didn't even argue with the farmer who reported them missing—a bus driver claims he saw a wolf with a chicken," says Lucash. "I just paid the farmer for the birds."

Such behavior wouldn't be too surprising in a not-quite-wild wolf. "We knew that the problem we were going to have with these transitional animals was that they were still too accustomed to people," says Lucash. "We can't get around that until we get pups born and raised in the open forest. We're just going to have to work real hard to get the animals through this transition until they start reproducing as real wild animals."

Wyoming Game and Fish Department veterinarian Tom Thorne agrees that keeping animals in a delineated area is the key to releasing them. Thorne is part of a joint venture between federal and state agencies to return black-footed ferrets to America's plains. "If they disperse to the four winds, they might as well be dead," he says. "Even if they escape getting killed, their chances of finding a mate and contributing to a stable population are zero. We want to keep them at the release site for as soon as possible, long enough so that they'll consider it their territory and stick around." A tall order, as Thorne and his colleagues have found out, when you're dealing with creatures as skittery, secretive, and nocturnal as ferrets.

These narrow, skinny two-pound animals are preyed on by coyotes, badgers, and owls, so it's a matter of survival for them not to show themselves. They spend much of their time underground in prairie dog burrows, and prairie dogs constitute virtually their entire diet. Earlier this century you were likely to find black-footed ferrets wherever there were prairie dogs. But farmers destroyed vast networks of burrows by plowing, and the U.S. government (responding to ranchers, who regarded prairie dogs as pests) systematically exterminated them. Ferrets died from eating strychnine-poisoned prairie dogs, and as prairie dog towns dwindled, ferret populations became increasingly isolated and fragmented. In fact, black-footed ferrets were believed extinct until a group turned up in Wyoming in the early 1980s. Beginning in 1985, the last 18 of these animals were rounded up to start a captive-breeding program.

The current plan is to reintroduce the animals to Wyoming, then to Montana and South Dakota, and eventually to the entire West. But bringing them back isn't proving easy. Just keeping track of the nocturnal animals is a challenge. Some are fitted with radio collars, but the collars are a burden to these small animals, and they are designed to disintegrate in about two or three weeks. After that biologists resort to "spotlighting"—sweeping lights across the prairie in the hope of glimpsing flashes of the ferrets' emerald green eyes—and searching the wintry, snow-covered plains for small footprints before the wind erases them.

In addition, the animals' vulnerability to predators has had biologists worried from the start. Had captivity blunted the ferrets' wariness? Could you train them to avoid predators they had never seen? "A ferret has one chance to learn about a predator: its first encounter," explains Thorne. "It either lives or dies. And the difference between living and dying is getting into a burrow fast." In the late 1980s, with captive breeding of black-footed ferrets barely begun, biologists with the U.S. Fish and Wildlife Service in Front Royal, Virginia, were already thinking about ways to improve the animals' chances. Using an unendangered ferret, the Siberian polecat, as a stand-in, they tried some rather fanciful experiments to gauge how well the animals would fare. A badger run over by a pickup truck in Wyoming was frozen and mailed to Virginia, explains Dean Biggins, who designed the experiment with fellow biologist Brian Miller (now at the National University of Mexico). The animal was stuffed by a taxidermist at the Smithsonian and mounted on a remote-control Radio Shack truck chassis.

Robo-badger, as he was called, was used to

pursue captive-bred Siberian ferrets living in a pen riddled with burrows to provide escape hatches. To drive home the danger of the situation, the researchers fired rubber bands at the animals with a toy-store gun. For good measure they occasionally dive-bombed the animals with a stuffed owl slung from an ingenious network of pulleys. Even Miller's Labrador, Rosa, was pressed into service, retrieving ferrets in her soft mouth before releasing them unharmed.

Still, when these animals were tracked into the wilderness of Colorado and Wyoming, they fared only somewhat better than those raised in cages. Though trained ferrets were more venturesome than untrained ones, they didn't live any longer—the record was 34 days. It was sobering news, because wild ferrets live for two to three years.

Thorne's team in Wyoming, meanwhile, has decided on a different tack. At the group's ferret breeding facility north of Cheyenne, they've added a series of 22-by-20-foot outdoor pens, each enclosed by a chain link fence sunk 8 feet into the ground. When prairie dogs were released inside them, they quickly established miniature prairie dog towns. These training pens allow families of black-footed ferrets to live, hunt prairie dogs, and raise their young in wildernesslike conditions.

When it comes time to release the animals, they're transported, barking and chattering, inside nest boxes to a windswept prairie called Shirley Basin, some 80 miles north of Laramie. There the team bolts the boxes to protective wire cages. Four-inch plastic tubing, big enough to accommodate a ferret but not a predator, provides safe passage from each cage to a nearby prairie dog burrow. After a ten-day acclimation period, a trapdoor springs open and the ferrets can wriggle along the tube to explore the surrounding plains and the labyrinthine tunnels beneath them. But they always have the security of the cage. If an owl swoops over them when they're on the open plain, they can duck into the tube and back to safety. It's an offer they seem to refuse. Ferrets have been observed returning to the cages for no more than a few days after the trapdoor was opened. Just in case, the team keeps food and water inside them for five days past the last sighting, but by then the animals are long gone, fending for themselves in the prairie dog town and beyond—far beyond.

Last fall 90 captive-bred ferrets were set loose in Shirley Basin. To the biologists' consternation, over half of them lit out for destinations unknown, for all intents and purposes lost forever. ("They suffer from some sort of wanderlust," shrugs Biggins.) Of the 37 ferrets that were radio-collared, at least 10 are known to have been killed, mostly by coyotes. "A hawk or eagle got one," says Biggins. "We found wing prints and tracks in the snow." But at least 16 ferrets are still known to be alive on the site.

The figures may look meager. But they're better than those for 1991, the first year of the release program, when only 4 of 49 ferrets made it through the winter. In fact, the reintroduction team is quite encouraged. Those four original survivors gave birth to six kits, more than doubling their number. If the latest batch does as well, the black-footed ferret has a fighting chance of making a comeback. Thorne, for one, is looking on the bright side. "Several days after the last release," he recalls with gratification, "one of our guys saw ferrets near their cage in Shirley Basin. One was barking at him from a burrow right next door. He was sticking close to his territory, warning the human to get away—just as he was supposed to do." ■

Questions

1. When the tamarinds were taken back to the rain forest what was the main problem they faced?

2. What are two characteristics the red wolf needs in order to adapt to the wilderness?

3. Why are the black–footed ferrets difficult to keep track of?

Answers are at the back of the book.

2

Predicting future events with accuracy helps to set science apart from most other human disciplines. But science can also predict which factors cannot be predicted. This understanding of unpredictability is the basis of chaos. Chaos, we find, characterizes the future of the solar system. Distance, energy, orbit size, orbit shape, and orbit inclination are just a few variables one must include in order to begin understanding chaos as it relates to the solar system. Four hundred years ago we did not know the motions of the planets. Now we have progressed to knowing not only the history of the solar system, but to the knowledge that we cannot predict how the solar system will evolve in the distant future.

Chaos in the Solar System

Neil de Grasse Tyson

The ability to predict future events with precision is what distinguishes science from almost all other human endeavors. Daily newspapers often give dates for upcoming phases of the moon or the time of tomorrow's sunrise. But they do not report news items of the future such as next Monday's plane crash or next Tuesday's closing prices on the New York Stock Exchange. The general public knows intuitively, in not explicitly, that science makes predictions, but it may surprise people to learn that science can also predict that something cannot be predicted. Unpredictability is the basis of chaos. And unpredictability characterizes the future evolution of the solar system.

A chaotic solar system would no doubt have upset the German astronomer Johannes Kepler, who is generally credited with the first predictive laws of physics, published in 1609 and 1619. Using a formula that he derived from planetary positions in the sky, he predicted the average distance between any planet and the sun by knowing the duration of the planet's year. In 1687, Isaac Newton published the *Principia*, which contains the law of universal gravitation from which Kepler's laws can be mathematically derived.

In spite of the immediate success of his new laws of gravity, Isaac Newton remained concerned that the solar system might one day fall into disarray. With characteristic prescience, Newton noted: "The Planets move one and the same way in Orbs concentric, some inconsiderable Irregularities excepted, which may have arisen from the mutual actions of...Planets upon one another, and which will be apt to increase, till the system wants a Reformation." Newton implied that God might occasionally be needed to step in and fix things. The celebrated French mathematician and dynamicist Pierre-Simon Laplace had the opposite view of the world. In his 1799 four-volume treatise *Mécanique céleste*, he declared that the universe was stable and fully predictable. Laplace later wrote, "[with] all the forces by which nature is animated...nothing [is] uncertain, and the future as the past would be present to [one's] eyes." When queried by Napoleon Bonaparte on the absence of any reference to God in his treatise, Laplace replied, "Sire, I have no need of that hypothesis."

The solar system does, indeed, look stable if all you have at your disposal is a pencil and paper—with or without God. But in the age of super-computers, where millions of computations per second are routine, solar system models can be followed for hundreds of millions of years. What thanks do we get for our deeper understanding of the universe?

Chaos—which reveals itself through the application of our well-tested physical laws in computer models of the solar system's future evolution. Today's leading solar system modelers include Scott Tremaine and his colleagues at the Canadian Institute of Theoretical Astrophysics and Jack Wisdom and his colleagues at the Massachusetts Institute of Technology.

Chaos has also reared its head in other disciplines, such as meteorology, predator-prey ecology, and in most other places where there are complex interacting systems. To understand chaos as it applies to the solar system, one must first recognize that the difference in location between two objects—their distance—is just one of many differences that can be calculated. Two objects can also differ in energy, orbit, size, orbit shape, and orbit inclination. It is therefore useful to broaden the concept of distance to include the separation of objects in these other variables as well. For example, two objects that are (at the moment) near each other in space may have very different orbit shapes. Our modified measure of distance would tell us that the two objects are widely separated.

A common test for solar system chaos begins with two computer models that are identical except for a small detail. For example, in one of the models, Earth much recoil slightly in its orbit as a result of being hit by a small meteor. We are now armed to ask a simple question: Over time, what happens to the "distance" between these two nearly identical models? It may remain constant, fluctuate, or increase. If the distance between the two models increases exponentially, then small changes to the system are extremely magnified over time, and the ability to predict future behavior based on well-known initial conditions is compromised. This is the hallmark of chaos. We owe much of our early understanding of the onset of chaos to Aleksander Mikhailovich Lyapunov (1857-1918), a Russian mathematician and mechanical engineer whose 1892 doctoral thesis, "The Stability of Motion," remains a classic to this day. (By the way, Lyapunov committed suicide during the political chaos that followed the Russian Revolution.)

It has been known since the work of Newton that the paths of two isolated objects in mutual orbit, such as a binary star system, can be solved exactly for all of time. No instabilities there. But as more objects are added to the dance card, orbits not only become more complex but also more sensitive to their initial conditions. In the solar system we have the sun, its nine planets, and sixty-plus satellites, along with innumerable asteroids and comets. While this may sound complicated, the story is not yet complete. Orbits in the solar system are further influenced by the sun's loss of 4 million tons of matter every second from the thermonuclear fusion in its core. (The matter is converted to energy that is subsequently released as light from the sun's surface.) The sun also loses mass from the continuously ejected stream of charged particles known as the solar wind. And the solar system is further subject to perturbing gravity from stars that occasionally pass by in their normal orbit around the galactic center.

To appreciate the task of the solar system dynamicist, consider that the equations of motion allow you to calculate, at any given instant, the net force of gravity exerted on an object by all other known objects in the solar system and beyond. Once you know the force on each object, you nudge them all (on the computer) in the direction they ought to go. But the force on each object in the solar system is now slightly different because everything has moved. You must therefore recompute all forces and nudge them again. This continues for the duration of the simulation, which in some cases involves billions of nudges. When you do these calculations or similar ones, you reveal that the solar system's behavior is chaotic: over time intervals of about 5 million years for the inner planets (Mercury, Venus, Earth, and Mars) and about 20 million years for the outer, "Jovian" planets (Jupiter, Saturn, Uranus and Neptune), arbitrarily small "distances" between initial conditions noticeably diverge. By 100 to 200 million years into the model, we have lost all ability to predict planet trajectories.

Yes, this is bad. Consider the following example: The recoil of Earth from the launch of a single space probe can influence our future in such a way that in about 200 million years, the position of Earth in its orbit around the sun will be shifted by nearly 60 degrees. Combine the effects of all past and future launches, and we simply do not know where Earth will be in its orbit so far in the future. By itself, this ignorance seem benign. But note that asteroids in one family of orbits can chaotically migrate to another family of orbits. If asteroids can migrate, and if Earth will one day be in some unpredictable place in its orbit, then there is limit to how far in the future we can reliably calculate the risk of a major asteroid impact and the global extinction that might ensue.

Should the probes we launch be made of lighter materials? Should we abandon the space program? Should we worry about loss of solar mass? Should we be concerned about the thousand tons of meteor dust per day that Earth accumulates as it plows through the debris of interplanetary space? Should we all gather on one side of Earth and leap into space together? None of the above, because even perfect information may not help. For example, the *Voyager* spacecrafts measured the masses of the Jovian planets with a high degree of precision, but some uncertainty remains. Simulations reveal that even if the masses of all planets were known exactly, our ability to predict the long-term evolution of the solar system would not improve.

A skeptical inquirer might worry that the unpredictability of a complex, dynamic system over long time intervals might be due to a computational round-off error or to some peculiar feature of the computer chip or computer program. If this suspicion were correct, then two-object systems might eventually show chaos in the computer models. But they don't. And if you pluck Uranus from the solar system model and repeat the orbit calculations for the Jovian Planets, then the chaos goes away. Another proof that chaos is not the result of a computer glitch comes from simulations of Pluto's orbit, which has the greatest eccentricity and the most extreme orbital tilt of any planet. Under the influence of the Jovian planets, Pluto actually exhibits well-behaved chaos; small distances between initial conditions of two models lead to an unpredictable, yet limited set of trajectories. Most importantly, however, different investigators using different computers and different computational methods have derived similar time intervals for the onset of chaos in the long-term evolution of the solar system.

Apart from our petty desire to avoid extinction, there are broader reasons why one might like to study the long-term behavior of the solar system. With a full evolutionary model, dynamicists can go backward in time to probe the system's history—when the planetary roll call might have been very different from that of today. For example, some planets that existed at the birth of the solar system (5 billion years ago), could have since been forcibly ejected. Jack-in-the-box planets are not idle speculation; there exists a small chance that among our current nine, the innermost planet, Mercury, will be ejected from the solar system or will collide with Venus in several million years.

In four centuries, we have gone from not knowing the motions of the planets to knowing that we cannot know the evolution of the solar system into the unlimited future. It is a bittersweet victory in our unending quest to understand the universe. ■

Questions

1. What is the basis of chaos and what does it characterize?

2. How does one understand chaos?

3. What is the hallmark of chaos?

Answers are at the back of the book.

3

We can "see" deep inside the earth with the help of seismic waves flowing from earthquakes. Studying the earth's mantle has helped geophysicists determine how the Andes and other major land forms have been shaped throughout time. Evidence of plate tectonics can be found by studying the elevation, width and bend of the Andes Mountains, some of which rise more than 20,000 feet, and by studying the effects of violent earthquakes. The subduction and convergence movement of moving plates, which forged most of the large features on the earth's surface, is caused by currents deep within the mantle. Plate motions were considered to be the best indicators of mantle motion, even though the causal forces behind plate motions were unclear. Wave splitting, a recent development using earthquake wave data, ascertains the orientation of mantle flow. By using this new development, scientists can now show that an active mantle is responsible for driving some plates and viscous rock deep within the earth, resulting in large-scale geological features on the North and South American continents.

The Andes' Deep Origins

Raymond M. Russo and Paul G. Silver

Peering into the mantle, geophysicists are now able to detect the currents of rock that shape the Andes and other major features of our planet.

On February 20, 1834, Charles Darwin was taking a noontime rest in a forest not far from the small Chilean coastal town of Valdivia when the ground began to shake. The motion continued for two minutes, and although he was able to stand up, the sensation of earth rocking beneath his feet made him feel giddy—as though he were back aboard the *Beagle*. Twelve days later, Darwin arrived in Concepcíon, farther up the coast near the epicenter of the quake. He was stunned by the utter devastation that had befallen he town. Only thirty-five lives had been lost, but nothing was left standing. "It is a bitter and humiliating thing to see works, which have cost men so much time and labour, overthrown in one minute; yet compassion for the inhabitants is almost instantly forgotten...In my opinion, we have scarcely beheld since leaving England, any other sight so deeply interesting."

Darwin also noted that shells of recent marine species had been uplifted with the land along much of the Pacific coast of South America. With the same insight that he brought to the biological world, he concluded that "the elevation on this western side of the continent has not been equable.... At several places the land has been lately, or still is, rising both insensibly and by sudden starts of a few feet during earthquake-shocks; this shows that these two kinds of movements are intimately connected together."

Both the lofty peaks of the Andes, many rising above twenty thousand feet, and the violent earthquakes that made such a strong impression on Darwin are spectacular evidence of plate tectonics. Since Darwin's time, the motions of the dozen or so thin, rigid plates that pave the earth's surface have been largely worked out. Recovering from a wound received early in World War I, Alfred Wegener, a German scientist and explorer, developed the idea that the continents were once assembled in a single landmass called Pangaea and were now drifting apart. For lack of proof, his work was not taken seriously, but in the years following World War II, surveys using a variety of new technologies gave geologists their first detailed look at the ocean floor—two thirds of the earth's surface that was previously uncharted. In a burst of creative insight in the mid-1960s, geologists realized that ocean crust was being created along the mid-Atlantic ridge, and as this new crust spread outward, it carried continents along with it.

Meanwhile, the Pacific Ocean floor was being continually renewed as new crust was being formed and older crust was destroyed. With the plate tectonic revolution, many of the earth's features suddenly made sense.

The earthquake Darwin experienced, for example, resulted from the convergence of two plates: in the slow-motion crash, the Nazca plate slid, with some resistance (hence the violent tremors), beneath the South American plate and was subducted, or forced down, into the mantle. The descending ocean plate, overridden by the lighter and thicker continental crust, disappeared along a deep trench in the ocean floor paralleling the Pacific coast of South America.

Although the theory of plate tectonics has explained a lot, the forces that drive the plate motions have remained unobservable, buried deep in the earth's interior. From very early on, geologists had assumed that the circulation of the mantle rock, flowing at a snail's pace in vast gyres, was intimately connected to plate motions, but the nature of the relationship remained murky.

In the case of the subducted Nazca plate, geologists assumed that it marked a region of sinking mantle and that at least some of the mantle was welded to the descending plate. The coupling of plate and mantle was presumed to be more or less complete in most cases: where the mantle went, so went the plates. Thus plate motions, and in particular the locations of subducting slabs, were thought to be our best indicators of mantle motions.

Within the last few years, however, a newly observed phenomenon called shear wave splitting has allowed us to determine the orientation of mantle flow using data from waves propagating outward from earthquakes. To test the technique on a continental scale, we decided to study the mantle beneath the Nazca subduction zone along the western edge of South America. We chose the coast because, first, it is a convenient place to put seismometers (many subduction zones are deep under the ocean). Second, the coast has plenty of earthquakes and therefore lots of earthquake data to "illuminate" the mantle below. And finally, because the motion of the two plates was particularly simple (the Nazca plate is moving essentially eastward at approximately 2.5 inches per year, and South America is moving westward at a little more than an inch per year), we expected the mantle flow to be simple, too. We were confident that the mantle beneath the Nazca plate would be flowing from west to east, and that it would be strongly coupled to the descending slab.

Some 350 seismograms later, we knew that the orientation of mantle flow beneath the descending Nazca plate was not what we had expected. We found a varied pattern of orientations, but most were *parallel* to the curvaceous western coast of South America and the subduction trench. We were flabbergasted. How could all those ideas on plate and mantle motions be wrong? Could the mantle be moving largely independently of the Nazca slab, often at right angles to the eastward motion of the oceanic plate? After double-checking our results and finding that they were correct, we realized that we would need a new model of the mantle flow and plate motions to explain what we were observing.

The westward motion of South America was the key to our new model. As the continent moves, it pushes the Nazca trench and the entire subduction zone westward before it. What if, as it is pushed westward, the descending Nazca slab is pushing the mantle out of the way like a continent-sized snowplow?

If true, this model implies that the mantle is not welded solidly to the descending slab, but is instead free to flow in any direction—as long as the way is clear. Because the descending slab represents an impermeable barrier, angling downward into the mantle to depths of 400 miles or more and extending the length of South America's western margin, we would expect the mantle to flow in any direction necessary to escape the slow westward push of South America.

But what we observed, mantle flow oriented parallel to the coast, indicates that the mantle under the slab does not move downward with the plowing slab. We suspect that it is prevented from descending deeper than 400 miles because an increase in mantle viscosity at this depth may prevent the flow from descending into stiffer mantle, or a pressure-induced change in the structure of mantle minerals, which occurs at this depth, may cause the top of the mantle to be much more buoyant than the denser, lower mantle.

Our measurements of seismic anisotropy reveal only the orientation of the mantle flow, not its direction. If we know, for example, that the orientation of the flow is north–south, we cannot be sure whether the direction of flow is north or south. But once again, the motion of South America resolves the problem. As the continent pushes westward, shoving the descending slab into the oncoming flow, the mantle backs up, especially near the middle of the coastline, far from the edges of the slab where the flow can escape. This pressure buildup should occur beneath the coastline at the border of Peru and Chile.

We suspect that the pressure that the flow exerts on the Nazca slab is transmitted to the leading edge of South America. And halfway down the coast, where the pressure is greatest, the Andes are the most deformed, in both horizontal shortening and vertical uplift. The great bend in the mountain chain is probably also a consequence of the great pressure; the leading edge of the continent is buckled inward in response to the impact of mantle flow. Because the flowing mantle is extremely viscous, the pressure beneath the Nazca slab may be transmitted backward to regions of the mantle beneath the unsubducted portion of the Nazca plate. The relatively shallow ocean floor off the coast of Peru is probably a result of the mantle billowing up, floating the crust higher.

From this point of high pressure in the mantle, the flow should diverge, heading northward to Colombia and southward to Patagonia, the tail of South America. And indeed this is what we observed; beneath the Caribbean and Scotia Sea plates that lie north and south of the continent, the mantle flow resumes its east–west orientation as it moves around South America.

The similarities between the Caribbean and Scotia Sea plates—in shape, area, and symmetry—have long intrigued geologists. In Alfred Wegener's third edition of *The Origin of Continents and Oceans*, he speculated that the Antilles and the South Sandwich Islands were moving westward more slowly than the westward drifting Central and South America: "Smaller portions of the blocks are left behind by the westward wandering of the larger blocks...The Lesser and Greater Antilles lag behind the movement of the Central American block, and similarly the so-called arc of the South Antilles [South Sandwich Islands] between Patagonia and West Antarctica."

An important point that Wegener did not consider is that all the subduction zones around the Pacific Ocean, the so-called ring of fire, are retreating toward the center of the Pacific Basin. The oceanic plates now in the basin will eventually be consumed by subduction, and the ocean itself will close. Seen in this light, the mantle flow around South America represents a transfer of mantle from a shrinking Pacific reservoir to a growing Atlantic reservoir. A decade ago, Walter Alvarez (the geologist famous for his impact crater theory of dinosaur extinction) postulated that such channelized flow was draining the Pacific mantle.

The westward motion of South America, "upstream" through the mantle flowing around it, is the best evidence we have of deep mantle flow driving the plate motions. But how does the mantle push the plate? Geophysicists have long suspected that South America (as well as other continents) has a peculiar deep "root," made of strong, cold mantle welded to the continental crust east of the Andes. We speculate that this root may actually extend deep enough to be swept up in, and pushed along by, deep mantle flow, rather like the keel of an iceberg. If this is true, then the deep mantle beneath South America must be flowing westward, pushing the continent and driving the Nazca slab backward. The Andes form because the leading edge of the continent is weaker and more easily deformed than both the mantle beneath the core of continent to the east and the Nazca plate and its underlying mantle to the west.

North America is probably propelled westward by the same deep mantle flow generated beneath the opening Atlantic. Wegener had postulated a connection between the motions of North and South America and the formation of the mountain chains running along their western coasts: "By the westward drift of the two Americas their anterior margin was folded together to form the mighty range of the Andes (which stretched from Alaska to Antarctica) as a result of the opposition of the ancient well-cooled and therefore resistant floor of the Pacific."

Wegener was remarkably prescient in relating the deformations of western North and South

America, even though his deformation mechanism, the resistance of the "floor of the Pacific," has since been superseded by the recognition of long-lived subduction zones. The connection between the westward motions of North and South America and the geology of the Rocky Mountains is clearly indicated in the similar shapes and great width of the Rockies and the Andes. Like the Andes, the Rockies are sharply bent. The range, which follows a northerly trend through New Mexico and Colorado, bends in Wyoming to take a northwest trend from Montana north. Also, the Rockies are widest and most deformed and uplifted in this central region. But the Rockies are a much older chain than the Andes and have since suffered several important tectonic episodes that have obscured the telltale embayment that may have existed along western North America during their formation. A jumble of exotic terrains (some of which resembled island chains such as the Philippines) were swept up and accreted to the western edge of North America as it advanced. The Basin and Range province, comprising much of the southwestern United States, formed as subduction ceased beneath western North America. Even later, the San Andreas fault system, a narrow corridor along which the Pacific and North American plates slide past each other, developed in California.

As seismic waves ripple through the earth, they give us new insight into the mantle forces that ultimately cause the tremors. Our initial observations of mantle flow beneath the subducted Nazca plate have already explained such intriguing features as the symmetry and motions of the Caribbean and Scotia Sea plates relative to South America, the bend in the Andean mountain chain (as well as its great elevation and width), and the anomalously shallow water depths of the Pacific Ocean over the Nazca plate. Most importantly, we can finally show that the motions of at least some plates are driven by an active mantle, and that flowing rock deep in the earth is responsible for the large-scale geological features we observe on the North and South American continents today. ■

Questions

1. What is the best evidence of deep mantle flow driving plate tectonics?

2. Why did the Andes form?

3. How is the connection between the westward motions of North and South America and the geology of the Rocky Mountains indicated?

Answers are at the back of the book.

4

A glass house named Biosphere 2 sits in the Sonoran desert 30 miles from Tucson, Arizona. It is a "greenhouse" designed to be self-sustaining and is almost completely sealed off from the outside world. This three-acre compound consists of nearly 4,000 introduced species of plants and animals. A tropical rainforest, marsh, desert, savannah, streams, agricultural area, and a miniature ocean with a coral reef gives Biosphere 2 the look of a Garden of Eden. A natural-gas power plant supplies energy as sunlight and electricity. The goal of the Biosphere 2 project is to explore scientific frontiers in ecotechnology in order to promote better management of Earth's resources, to act as a model for colonizing space, and to motivate the human spirit. Eight "Biospherians" spent two years voluntarily confined within this futuristic glass and steel greenhouse. During their two years, the Biospherians became critically conscious of their intrinsic connections and complete reliance on the delicate ecosystems of Biosphere 2. Sadly, we take these ecosystems for granted—as though we can continuously pollute and overpopulate the world. The message of Biosphere is that we have only one home, and it is fast becoming overpopulated. Colonizing space is nice to contemplate—but first we must aim our goals to preserving Earth.

The Real Message from Biosphere 2

John C. Avise

Last September, eight gaunt but triumphant Biospherians emerged through the airlock doors of Biosphere 2 after two years under public scrutiny and sealed glass (Alling & Nelson 1993). Their reentry into Biosphere 1 (Earth) marked completion of the first in a century-long series of planned missions, the stated objectives of which are to explore scientific frontiers in ecotechnology (for better husbandry of the plant's resources and as a model for colonizing space) and in general to inspire the human spirit (Allen 1991). The latter goal already may have been achieved. Aficionados see the endeavor as audacious and visionary—"the most exciting venture undertaken in the U.S. since President Kennedy launched us towards the moon" (see the previous reference). And, unlike NASA's lunar mission, this $150 million program was launched entirely from private venture capital!

For those who don't already know, Biosphere 2 is the futuristic glass and steel "greenhouse" nestled in Arizona's Sonoran desert, about 30 miles north of Tucson. Engineered to be a self-sustaining mesocosm, almost completely sealed off from atmospheric or other material exchange with the outside world, the graceful three-acre enclosure houses nearly 4000 introduced species of plants and animals in a Garden-of-Eden-like setting of tropical rainforest, marsh, desert, savannah, streams, agricultural area, and even a miniature ocean complete with coral reef. Biosphere 2 receives energy as sunlight and as electricity (from an adjacent natural-gas power plant) that drives a vast "technosphere" of pumps, sensors, scrubbers, air-cooling systems, and other electronic and engineering wizardry designed to keep the environmental systems within boundaries suitable for life.

I recently returned from a second visit to Biosphere 2 (as an independent researcher), and once again my mind is aspin with ambivalent impressions. There is the commercial side—on adjacent grounds you can purchase biomeburgers, habitat hotdogs, and planetary pizzas, or browse gift shops and bookstores. There is the mystical side, exemplified by the many evocative sculptures with names of Indian Gods fashioned of stainless steel salvaged from the Los Alamos atomic bomb project. There is the educational side, where thought-provoking films and tours explain ecosystem functions and their relevance to the design of space modules. There are the many ecotechnological paradoxes of Biosphere 2 itself, where earthy smells of compost and forest contrast with the electronic sterility of the computer control room and where the Biospherians' simple agrarian lifestyle seems in opposition to their sophisticated telecommunications with the international press. And then there is the scientific side, a focus of much controversy and media attention. Whether sound basic research eventually can find a good home in Biosphere 2 remains to be seen (Watson 1993), but I am optimistic.

Overriding scientific lessons from Biosphere 2 already may be available. To many of us, healthy ecosystems and biodiversity have inestimable aesthetic value, but such philosophical orientations are difficult to translate into the kinds of economic terms that carry weight with business or industrial interests. Some far-thinking economists have sought to attach dollar values to natural ecosystems by virtue of the fundamental life-support services rendered (e.g. atmospheric regulation by rainforests and oceans, water purification by marshes, groundwater storage by aquifers, soil generation and maintenance by decomposers), but such attempts are almost hopelessly complicated by the vast range of spatial and temporal scales over which the monetary valuations might be tabulated. However, thanks to the controlled experiment of Biosphere 2 we now have a more explicit ledger.

The cost of the man-made technosphere that (marginally) regulated life-support systems for eight Biospherians over two years was about $150 million, or $9,000,000 per person per year. These services are provided to the rest of us more-or-less cost-free by natural processes, but if we were being charged, the total invoice for all Earthospherians would come to an astronomical three quintillion dollars for the current generation alone! The sad irony is that, as a species, we blithely take these ecosystem services for granted, acting as though we can endlessly befoul and overpopulate our planet.

During their two years of voluntary incarceration the Biospherians became acutely aware of their intimate connections with, and complete dependence upon, the fragile ecosystems within Biosphere 2: "It seemed as though we had touched every aspect of our world; we interacted with molecules and with trees, we knew our environment's boundaries and its subtleties" (Alling & Nelson 1993). The Biospherians would never have tolerated in their small household the kinds of practices that are so widespread in our broader world—massive deforestation, water and atmospheric pollution, the dumping of toxic chemicals, or overexploitation of renewable and nonrenewable resources. Nor would human population growth within Biosphere 2 have been tolerable—both oxygen and food supplies already were stretched to the very limits, to the point where supplemental oxygen had to be injected at the end of year one, and the scanty food stores had to be placed under lock-and-key to prevent recurring incidences of theft by the hungry Biospherians (see previous reference). Clearly, the facility was close to if not well beyond human carrying capacity, even in the short term, and even with massive energy subsidies from the outside.

Exactly how many people the earth can hold remains uncertain (Cohen 1992), but many signs indicate that we are rapidly approaching achievable limits. Indeed, if carrying capacity is defined (as it often is) as the maximum population that can be supported without degrading the environment, then the earth's carrying capacity already has been exceeded. Ozone depletion and atmospheric pollution are global concerns, as are losses of ground-water supplies and usable surface-waters, soils, fossil fuels, and species. Massive hunger, starvation, and conflicts over limited resources are recurring themes in many regions of the world. Current population densities over vast areas are not grossly different from those in the crowded Biosphere 23. For example, across the nearly four million square miles of Europe, densities already *average* nearly 0.3 people per acre, more than 1/10th the density of Biospherians inside Biosphere 2. Astonishingly, our species currently shows a net increase of more than 10,000 people every hour, a quarter million people each day (Meffe et al. 1993), and within our children's lifetimes the global population is projected by the United Nations to quadruple under current fertility rates. How much farther the earth's life support systems can be pushed remains to be seen, but all of us are the unwitting guinea pigs in this reckless and utterly pointless experiment with global carrying capacity. Unlike the inhabitants of Biosphere 2, we have no outside source of rescue or escape. We can only save ourselves, through immediate and humane efforts at population control.

Herein lies the real message from Biosphere 2. It may be fun and even inspirational to dream of colonizing other planets, but the harsh reality is that we have but one home, and it is getting untenably crowded. Whether based on ethical or purely utilitarian considerations, human societies must learn to properly value our Earth, and quickly. Like the astronauts' views from space, Biosphere 2 should give us a novel perspective and renewed appreciation of Biosphere 1.

Literature Cited

Ailing, A., and M. Nelson. 1993. Life under glass. The Biosphere Press, Oracle, Arizona.

Allen, J. 1991. Biosphere 2: The human experiment. Penguin Books, New York.

Watson, T. 1993. Can basic research ever find a good home in Biosphere 2? *Science* **259**:1688–1689.

Cohen, J.E. 1992. How many people can Earth hold? *Discover* **13**:114–119.

Meffe, G.K., A.H. Ehrlich, and D. Ehrenfeld. 1993. Human population control: The missing agenda. *Conservation Biology* **7**:1–3. ■

Questions

1. Why is it complicated to attach dollar values to natural ecosystems by virtue of the fundamental life-support services rendered?
2. What practices would the Biospherians not tolerate in their home?
3. Define carrying capacity.

Answers are at the back of the book.

5

The world's supply of national resources is diminishing. Now we must conserve what we have and find other ways of meeting our needs with a minimum of materials and energy usage. During the last 50 years, the production of raw materials has brought extreme ecological destruction. Industrial logging has more than doubled since 1950, and is responsible for the destruction of primary rain forests in Central Africa and Southeast Asia. Overall, nearly one fifth of the earth's forested area has been cleared. The U.S., in particular, has an unquenchable thirst for energy. This thirst greatly attributes to global warming, acid rain, the flooding of valleys, and the destruction of rivers for hydroelectric dams. Collection systems need to be developed to collect waste and transform it into new products. Such systems can work if public policies encourages sharing, maintains the value of materials, and promotes the design of efficient goods and services. If 60 percent of all materials were recycled it would be the equivalent of 315,000,000 barrels of oil a year.

Mankind Must Conserve Sustainable Materials

John E. Young and Aaron Sachs

As the supply of natural resources dwindles, the world must focus on meeting human needs with a minimum of materials and energy usage.

The culture of consumption that has spread from North America to Western Europe, Japan, and a wealthy few in developing countries has brought with it an unprecedented appetite for physical goods and the materials from which they are made. People in industrial countries account for 20% of global population, yet consume 86% of the world's aluminum, 81% of its paper, 80% of its iron and steel, and 76% of its timber.

Sophisticated technologies have let extractive industries produce these vast quantities of raw materials and have helped to keep most materials prices in decline. However, the growing scale of those industries also has exacted an ever-increasing cost. Raw materials production has brought about unparalleled ecological destruction during the last half-century.

The environmental costs of waste disposal, ranging from toxic incinerator emissions to the poisoning of groundwater by landfills, have been documented with increasing frequency. Even greater damage is caused by the initial extraction and processing of raw materials by an immense complex of mines, smelters, petroleum refineries, chemical plants, logging operations, and pulp mills. Just four primary production industries—paper, plastics, chemicals, and metals—account for 71% of the toxic emissions from all U.S. manufacturing. The search for virgin resources increasingly has collided with the few indigenous peoples who had remained relatively undisturbed by the outside world.

Though not many of the world's mostly city-dwelling consumer class comprehend the impacts and scale of the extractive economy that supports their lifestyles, the production of virgin materials alters the global landscape at rates that rival the forces of nature. Mining moves more soil and rock—an estimated 28,000,000,000 tons per year—than is carried to the seas by the world's rivers. Mining operations often result in increased erosion and siltation of nearby lakes and streams, as well as acid drainage and metal contamination by ores containing sulfur compounds. Entire mountains, valleys, and rivers have been ruined by mining. In the U.S., 59 former mineral operations are slated for remediation under the Federal Superfund hazardous-waste

Reprinted with permission from *USA Today* magazine.

cleanup program, at a cost of billions of dollars.

Cutting wood for materials plays a major role in global deforestation, which has accelerated dramatically in recent decades. Since 1950, nearly one-fifth of the Earth's forested area has been cleared. Industrial logging has more than doubled since 1950, and is particularly culpable in the destruction of primary rain forests in Central Africa and Southeast Asia. Production of agricultural materials has dramatic environmental impacts as well. In the former Soviet republics of Kazakhstan and Uzbekistan, for instance, decades of irrigated cotton production have contaminated large areas of farmland with toxic chemicals and salt.

The chemical industry has become a major source of materials, including plastics, which increasingly have been substituted for heavier materials, and synthetic fibers, which have become crucial to the textile industry. The impacts of chemical production—from hazardous-waste dump sites such as Love Canal to industrial accidents like the release of dioxin from a Seveso, Italy, plant in 1977—generally are more familiar than those from mining, logging, and agriculture, since chemical facilities usually are located closer to urban areas.

Raw materials industries are among the planet's biggest consumers of energy. Mining and smelting alone take an estimated five to 10% of global energy use each year. Five primary materials industries—paper, steel, aluminum, plastic and container glass—account for 31% of U.S. manufacturing energy use. This thirst for energy contributes significantly to such problems as global warming, acid rain, and the flooding of valleys and destruction of rivers for hydroelectric dams.

Despite the environmental impacts of the materials economy, the principal subject of debate over materials policy in the last several decades has been how soon Earth is likely to run out of nonrenewable resources. Yet, the ecosystems that provide renewable resources could collapse long before that point is reached.

Since the 1970s, growth in industrial nations' raw materials consumption has slowed. Some observers believe that these countries have reached a consumption plateau, for much of their materials-intensive infrastructure—roads and buildings—already is in place, and markets for cars, appliances, and other bulky goods largely are saturated. The plateau they sit on is a lofty one, though, and the consumer culture still is going strong.

Materials use has reached extraordinary levels in industrial countries because of an outdated global economic framework that depresses virgin materials' prices and, most important, fails to account for the environmental costs of their extraction and processing. Prices have continued to fall even as ecological expenses of the global materials economy have risen sharply. During the past decade, almost every major commodity has gotten significantly cheaper throughout the world—a trend that, in turn, allowed consumption rates to continue their steady growth.

International trade rules and the policies of industrial nations tend to reinforce materials consumption patterns that date back to the colonial era, when empires were assembled to secure access to raw materials for manufacturing industries in home countries. The development assistance policies of former colonial powers tend to favor the production and export of primary commodities, which they often still receive in large quantities from the countries they once ruled. World Bank and International Monetary Fund planners generally advise commodity-exporting developing nations—many of which are deeply in debt—to invest in those sectors to gain foreign exchange. Such policies, combined with tariffs that are lower for primary commodities than for processed intermediates or manufactured goods, have tended to depress prices of primary material commodities as compared with recovered materials.

At the other end of the cycle, industrial countries commonly subsidize waste disposal as well. In the U.S., where national policy officially favors waste reduction, reuse, and recycling over landfilling and incineration, actual practice has been the reverse. Local communities have spent billions of dollars to finance construction of disposal facilities, while cheaper, more environmentally sound waste management options have received little funding. A large share of these waste disposal costs are hidden in property taxes or utility assessments, rather than being paid for directly per unit of waste. Thus, there is little incentive not to throw things away.

Favoring disposal over waste reduction, re-use, and recycling squanders not only materials, but the large amounts of energy embodied in products that are buried or burned. A 1992 study of recycling and incineration found that, while significant amounts of energy can be recovered through burning, three to five times more can be saved by recycling municipal solid waste. Increasing the recovery of materials in U.S. solid waste so that at least 60% of all materials are recycled could save the equivalent of 315,000,000 barrels of oil a year.

Preserving the natural resource base will require the creation of an economy that produces much less waste and can function with relatively small inputs of virgin materials. Sooner or later, of course, the over-all efficiency of the system will have to improve on a massive scale; all the goods and services the economy produces will have to be designed to need far fewer materials. On a more immediate level, though, it is necessary to look at "wastes" and secondary materials in an entirely different light. The throwaway culture of "convenience" and planned obsolescence must be discarded in favor of an approach that seeks value in products even after people think they have finished using them.

The practical consequences of this attitude will be complex and varied. Perhaps most important, entrepreneurial and employment opportunities would grow rapidly in the recovery and reprocessing of used materials. A wide variety of items—from bottles to shipping containers—could be reused dozens of times, then collected for remanufacturing. Car owners might bring their tires to a local auto parts dealer to get retreaded and, later perhaps, melted down into completely different products. Composted kitchen, yard and, sewage wastes would be plowed back into gardens and farms. Recycled-paper mills would outnumber those equipped only for virgin fiber, and smelters fed by recycled metals would replace a major share of mining operation. In general, cities—where used resources, factories, and labor are concentrated—would become a more important source of materials than rural mines or forests.

Bringing about change on this scale is going to require more than today's incremental increases in governments' environmental budgets, curbside pickup of newspapers and the occasional trip to the community bottle bank. The job demands an infusion of capital, design skill, imagination, and public commitment comparable to America's economic mobilization during World War II. Like that process, this one will have to proceed from public policies, but its principal players will be industrialists, financiers, engineers, designers, and thousands of small businesses. In the long run, efficient use of materials should mean not only less environmental damage, but also a more stable economy, better long-term investment opportunities, and more skilled jobs, especially in design industries.

The most obvious place to start is with the current subsidization of virgin materials extraction. Raw material production should be taxed, not subsidized. A reformed tax system could force industries to cover the full environmental costs of their activities, instead of leaving the bill for the public to pay. By raising prices to more realistic levels, such a system would provide strong incentives not to degrade the natural resource base in the first place. Market forces need to be aligned for, rather than against materials efficiency.

A related policy could force households and businesses to pay the full cost of disposing of their waste—with the clear understanding that a more efficient economy would make it well worth their while.

Truly taking responsibility for garbage will involve far more than just paying a little extra for its disposal. The ultimate goal is to develop comprehensive systems for collecting waste and transforming it into new products, which will be possible only if many consumer goods are redesigned to be reused and recycled easily.

Recovering secondary materials and re-integrating them into the economy will be crucial in the struggle to reduce the need for virgin resources. Over the long term, though, it will be necessary to make basic design changes in the materials economy to eliminate materials needs and wastes at the source.

Two decades ago, when the world faced an energy crisis, skeptics scoffed at the idea that efficiency was the key to a sustainable energy policy. Since then, new lighting, heating, cooling, insulation, and manufacturing technologies have made it possible to cut energy use by three-fourths or more in many applications. The improved technologies often are cheap enough to make energy efficiency a better investment than energy production.

Houses can be designed to save materials without sacrificing comfort. Even the boards they are made of could be produced more efficiently. In recent years, managers of industrial sawmills, frustrated at how much wood was being lost as sawdust, determined that they could realize considerable savings simply by using thinner blades to saw logs; the thinner blades cut just as well as the originals but left more of the wood intact. By combining similar technologies already available—ranging from two-sided copying in offices to the adoption of efficient architectural techniques—the U.S. could cut its wood consumption in half.

Efficiency policies will have to cover a wide range of issues, but on the most basic level they all need to spark smarter design. Three principles may help designers, architects, engineers, planners, and builders work together to make that happen.

The first is to promote sharing. For years, consumers have obtained reading materials from free public libraries instead of buying increasingly expensive books that they probably would read only once. Many people likely would welcome the opportunity to apply the same concept of sharing and re-use to thousands of other everyday items—power tools, bottles and jars, cars, or computer data, for example.

A second principle is to maintain the value added to materials. Extraction, processing, refining, and manufacturing all add value to a raw material. These processes also have major environmental costs. If a computer becomes worthless a few months after its purchase because a much better, cheaper model has hit the market, the economy has wasted all the effort and environmental damage that went into the device's manufacture. The item would lose value much more slowly if it were designed specifically to be repaired or upgraded easily. The more durable a product is, the less frequently the cycle of processing or reprocessing has to start over again.

The third is to design goods and services in context. A product design is most likely to be materials- and-energy efficient if it is considered as part of the entire system in which it functions. It often is more efficient to substitute a broad systemic change for an individual product.

Synergistic gains between components of design are not realized simply by plugging in energy-efficient technologies or materials. They emerge only when design professionals work together from beginning to end—as they did, for example, when the National Audubon Society built its new headquarters in New York City. Audubon achieved massive improvements in lighting, heating, cooling, ventilation, and over-all indoor air quality. The architects drew up floor plans to take full advantage of natural light; the contractor installed windows that let just the right amount of light and heat pass through them, the lighting technician knew, accordingly, that the building would need fewer lamp fixtures; and the interior designer arranged surfaces and finishes to get the most out of the lamps.

Such integrated design costs perhaps three times as much as conventional design. Nevertheless, according to Amory Lovins, whose Rocky Mountain Institute has done pioneering studies on the subject, the resulting efficiency improvements can yield as much as 25% more floor space in a building of the same size. The extra expense may be recovered immediately in materials savings—fewer ducts will be necessary, for instance, if climate-control systems are smaller—and a good design would yield substantial energy-saving dividends over the life of the structure.

This truly thoughtful design no doubt would be more common if society rewarded it more directly. An engineer's commission usually is based on a percentage of the overall project budget—a practice that in many cases rewards oversizing. Taking the opposite approach, the Canadian utility Ontario Hydro recently announced that it would reward design that met certain energy-efficiency standards with a rebate equivalent to three years' energy savings, to be shared by the developer, architect, and consulting engineers. By basing the rebate on the finished product—the building's actual energy performance—the utility was adding an incentive for the design professionals to stay involved and ensure that their ideas were executed properly during the structure's construction, operation and maintenance. Although materials efficiency is harder to measure than energy efficiency, similar methods of compensating designers for work that reduces materials intensiveness could be just as effective.

Even with such incentives, smarter design will remain difficult until information systems are in place that give fuller descriptions of products and materials. The "green labels" now seen in several countries are intended to encourage purchases of environmentally preferable products, but they provide little detailed data and are directed primarily at consumers. More promising would be in-depth green labeling for designers and builders. Information on a material's origin, its capacity for re-use and recycling, the environmental costs of its production, etc., will have to become an essential part of its design specifications.

Systematically reworking materials specifications to include environmental information would be a step in the right direction. Accomplishing this reform on a broad scale, though, will require a much clearer understanding of how materials production and use actually affect the environment. Currently, detailed information on that topic is just as scarce as information about waste. Yet, materials-efficient design ultimately will be impossible without it. Data must be developed at every level of society, from corporate materials audits—which should help firms identify opportunities for efficiency improvements—up to national and international accounting of materials flows. These statistics also need to be linked with data on the amount of energy, pollution, and economic activity associated with materials production and use.

There have been at least a few promising initiative in this area. In the U.S., the Bureau of Mines has started compiling limited, but extremely useful, information on materials production, consumption, and recycling. The eventual goal is to track comprehensively the quantities of materials flowing into and out of the American economy, allowing progress in materials efficiency to be measured. Similarly, the Department of Energy has begun to collect more detailed energy-use statistics. Combining the two data sets undoubtedly will uncover valuable opportunities for saving energy through more efficient materials use.

Data on pollution and hazardous waste generation from production processes, exemplified by the information collected annually in the U.S. Toxics Release Inventory (TRI), also has been useful. Unique in the world, the Environmental Protection Agency's TRI lists the reported output of several hundred toxic substances by American manufacturers. Data are available by the specific facility or company, by geographical area or industrial sector, or by chemical. The TRI has its flaws—including limited coverage of industries and chemicals, and poor quality control on its data—but it is a good starting point for the sort of comprehensive system that is needed. It would be very useful if such a system included data on raw materials that flow into industrial facilities—a reform Massachusetts has implemented on a limited basis through its toxics use reduction law.

In the long run, materials-data collection efforts—like energy-use tracking—make sound economic sense, since they could inspire efficiency improvements that would far outweigh the cost of amassing the information. They could also help in making materials choices by facilitating quick assessments of the energy use, pollution, jobs, and waste associated with production of a given material or product—a virtually impossible task today. ■

Questions

1. Name five primary materials industries and the problems they contribute to.

2. Explain what would need to be done in order to take responsibility for garbage.

3. Why has materials use in industrial countries reached extraordinary levels?

Answers are at the back of the book.

6

Cowbirds are pushing songbirds to extinction. As a result conservationists are debating whether or not killing the cowbirds is the solution. Cowbirds reproduce at an incredible rate. They do not have to feed their young and as a result, a single female can lay as many as 50 eggs in one breeding season by co-opting the nests of a hundred or so different species of birds. These species are raising the cowbirds' eggs and not enough of their own. This is pushing many species towards extinction, including: Bell's vireo of California, the black–capped vireo of Texas, and the Kirtland's warbler of Michigan. This has conservationists advocating cowbird extermination. Even though there will always be disagreements between researchers with this way of dealing with the cowbird, most agree that the restoration of the habitat is the only long-term solution for the songbird.

Taking Back the Nest

Daniel Dunaief

As cowbirds push some songbirds toward extinction, conservationists are debating an uncomfortable solution: killing the pests.

Historically speaking, the brown-headed cowbird should be called the buffalo bird. This resourceful parasite, which lays its eggs in the nests of other birds, once followed herds of buffalo around the Great Plains, feeding on insects and seeds in the soil churned up by the grazing animals. As Europeans settled North America, however, they replaced forests with pastures full of cows—that is, with cowbird territory. Since cowbirds reproduce ferociously—not having to feed her young, a single female can lay as many as 50 eggs in a breeding season—they soon fanned out across the continent. Today their eggs are routinely found in the nests of a hundred or so species of birds.

Some of these species are in trouble as a result: they're raising too many baby cowbirds and not enough young of their own. Among the songbirds that have been pushed closer to extinction by cowbirds are the least Bell's vireo of California, the black-capped vireo of Texas, and the Kirtland's warbler of Michigan. The situation has forced conservationists into the uncomfortable position of advocating cowbird extermination programs. "While we don't like killing cowbirds," says Jane Griffith, a biological consultant, "we do like to hear the songs of endangered species."

On a local level, at least, such programs seem to help. An example is the once that Griffith and her husband, John Griffith, have worked on at Camp Pendleton, a Marine base in Oceanside, California, that is one of the last refuges of the least Bell's vireo. In the early 1980s, half the vireo nests there were parasitized. By 1994, after some 4,800 cowbirds had been trapped and gassed with carbon monoxide, parasitism had declined to 1 percent. More important, the number of male vireos had increased more than fifteenfold, from 27 in 1981 to 420 in 1993. Similar success has been reported by a cowbird control program at Ford Hood, Texas, which has become a retreat for the black-capped vireo.

Meanwhile, a new threat to songbirds has surfaced—an invasion of another species of cowbird. Since 1985 the shiny cowbird of South America has been sighted in Florida. According to Alexander Cruz, a biologist at the University of Colorado in Boulder, it hasn't yet been caught parasitizing nests, but that's just a matter of time. Cruz says the Cuban yellow warbler and the Florida prairie warbler may be particularly vulnerable.

Florida is the cowbird frontier: as the shiny

cowbird spreads into the state from the south, the brown-headed cowbird is invading from the north. No one knows what's going to happen as the two species converge. They may get along just fine—which would be bad news for songbirds. The shiny cowbird has a longer breeding season, and so it may be able to use the nests of early- and late-breeding songbirds that escape its brown-headed cousin. It also seems to favor coastal regions, whereas the brown-headed cowbird is happy inland as well. Cruz thinks the shiny cowbird is likely to spread along the Atlantic and Gulf coasts.

Some conservation biologists, including the Griffiths, believe that more radical steps must be taken to control cowbirds. They favor killing brown-headed cowbirds not only in places where the birds are parasitizing the nests of endangered species but also at their winter roosting grounds in the southern third of the country. Cowbirds from all over congregate there by the millions and would be easy to kill in large numbers.

But other researchers who oppose such radical measures point out that cowbirds are not the primary threat to endangered songbirds—humans are. The least Bell's vireo, for example, has lost 95 percent of its habitat—trees and bushes along southwestern riverbanks—to farms and other human uses. Camp Pendleton is one of the few places where it can breed in peace, and in that confined space it is particularly vulnerable to cowbirds.

Restoring habitat, most researchers agree, is the only long-term solution to the songbirds' woes. It would help them directly, by giving them breeding space, and indirectly, by taking that space from cowbirds. In New England, for instance, the farms established by settlers have been reverting to forest for the past century or so—and there the cowbird population has declined substantially. ■

Questions

1. What state is considered the cowbird frontier?

2. Some researchers feel that the cowbird is not the primary threat for the songbird. Who or what do they is the primary threat?

3. What is the long-term solution for the songbird and why?

Answers are at the back of the book.

7

Deforestation, soil erosion, desertification, wetland degradation, and insect infestation are some of the major factors that are indicative of how severe environmental degradation is in sub-Saharan Africa. As poverty runs rampant throughout the area, the inhabitants of Africa have to overexploit natural resources, therefore causing environmental degradation in order to just survive. Three elements that increase degradation in this area are demographics, foreign debt, and the absence of democracy. Population has been a source of contention as well. During the last 25 years, social services, especially in education and healthcare, have led to a decrease in infant mortality. However, these services have also led to an increase in population. The developmental strategies attempted in these countries not only caused destruction to the environment at large, but did not improve the average standard of living. Furthermore, even though sub-Saharan Africa has a large natural resource base, degradation is still prevalent. But perhaps the most significant element is that the information on the environment and its degradation remains insufficient. Hope for sub-Saharan Africa lies with institutional development. Institutional development can reduce poverty and conserve the environment in the following ways: encouraging democracy, augmenting human rights, and enlarging the information base so that environmental concerns can be decided upon. If these institutional developments can be implemented, the people of sub-Saharan Africa can turn the economic hardships of their countries around and improve the attributes of their environment.

The Environmental Challenges in Sub-Saharan Africa

Akin L. Mabogunje

Sub-Saharan Africa suffers from some serious environmental problems, including deforestation, soil erosion, desertification, wetland degradation, and insect infestation. Efforts to deal with these problems, however, have been handicapped by a real failure to understand their nature and possible remedies. Conventional wisdom views the people of this region as highly irresponsible toward the environment and looks to the international community to save them from themselves. It tends to blame all of the region's environmental problems on rapid population growth and poverty. Yet, there is no conclusive evidence that Africans have been particularly oblivious to the quality of the environment, nor has the international community shown any genuine concern for it until recently. Clearly, protecting the environment of sub-Saharan Africa is an issue that needs to be examined more carefully and incorporated into an overall strategy of sustainable economic development.

Formulating such a strategy will not be easy: In the closing years of the 20th century, virtually every country in this region is slipping on almost every index of development. The heady post-independence period of the 1960s and early 1970s, when development was considered simply a matter of following a plan formulated by Western experts, has now been succeeded by a time of fiscal crises and international marginalization. The region now finds itself afflicted by the consequences of inappropriate policies, as well as by almost endemic political instability, an inability to manage its economies effectively, and an increasingly hostile external economic milieu. As simple survival has become more problematical, it has become increasingly difficult to avoid overexploiting natural resources and degrading the environment. Analysts are now concerned that this will compromise the prospects for sustainable development in the near future.[1]

To understand the full dimensions of the prob-

***Environment*, Vol. 37, No. 4, pp. 4–35, May 1995. Reprinted with permission of the Helen Dwight Reid Educational Foundation. Published by Heldref Publications, 1319 Eighteenth St., NW., Washington, DC 20036-1802.**

lem, it will first be necessary to examine the factors that predispose sub-Saharan Africa to serious environmental degradation. This will permit a detailed investigation of the environmental problems caused by humans in both rural and urban areas, along with a suggestive comparison between those problems and ones caused solely by nature. It will then be possible to look at the question of environmental protection in terms of sustainable development in the region and to suggest the roles that the state and international assistance ought to play. The present situation offers an important opportunity to redirect development strategy in ways that will not only improve the social and economic well-being of people in this region but also enhance the quality of the environment in which they live.

Factors Predisposing to Environmental Degradation

There are three factors that strongly increase the threat of environmental degradation in sub-Saharan Africa: its demographics, its heavy burden of foreign debt, and the absence of democracy. Throughout the region, the end of the colonial period saw a tremendous expansion of social services, especially in the areas of education and healthcare. This led to a sharp decline in infant mortality and to a rapid increase in population. During the last 25 years, annual growth rates of 2.5 to 3.5 percent have caused the population of sub-Saharan Africa to double (to 570 million); at the current rate of increase, it should double again in the next 25 years.[2]

An increase of this magnitude within a relatively short time span implies a rising proportion of children in the population and thus a heavier burden on those who must care for them. This has led to mass migration to the cities (particularly by adult males) and other efforts to supplement family income through nonfarm employment. As a result, there has been less time for farm work, and more labor-saving but environmentally harmful shortcuts are being taken. In forested areas, for instance, cleared land is used continuously, even though allowing it to lie fallow from time to time would result in greater productivity and less degradation. In dryland regions, cultivation has been extended into marginal lands that are more easily cleared and cultivated.

Turning to the second factor, countries in sub-Saharan Africa incurred large foreign debts in their efforts to industrialize and to provide their rapidly growing populations with modern social services. Most of these loans have been long-term ones from official sources and on concessional terms; as the need for borrowing has become more urgent, however, countries have turned increasingly to private, short-term loans at market rates. Thus, while in 1970 the region's total official debt (excluding that of South Africa and Namibia) was slightly more than $5 billion (U.S. dollars), by 1990 it had risen to nearly $140 billion. Total private debt, which was zero in 1970, was more than $20 billion in 1990. (With other external loans, the total indebtedness of the region was more than $171 billion by 1990.[3])

The problem, however, lies not so much in the rising level of debt as in the region's dwindling ability to service it. High dependence on the export of primary products left sub-Saharan African countries vulnerable to the long decline of commodity prices that began in the late 1970s. The total value of the region's agricultural exports has fallen dramatically, with the decline averaging 0.8 percent a year from 1975 to 1980, 2.9 percent a year from 1980 to 1985, and 2.5 percent a year from 1986 to 1988. (For some countries the decline has been even more pronounced.[4]) As a result, the burden of debt has risen markedly for most countries in the region. Between 1980 and 1989, the total external debt rose from 27 percent to 97 percent of gross national product and from 97 percent to 362 percent of exports.[5]

Not unexpectedly, most countries in sub-Saharan Africa have had to undergo major structural adjustments. This has entailed not only a drastic compression of imports and a sharp devaluation of national currencies but also the retrenchment of a sizable portion of the wage- and salary-earning population. As living conditions deteriorated, more people turned to survival agriculture, both in urban and rural areas. At the same time, sharply rising prices for imported energy products forced many families to fall back on wood and charcoal for their domestic energy needs. Clearly, these developments put acute strain on the environment everywhere in the region.

The performance of most African governments in implementing the reforms necessary to turn their economies around has also been a source of serious concern. The international community spent the years immediately following independence rationalizing (and sometimes applauding) the necessity for authoritarian one-party or military rule. Over time, these regimes have become inordinately corrupt and have managed the countries' economies without due concern for transparency and accountability. In most countries, this has led to a high level of political instability and social alienation that has impaired both development efforts and environmental protection. There is a growing realization that economic reforms cannot be achieved without a much greater degree of decentralization and democratization in the political process.

Much of the debate about sustainable development in sub-Saharan Africa has focused on the region's severe poverty. There is no question that poverty has become widespread. The World Bank estimates that between 1985 and 2000, the number of persons living below the poverty line will rise from 180 million to 265 million.[6] By 1990, the combination of rapid population growth and an economy in crisis had lowered per capita gross national product to $340, making this region one of the least developed in the world.

For neo-Malthusians, this poverty stems directly from overpopulation; in their view, the two will inevitably lead to an increase in land fragmentation, overutilization of agricultural and grazing land, more frequent famines, lower life expectancy, and considerable environmental degradation.[7] By contrast, the renowned agricultural economist, Ester Boserup, and others argue that population growth need not result in such dire consequences. In their view, population growth can promote more intensive agricultural practices and induce more favorable attitudes toward technological and organizational innovation that will not only increase productivity but improve environmental quality as well.[8]

Two considerations suggest that the second view is more applicable to the situation prevailing in sub-Saharan Africa. First, over the period 1600 to 1900, this region lost a large part of its population to internecine warfare and the slave trade. As a result, by 1900 the region was more sparsely populated than it had been earlier. Second, at 23 persons per square kilometer, the region's current population density is still low compared to that of Asia or Europe.

This is not to imply that there is no cause for concern about the environment in sub-Saharan Africa. One needs to keep a sense of perspective in addressing the question, however. The proper focus is on the region's poverty per se (as opposed to its population growth) and on the impact this has on the environment in both rural and urban areas.

Poverty and the Rural Environment

Despite its pervasive poverty, sub-Saharan Africa has substantial natural recourses in its rural areas, including forests and grasslands, wetlands, cultivable soils, and other biological resources.[9] Although only 40 percent of the total land area is under cultivation or used for pasture, much of it is threatened by one form of damage or another.

Three types of environmental damage are occurring in sub-Saharan Africa: deforestation, degradation, and fragmentation. Deforestation is defined as "the temporary or permanent clearance of forest for agriculture or other uses, resulting in the permanent depletion of the crown cover of trees to less than 10 percent."[10] Degradation, on the other hand, refers to the temporary or permanent deterioration in the density or structure of vegetation cover or species composition.[11] It results from the removal of plants and trees important in the life cycle of other species, from erosion, and from other adverse changes in the local environment. Fragmentation arises from road construction and similar human intrusions in forest areas; it leaves forest edges vulnerable to increased degradation through changes in microclimates, loss of native species and the invasion of alien species, and further disturbances by human beings.

While there is no doubt that all three processes are taking an increasingly heavy toll on the forest and woodland areas of sub-Saharan Africa, there is considerable controversy over the exact rate at which this is occurring. Estimates based on the subjective judgment of experts or on data from low-resolution sensors on weather satellites are generally higher than those based on the more accurate data from high-resolution sensors on the Landsat and Spot satellites.[12] Until more of the latter data are available, the actual extent of deforestation will remain uncertain.

The United Nations Food and Agriculture Organization (FAO) estimates that forested land was converted to agricultural uses at increasing rates over the period 1981 to 1990 and that such changes accounted for 25 percent of the changes in forest cover during the period.[13] These changes were concentrated in the moist and dry forest lowland areas, where the average annual conversion rate was higher than in tropical rain forest areas. Except for the dry forest areas, however, conversion rates in sub-Saharan Africa are lower than those in Latin America and the Asia-Pacific region.

Degradation and fragmentation involve a much larger area than deforestation and pose a greater threat to the diversity of plant and animal life. Selective logging and the failure to pursue a systematic program for forest regeneration (either natural or artificial) are the two major factors promoting rapid degradation of the forest and woodland environment. Owing to the desperate need for foreign exchange, the rate of logging in sub-Saharan Africa rose more than 34 percent between 1979 and 1991, compared with a global average of only 19 percent. Similarly, the lack of foreign exchange to purchase petroleum has led to a rapid rise in the production of fuelwood and charcoal.[14]

There are no firm estimates of the harm resulting from degradation and fragmentation in the region. Two factors suggest that it is considerable, however: First, the ratio of forest regeneration to deforestation is as low as 1:32 in sub-Saharan Africa, compared with ratios of 1:2 in Asia and 1:6 in Latin America.[15] Second, one out of every six species in the tropical moist forests has some economically valuable, nontimber use.[16]

Nowhere is this loss of biological resources more marked than in the region's wetlands. These wetlands include river floodplains, freshwater swamps and lakes, and coastal and estuarine environments. They provide a number of valuable resources, including wood; foraging, hunting, and fishing opportunities; and land for crops and pasture. They also contribute significantly to aquifer recharge and flood control, as well as providing habitat for migratory birds and other organisms.

Degradation of these wetlands is due not to population growth or poverty but to modern development, principally the construction of major dams on important rivers. These dams control water flow over much of the rivers' length and impair the agricultural value of wetlands both by lowering water quality and by altering the extent and timing of floods.[17] In Nigeria's Benue floodplain, for instance, the reduction in flooding caused by the Lagdo Dam led to a 50 percent reduction in the area used for environmentally friendly flood-recession sorghum farming in the dry years of the mid-1980s.[18] Similarly, the Diama Barrage on the Senegal River, built to prevent incursion of salt water during periods of low river flow, is expected to cause the loss of some 7,000 tons of shrimp and fish, while Manantali Dam is expected to greatly reduce the fish catch in that river.

If wetlands are being degraded by inappropriate development, grasslands and other relatively dry areas are being degraded by both rapid population growth and inappropriate technology and land-use practices. Recent studies, however, highlight the resilience of dryland ecosystems and caution against confusing natural changes due to recurring droughts with the long-term degradation caused by human activities.[19] They also argue against the simplistic application of the general concepts of overgrazing and carrying capacity in dryland environments.

This is not to say that such factors as overgrazing, overcultivation, and excessive use of wood for fuel have not contributed to the degradation of drylands. Rather, it is to stress that degradation occurs only where these activities lead to detrimental changes to the soil system itself as well as to plant cover. Damage to the soil system results either from erosion or from physical and chemical changes in the soil itself. Erosion by wind or water is a serious problem in dryland areas because the naturally thin soil and its slow rate of formation make recovery difficult. Such erosion accounts for nearly 86 percent of the total degradation of dryland areas in sub-Saharan Africa.[20] The remainder is primarily due to the loss of nutrients from excessive cultivation and lack of fertilization.

Estimates based on the GLASOD (Global Assessment of Soil Degradation) approach indicate that by 1992 nearly 320 million hectares of drylands in sub-Saharan Africa had degraded soils, ranging from light and moderate (77 percent) to strong and extreme (23 percent).[21] These estimates, however, represent a considerable (almost two-thirds) reduction in the area previously thought to be suffering desertification as a result of human activities. Improved monitoring capabilities are making it increasingly clear that climatic variations are responsible for much of the degradation of soils in the region.

Even so, the loss of biological diversity—due to habitat destruction, the introduction of exotic species, overharvesting, pollution, and other activities that affect natural ecosystems—is a growing problem in sub-Saharan Africa. This is especially significant because biodiversity is greatest in the tropics. According to one estimate, between 40 and 90 percent of all plant and animal species are found in tropical forests.[22] Based on current trends in habitat destruction, as many as 11 percent of total species may become extinct during every 10-year period from 1975 to 2015.[23]

Despite international agreements such as the Convention on International Trade in Endangered Species of Flora and Fauna (CITES), hunting of elephants, rhinoceroses, and alligators is still a major problem in some African countries. The situation is even more serious in the case of birds, because widespread pollution is destroying their habitats, often in imperceptible ways. Tragically, although most endangered species are technically under government protection, as a practical matter they are resources free for the taking.

Poverty and Urban Environments

Although sub-Saharan Africa is the least urbanized region in the world, its urban population is growing quite rapidly. In 1965, urban areas accounted for only 14 percent of the total population of the region; by 1990, however, such areas accounted for 29 percent, and by 2020, the figure should be more than 50 percent. Already, there are 27 metropolitan areas with populations greater than 1 million and 1 (Lagos) with a population of at least 10 million.[24]

Consequently, even in urban areas, the widespread poverty exerts a strong negative impact on air, water, and land resources. The ongoing economic crisis has intensified the level of air pollution in most countries. Most households can no longer afford to use petroleum or gas products for fuel and are relying increasingly on charcoal or wood. This has greatly increased the amount of carbon dioxide generated by cities in the region. While in 1991 the region accounted for just 2 percent of global carbon dioxide emissions from industrial processes, its total contribution to global carbon dioxide emissions (from both urban and rural areas) was 19 percent.[25] Also contributing to the problem is the increased use of substandard industrial equipment and motor vehicles, made necessary because the region lacks the funds to invest in more environmentally responsible devices. Thus, three countries in the region—South Africa, Zaire, and Nigeria—are now ranked among the top 50 countries in terms of their 1991 contributions to global greenhouse emissions.[26]

Air pollution, both indoor and outdoor, exposes the population to serious health hazards, especially from suspended particulate matter and lead. Most sub-Saharan African cities do not yet suffer from serious outdoor pollution. Nonetheless, the increased use of wood and charcoal in household kitchens is exacerbating indoor air pollution and heightening the risk of acute respiratory infections, particularly among infants and children. In some areas, lead pollution from substandard vehicles is also starting to increase the risks of hypertension, heart attacks, and strokes.

Access to clean water is also a major problem throughout sub-Saharan Africa. Although the region as a whole has large water resources, a number of countries in the drier areas have experienced serious shortages. Within these countries, some 35 percent of the urban population has no source of drinking water within 200 meters of their homes. For 13 of the 18 countries for which data are available, the proportion of the population with ready access to water has declined since the 1970s[27]; in countries without data, the situation is probably worse.

Safe water, however, is a premium everywhere in the region. Water pollution, largely from human waste, has become a serious health hazard because the economic crisis is preventing most countries from providing adequate water treatment. Diseases such as typhoid, cholera, and diarrhea are spread by drinking contaminated water, while bilharzia, guinea worm, roundworm, and schistosomiasis are spread by bathing in it. Water pollution thus exacts a tremendous toll on the population of sub-Saharan Africa, raising infant mortality rates and impairing the health of all age groups.[28]

Degradation of land resources, mainly from improper disposal of solid and toxic wastes, is another problem. Although the volume of such wastes is much lower than in industrial countries, most of the region lacks even rudimentary collection and disposal facilities. Refuse is simply dumped along roads and other public places or into waterways, contributing to the spread of disease. Although toxic wastes are not yet widespread and exposure is fairly localized, there is fear of surreptitious trade in and dumping of such wastes in some countries.

Natural Disasters

In addition to human-induced degradation, geophysical events (such as droughts, floods, tornadoes, windstorms, and landslides) and biological events (such as locust and pest invasions) greatly affect the environment and the well-being of people in sub-Saharan Africa. The two most important geophysical events are probably drought and floods. Drought is defined as a period of two or more years during which rainfall is well below average.[29] In ecological terms, it is simply a dry period to which an ecosystem may be adapted and from which it often recovers quickly. Drought should be distinguished from dryland degradation, which, as pointed out earlier, is brought about by inappropriate land-use practices under delicate environmental conditions.

Unlike dryland degradation, however, drought inflects acute distress on human beings and animals, forcing mass migrations from the affected areas. Given the region's poverty and its inability to invest in new techniques, plant strains, storage facilities, and so on, the capacity to deal with drought is severely limited. As a result, many countries have become dependent on international assistance; this is especially true for small countries, but even large ones like Ethiopia (where a drought occurred in the midst of a prolonged civil war) have needed significant help. However, once normal rainfall resumes, recovery takes place quickly and people tend to return to their native areas.

Floods, on the other hand, stem from periods of heavy rainfall, either in the immediate locality or upstream of it. They are most common in river valleys and floodplains. In rural areas, their effects can be beneficial as well as harmful. Although flash floods destroy crops, livestock, and settlements, they provide ideal conditions for certain fish and for cultivating crops such as paddy rice, millet, sorghum, and vegetables. In urban areas, however, especially where there has been indiscriminate building on floodplains or where channels are blocked, floods pose a real danger to life and property. In Ibadan, Nigeria, for instance, the flood of 31 August 1980 claimed about 200 lives, displaced about 5,000 persons, and damaged property worth millions of dollars.[30]

Insects are the most important biological hazard in sub-Saharan Africa. According to Thomas Odhiambo, the leading African entomologist, "The insect world in tropical Africa is a rich and diverse one."[31] Although some species confer benefits such as pollinating trees and other plants, many others are pests to plants or serve as vectors in the transmission of disease. Serious study of insects began only recently. The initial emphasis has been on combating plant pests through heavy pesticide use—with all the deleterious environmental consequences this implies. Fortunately, most farmers in the region cannot afford to use pesticides to any great degree.

As the entomology of the region becomes better understood, there is growing appreciation of the potential for biological control of pests. The best example so far is the cassava mealybug program. Mealybugs were inadvertently introduced into sub-Saharan Africa from South America in the early 1970s. Within less than a decade, they had cut cassava yields by two-thirds in most parts of the region. However, biologists eventually found a natural enemy of the mealybug; specially bred in laboratories and distributed throughout the cassava-growing areas, it has brought losses substantially under control. Thus, this program not only saved a staple on which so many people depend but also prevented major harm to the environment.

The Potential for Sustainable Development

Three points stand out clearly from this review of environmental challenges in the rapidly growing but poor countries of sub-Saharan Africa. First, the development strategy pursued in most of these countries has wrought serious havoc on the environment without necessarily improving the average person's standard of living. Second, this has taken place despite the region's relatively ample natural resources. Third, and perhaps most important, knowledge about the region's environment and its degradation remains inadequate.

Nowhere is this last point more true than in the attempt to explain environmental degradation in terms of population growth. This Malthusian argument depends on there being a "carrying capacity" beyond which the environment will inevitably suffer. But as already pointed out, in most of sub-Saharan Africa the population density is relatively low. Furthermore, some prime agricultural lands are clearly "undersettled," while areas less suited to agriculture are densely populated.[32]

A recent study of the relationships among population growth and density, the intensification of agriculture, and the implications for sustainability offers some useful insights on this issue.[33] The study focused on 10 areas with relatively dense populations (ranging from 150 to more than 1,000 persons per square kilometer). Five of these were in East Africa (in Kenya, Rwanda, Uganda, and Tanzania), while the remaining five were in West Africa, mainly Nigeria. In all of these areas, the study found that "contrary to much conventional wisdom that portrays the African smallholders as wrecking their physical resources, particularly in the face of land-intensive conditions...farmers...made considerable investments in resource-based capital, thereby protecting their farms from major environmental deterioration and the negative impacts of intensification and production that usually follow."[34]

Similar conclusions have been reached regarding other aspects of the population-environment equation in sub-Saharan Africa. Contrary to conventional wisdom, detailed field investigations in Nigeria have found that the rising demand for fuelwood has not led to greater deforestation or desertification. Far from "overcutting their trees," farmers have been maintaining their tree stocks by planting and by protecting spontaneous seedlings. The area studied showed "a 2.3 percent per annum increase in tree density between 1972 and 1981, in the wake of the disastrous drought of the late 1960s and early 1970s when pressure on woody vegetation from human and natural sources must have been very intense."[35] Field investigations in Uganda and Mali drew similar conclusions.[36]

This is not to imply that there have been no instances of severe environmental degradation. These have only occurred under three special circumstances, however: where the population density was greater than 500 per square kilometer; where the area itself was physically or biologically vulnerable; and where socioeconomic conditions impeded the implementation of conservation measures. Indeed, decreases in well-being (indicated by reduced food availability) are attributable not to rapid population growth but to the persistence of customary land tenure arrangements, misguided macroeconomic policies, and inadequate infrastructure. According to the World Commission on Environment and Development, chaired by Gro Harlem Brundtland in 1987, sustainable development is "a process of change in which the exploitation of resources, the direction of investments, the orientation of technological development, and institutional change are *in harmony* [emphasis added] and enhance both current and future potential to meet human needs and aspirations."[37] Included in the concept of harmony, of course, is the access of producers to the various factors of production, especially land.

The problem of land tenure, like many of the other problems besetting the development process in sub-Saharan Africa, probably stems from the region's incomplete transition from one mode of production to another. Colonialism attempted to shift the economies of these countries from a precapitalist mode of production (based largely on kinship relations) into a global capitalist mode (based on "commoditized" factors of production whose prices were subject to the forces of supply and demand in a self-regulating market). Though praiseworthy in many ways, these efforts failed signally in the one major area where they could have made a real difference: the patterns of land ownership in rural areas. By and large, colonial administrators left the traditional patterns intact, thus introducing a major contradiction into the development process.

While capitalism requires well-established individual property rights, most smallholders in sub-Saharan Africa have no such rights, even though they have longstanding rights to the use of communal land. Smallholders thus have no "economic assets" in the conventional sense of the term.[38] Second, they have no real collateral against which to borrow and thus no access to the credit they need to invest in improved farm infrastructure, new cultigens, and modern technologies generally. When one considers the heavy investment that went into producing the polders of the Netherlands or the wheatlands of North America, the disability under which African farmers labor becomes readily apparent.

Consequently, much as colonial and post-colonial governments tried to make farmers more market oriented, the fact that one of their major inputs lies outside the market system has always limited the success of this effort. In many cases, farmers have chosen simply to "opt out" of the system, especially now that governments make little attempt to ensure that they receive fair prices for their output.[39] The unnecessary liabilities under which farmers labor probably account for a large part of the poverty in sub-Saharan Africa.

Other aspects of the macroeconomic policies pursued by most African governments simply served to deepen the poverty under which the majority of their rural populations labored and exacerbated the negative impact of their activities on the environment. Widening budget deficits eroded the value of national currencies, fueled inflation, undermined peoples' real income, and encouraged excessive exploitation of natural resources to maintain even a subsistence level of existence.

The Roles of the State and International Assistance

It is clear that the environmental challenges in sub-Saharan Africa are more complex than the simple model linking environmental degradation to population growth and inappropriate macroeconomic policies indicates. Because of this complexity, no easy solutions are available. But whatever policies are adopted, to succeed they must increase peoples' interest in protecting the environment by involving them directly in the process; reduce the incidence of poverty to reduce the pressure on natural resources; and show people how a high level of resource use can go hand-in-hand with the maintenance of environmental quality.

The state can play an important role in promoting sustainable development and improvement of the environment. By setting the correct investment priorities, it can provide needed infrastructure, services, and education. In urban areas, it should focus on providing safe water, collecting and disposing of solid wastes, and improving the physical layout of congested places; in the rural areas, it should focus on health education and basic sanitation.

Regulatory measures, however, may be more important than public investment. In this regard, the state should set environmental standards that are realistic in terms of the country's particular socio-economic circumstances. For example, setting strict standards for indoor air pollution when most people cannot afford less-polluting energy sources simply makes enforcement impossible. Regulatory measures should also aim to remove those distortions in the economy that tend to penalize producers and/or promote overconsumption. Important examples include underpricing agricultural commodities and subsidizing public goods and services, both of which favor the urban population at the expense of the rural population. Such distortions, of course, are partly responsible for the economic collapse of many countries in sub-Saharan Africa. Although structural adjustments now taking place may improve matters, the governments' lack of commitment has left the situation far from satisfactory.

Conservation measures have been important in protecting most natural resources from excessive use or degradation. Through its power of eminent domain, the state has been able to set aside sizable tracts of land to protect watersheds, prevent soil erosion, allow natural regeneration to take place, and preserve habitats, species, and biodiversity. As of 1993, there were 663 public reserves or parks in sub-Saharan Africa, totally 125.2 million hectares.[40] This, however, is no more than 4.6 percent of the total land area of the region, much less than the 6 percent for the world as a whole. Moreover, the 1992 Caracas Action Plan of the World Parks Congress set a goal of protecting at least 10 percent of each of the world's major biomes; sub-Saharan Africa currently falls far short of this standard.

Simply setting land aside, however, does not mean being able to manage it properly. Many governments in the region lack the staff or financial resources to administer their protected areas, much less invest in new ones. Innovative strategies, such as involving private groups and nongovernmental organizations, are being considered and may provide another option for conservation management. Such groups are believed to be better able to raise funds to purchase land, to support conservation activities in existing parks and reserves, to incorporate the local population in management decisions, and to negotiate land-use disputes within and between communities.

Important as public investment, regulation, and conservation are, however, it is institutional development that offers the most hope for alleviating poverty and protecting the environment. Three aspects of institutional development are paramount: promoting democracy, expanding individual property rights, and increasing the knowledge base.

Decentralization and democratization must go down to the community level and must entail not only giving people a voice in decisions but also ensuring that they can raise the revenues necessary to translate their desires into reality. It is this that will promote transparency and accountability in government and foster a proprietary interest in the quality of the environment.[41]

The importance of expanding property rights was made clear earlier. Although it is often claimed that land tenure in sub-Saharan Africa is so complex that nothing can be done about it, it is difficult to believe that meaningful reforms cannot be introduced. The most serious mistake that many governments have made, however, was to resort to nationalization.[42] From a conservation standpoint, nationalization often fails to distinguish between traditional communal property systems (which promote sound management of natural resources) and open-access systems (which result in excessive exploitation). When land and water have been nationalized and sound management practices disturbed, the environmental consequences have often been very severe.[43]

Nationalism has also led governments to give short shrift to titling and registration. Yet, until such procedures clearly define rights to land (on either a freehold or a leasehold basis), much of the region's natural resources are bound to be treated as common property and therefore suffer degradation and "the tragedy of the commons."

The third aspect of institutional development relates to the knowledge available for making decisions on environmental matters. People in sub-Saharan Africa have been adapting to the region's various environments for thousands of years. In the process, they have accumulated valuable information that should be incorporated into more formal analyses of sustainable development.[44] Along with such knowledge, of course, must go the collection and analysis of field data by modern techniques. This is necessary to correct the "hallowed but mistaken" notions of conventional wisdom and to give governments in the region better appreciation of the causes and effects of environmental damage as well as the costs and benefits of different policy options. In this regard, independent commissions provide a useful way for governments to draw on technical expertise both within and outside of their countries; they can also be instrumental in bringing the results of advanced research to bear on local problems.

As mentioned earlier, current knowledge of the ecology of tropical forests and grasslands is still rudimentary. The rich biological resources of these areas—and the ways in which humans relate to them—have yet to receive as much study as they deserve. Given the shortage of funds and trained personnel in most sub-Saharan African countries, this is an area where bilateral and multilateral assistance could make a real difference. The Convention on Biological Diversity, signed by 153 countries at the 1992 United Nations Conference on Environment and Development (UNCED) in Rio de Janeiro, is correct in insisting that tropical countries be compensated for protecting biological diversity from which others benefit.[45] If such compensation became the order of the day, some of it should be used to finance further study of tropical ecosystems.

Poor countries in sub-Saharan Africa could also use international assistance in reforming their environmental laws and in selecting optimal strategies for environmental management. Not enough emphasis has been given to the role of law in alleviating poverty and protecting the environment. Particularly in the area of pollution charges, the experience of developed countries could be invaluable. But countries in the tropics that are being asked to protect biodiversity and genetic resources partly for the benefit of others also need technical assistance in legally defining and protecting their rights regarding these resources.

Consequently, management strategies must go beyond assessing the impact of individual projects, as this tends to address the symptoms rather than the root causes of environmental problems. Such strategies must pay greater attention to broader issues and recognize intersectoral links and intergenerational concerns. This would entail integrating natural resource management with national economic planning as well as tailoring international assistance to specific aspects of resource conservation. To implement such strategies, African countries must strive to secure broad consensus and support, both nationally and internationally.

Conclusion

In the closing years of the 20th century, most countries in sub-Saharan Africa find themselves almost returning to the drawing board. Three decades of trying to drive their economies according to Western models have left them prostrate, their people wallowing in poverty, and their environment exposed to many hazards. More importantly, the international indebtedness of these countries and their present unattractiveness to foreign investors are forcing them to rethink the whole question of development.

The next 25 years thus offer real opportunities for improvement, beginning with population control. At the household level, the economic crisis is inducing a reassessment of the viability of large families; at the governmental level, political inertia and indifference to family planning programs are being replace by more effort and initiative. Already, fertility has begun to decline in some countries, such as Zimbabwe, Kenya, and Cameroon.[46] Although the future remains uncertain, there is every likelihood that this trend will spread across the region.

A decline in fertility, however, will not completely eliminate the momentum that has built up in the years of rapid population growth. The number of people will continue to increase, raising the population density all over the region. But as already emphasized, there is growing evidence that the African environment is more resilient than conventionally thought and can probably support a higher level of population and more intensive agriculture. Technological innovations and institutional developments are thus more important to maintaining a sound environment in sub-Saharan Africa than are efforts to reduce population pressure. If the focus of development shifts from mere economic growth to eradicating the widespread poverty, the people as a whole can play a more decisive role, not only in turning the economic fortunes of their countries around but also in enhancing the quality of the environment.

Notes

1. World Bank, *Sub-Saharan Africa: From Crisis to Sustainable Development: A Long Term Perspective Study* (Washington, D.C., 1989), 22.
2. United Nations Centre for Human Settlements, *Global Report on Human Settlements 1986* (New York: Oxford University Press, 1987), table 1.
3. World Bank, *World Development Report 1992* (New York: Oxford University Press, 1992), 258–59.
4. W.B. Morgan and J.A. Solarz, "Agricultural Crisis in Sub-Saharan Africa: Development Constraints and Policy Problems," *Geographical Journal* 160, no. 1 (1994); 57–73.
5. World Bank, note 3 above, pages 250–51.
6. World Bank, *World Development Report 1990* (New York: Oxford University Press, 1990), 139.
7. A.S. MacDonald, *Nowhere to Go But Down? Peasant Farming and the International Development Game* (London: Unwin Hyman, 1989).
8. E. Boserup, *The Conditions of Agricultural Growth* (Chicago: Aldine, 1965); and E. Boserup, *Population and Technological Change: A Study of Long-Term Trends* (Chicago: University of Chicago Press, 1981). See also R.W. Kates and

V. Haarmann, "Where the Poor Live: Are the Assumptions Correct?" *Environment*, May 1992, 4.
9. World Bank, note 3 above, page 201.
10. A. Grainger, "Quantifying Changes in Forest Cover in the Humid Tropics: Overcoming Current Limitations," *Journal of World Forest Resource Management* 1 (1984): 3–63.
11. World Resources Institute, *World Resources 1994–95* (New York: Oxford University Press, 1994), 133. See also A. Grainger, *Controlling Tropical Deforestation* (London: Earthscan Publications, 1993).
12. A. Grainger, "Rates of Deforestation in the Humid Tropics: Estimates and Measurements," *Geographical Journal* 159, no. 1 (1993): 33–44.
13. Food and Agriculture Organization, *Forest Resources Assessment 1990: Tropical Countries*, FAO Forestry Paper 112 (Rome, 1993), 38.
14. World Resources Institute, note 11 above, page 310.
15. A.D. Jones, "Species Conservation in Managed Tropical Forests," in T.C. Whitmore and J.A. Sayers, eds., *Tropical Deforestation and Species Extinction* (London: Chapman and Hall, 1992), 3.
16. J. Davidson, *Economic Use of Tropical Moist Forests*, Commission on Ecology Paper No. 9 (Gland, Switzerland: International Union for Conservation of Nature and Natural Resources, 1985), 9.
17. B.R. Davies, "Stream Regulation in Africa," in J.V. Ward and J.A. Stanford, eds.., *The Ecology of Regulated Streams* (New York: Plenum Press, 1979). For more information on one such project in Botswana, see the review by C.W. Howe of *The IUCN Review of the Southern Okavango Integrated Water Development Project, Environment*, January/February 1994, 25.
18. C.A. Drijver and M. Marchand, *Taming the Floods: Environmental Aspects of Floodplain Development in Africa* (Leiden, the Netherlands: University of Leiden, 1985).
19. A. Warren and M. Khogali, *Assessment of Desertification and Drought in the Sudano-Sahelian Region 1985–1991* (New York: United Nations Sudano-Sahelian Office and United Nations Development Programme, 1993).
20. D.S.G. Thomas, "Sandstorm in a Teacup? Understanding Desertification," *Geographical Journal* 159, no. 3 (1993): 318–31. See also United National Environment Programme, *World Atlas of Desertification* (London: Edward Arnold, 1992).
21. D.S.G. Thomas, note 20 above, page 328.
22. P.H. Raven, "Biological Resources and Global Stability," in S. Kawano, J.H. Connell, and T. Hidaka, eds., *Evolution and Coadaptation in Biotic Communities* (Tokyo: University of Tokyo, 1988), 16–23.
23. W.V. Reid, "How Many Species Will There Be?" in T.C. Whitmore and J.A. Sayers, note 15 above, table 3.2.
24. United Nations Centre for Human Settlements, note 2 above, table 6. See also G. McGranahan and J. Songsore, "Wealth, Health, and the Urban Household: Weighing Environmental Burdens in Accra, Jakarta, and Sao Paulo," *Environment*, July/August 1994, 4.
25. World Resources Institute, note 11 above, pages 362–65. See also K.R. Smith, "Air Pollution: Assessing Total Exposure in Developing Countries," *Environment*, December 1988, 16.
26. World Resources Institute, note 11 above, page 201.
27. United Nations Development Programme, Environment & Natural Resources Group, *The Urban Environment in Developing Countries* (New York: United Nations, 1992), 29.
28. D. Satterthwaite, "The Impact on Health of Urban Environments," *Environment and Urbanization* 5, no. 2 (1993): 87–111. For another side of this, see J. Briscoe, "When the Cup is Half Full: Improving Water and Sanitation Services in the Developing World," *Environment*, May 1993, 6.
29. D. Wilhite and M. Glantz, "Understanding the Drought Phenomenon: The Role of Definitions," in D. Wilhite and W. Easterling, eds., *Planning for Drought* (London: Westview Press, 1987), 1130.
30. A.B. Oguntola and J.S. Oguntoyinbo, "Urban Flooding in Ibadan: A Diagnosis of the Problem" *Urban Ecology* 7 (1982): 39–46.

31. T.R. Odhiambo, "Insect Pests," in E.S. Ayensu and J. Marton-Lefevre, eds., *Proceedings of the Symposium on the State of Biology in Africa* (Washington, D.C.: International Biosciences Network, ICSU [International Council of Scientific Unions]), 115.
32. A.J. Dommen, *Innovation in African Agriculture* (Boulder, Colo.: Westview Press, 1988), 115–17.
33. B.L. Turner II, G. Hyden, and R. Kates, eds., *Population Growth and Agricultural Change in Africa* (Gainesville, Fla.: University Press of Florida, 1993).
34. Ibid., page 422. See also M. Mortimore and M. Tiffen, "Population Growth and a Sustainable Environment: The Machakos Story," *Environment*, October 1994, 10.
35. R.A. Cline-Cole, H.A.C. Main, and J.E. Nichol, "On Fuelwood Consumption, Population Dynamics and Deforestation in Africa," *World Development* 18 (1990): 522–23. The conventional view is presented in E. Eckholm, G. Foley G. Barnard, and L. Timberlake, *Firewood: The Energy Crisis that Won't Go Away* (London: Earthscan, 1994); and E. Eckholm and L. Brown, *Spreading Deserts: The Hand of Man*, Worldwatch Paper 13 (Washington, D.C.: Worldwatch Institute, 1977).
36. H.P. Andersen, "Land Use Intensification and Landscape Ecological Changes in Budondo Sub-County, Uganda," (Master's thesis, University of Oslo, 1993); and T.A. Benjaminsen, "Fuelwood and Desertification: Sahel Orthodoxies Discussed on the Basis of Field Data from the Gourma Region of Mali," *Geoforum* 24, no. 4 (1993): 397–409.
37. World Commission on Environment and Development, *Our Common Future* (New York: Oxford University Press, 1987), 46.
38. World Bank, note 6 above, pages 31–32.
39. A.O. Hirschman, *Exit, Voice and Loyalty* (Cambridge, Mass.: Harvard University Press, 1990); see also G. Hyden, *No Shortcuts in Progress: African Development Management in Perspective* (London: Heineman, 1983).
40. World Resources Institute, note 11 above, page 316.
41. See R.L. Paarlberg, "The Politics of Agricultural Resource Abuse," *Environment*, October 1994, 6.
42. A.L. Mabogunje, *Perspective on Urban Land and Urban Management Policies in Sub-Saharan Africa*, Technical Paper No. 196 (Washington, D.C.: World Bank, 1992), 15–22.
43. World Bank, note 3 above, page 12.
44. P. Richards, *Indigenous Agricultural Revolution: Ecology and Food Production in West Africa* (London: Hutchinson, 1985).
45. J.A. Tobey, "Toward a Global Effort to Protect the Earth's biological Diversity," *World Development* 12, no. 12 (1993): 1931–45.
46. World Resources Institute, note 11 above, page 30. See also the articles in the January/February 1995 issue of *Environment*. ■

Questions

1. Name three types of environmental damages in sub-Saharan Africa.

2. Based on current statistics what percent of species in sub-Saharan Africa may become extinct from 1975 to 2015?

3. What three countries are ranked in the top 50 countries in terms of global greenhouse emissions?

Answers are at the back of the book.

8

In September of 1994, the third United Nations population conference was held. It was called the International Conference on Population and Development (ICPD). Its approach was gender-sensitive and covered such areas as sustainable human development, family planning in regards to population control, socioeconomic development, education, the environment, and migration. The conference attracted participants from over 150 countries who were part of the governmental delegations, international agencies, academia, feminists, environmentalists, and the press. During the conference, the official delegations negotiated a 16-chapter, 113-page World Programme of Action (WPOA) that set guidelines for attaining population stabilization, education of females, and the setting up of a universal high-quality health care system that would improve child survival and instigate reproductive health care services. The final agreements focused on sustainable development and gives priority to reproductive health, women's empowerment, and reproductive rights. Of the $17 billion allotted for population endeavors, $10 billion is designated for family planning. Overall, the Cairo conference provided a model for the management of global issues.

Women, Politics, and Global Management

Lincoln C. Chen, Winifred M. Fitzgerald, and Lisa Bates

The third United Nations population conference, held in Cairo last September, was a watershed global event. The International Conference on Population and Development (ICPD) succeeded in both shifting concern about world demographics into a gender-sensitive, peoples-centered approach of sustainable human development and propelling sensitive and ideologically charged population issues into the public domain.[1] Even more important, ICPD was a landmark in the management of complex global problems, such as population, through an imperfect international institutional machinery. The international consensus in Cairo, summarized in a World Programme of Action, was, despite fierce opposition to abortion and other sensitive subjects, a monumental achievement.

ICPD was the third decennial UN-sponsored world population conference. The first, held in Bucharest in 1974, saw the North and the South become polarized over the importance of demographics relative to other development concerns.[2] Led by the United States, Northern countries proposed vigorous family planning (contraception) programs to control rapid population growth, while many Southern governments, led by China and India, argued that, to attain global equity, higher priority should be given to socioeconomic development. Southern countries rejected the Northern assertions and insisted that the key to slowing population growth was through a more equitable distribution of resources between North and South. The South's assertions were captured in the phrase "development is the best contraceptive."

By the second conference, held in 1984 in Mexico City, the North-South dynamics had changed dramatically. After a decade of the fastest population growth in their histories, most Southern countries had adopted antinatalist policies, usually in the context of health and social development objectives. The United States had also switched its position.

Environment, Vol. 37, No. 1, pp. 5–33, January/February 1995. Reprinted with permission of the Helen Dwight Reid Educational Foundation. Published by Heldref Publications, 1319 Eighteenth St., NW., Washington, DC 20036-1802.

Under domestic political pressure from conservative and religious groups, the U.S. delegation opposed abortion and stated that demographic factors were "neutral," arguing that private markets would solve many population problems. In support of its position, the United States government withdrew financial support to several international organizations, such as the International Planned Parenthood Federation (IPPF) and the United Nations Population Fund.[3]

Given this history, few could have predicted the outcome at Cairo. Ten years after Mexico City, a broad and inclusive women-centered approach to population policies as part of sustainable human development was able to attract remarkable consensus in Cairo. Southern countries, after a "lost decade" of development associated with indebtedness and structural adjustment programs, were concerned over the dwindling concessional resource transfers from the North. They continued to view population both as an independent problem and as an important component of socio-economic development.[4] As for the United States, the Clinton administration was more supportive of environmental and women's issues than the previous administration and had reversed the Mexico City Policy after coming to office.[5] For Cairo, the United States supported a global consensus through active engagement with nongovernmental organizations (NGOs) and women's health groups.[6]

ICPD was structured along two parallel processes: official negotiations over the World Programme of Action and the NGO Forum for nonofficial participants. More than 1,500 organizations were represented by the estimated 10,000 participants who flocked to Cairo for the nine-day meetings.[7] The simultaneous processes drew participants from more than 150 countries who were part of the governmental delegations, international agencies, NGOs, academia, and the press. Although the battle over abortion received significant attention from the media, many other population-related issues were addressed, such as development, education, health, family planning, the environment, and migration.

At Cairo, certain North-South tensions evident at the Bucharest and Mexico City conferences appeared in the context of only a few issues, such as international migration and volume of foreign aid commitments. Furthermore, the vigorous dissent of the Holy See and extremist Islamic groups galvanized participants to leap across traditional divisions and to nurture common ground among most official and NGO groups.

During ICPD, official delegations negotiated the 16-chapter, 113-page World Programme of Action (WPOA)[8] that sets out recommendations for achieving population stabilization, educating girls and women, improving child survival and health, and providing universal access to high-quality reproductive health care services. As a UN statement about population problems and approaches, WPOA is not a treaty or a convention and thus has no legal force. However, it can become a powerful instrument for promoting national compliance through actions by citizenry and NGOs to hold governments accountable for pledges made and for attaining international standards.[9]

Before the negotiations at Cairo, the participating governments had agreed on about 92 percent of the WPOA text during three UN preparatory committee (PrepCom) meetings.[10] Left for finalization at Cairo were the more contentious issues, such as reproductive rights, sexuality and family structures, and abortion. Negotiations over these were not easy; indeed, there were times when consensus appeared beyond reach. Among the difficulties hampering the negotiations was the simultaneous use of the five official UN languages. Translating ambiguities and subtle nuances was a tedious and sometimes troublesome chore that occasionally proved helpful in reaching compromises on acceptable language.

A Leap to Women's Priorities

The final World Programme of Action that was agreed on at Cairo is a major departure from previous mainstream discourse on population and development issues; it shifted the focus from population control and demographic targets to sustainable development that gives priority to reproductive health, women's empowerment, and reproductive rights.[11] "Reproductive health" is a term that encompasses health issues associated with human sexuality and reproduction: safe and effective contraception, maternal and child health and nutrition, safe obstetrical practices (safe motherhood), and control of reproductive tract infections, including sexually transmitted diseases and HIV/AIDS.[12] "Women's empowerment" refers to improving the status of women and achieving gender equity through education and economic opportunities, thereby allowing women to exercise greater control over their reproductive and productive lives. "Reproductive rights" can be viewed in terms of the power and resources that enable individuals and couples to make informed and safe decisions about their reproductive health.[13]

Moving beyond the customary emphasis on family planning, the Cairo agenda focused on an expanded range of high-quality reproductive health services responsive to the needs of clients, male participation in family planning, and male responsibility in sexual health and childrearing. Other social concerns were adolescent sexuality and violence against women.

The consensus achieved on advancing the health and safety of human reproduction and on promoting gender equity and empowerment was remarkable in light of the controversy over abortion, adolescent sexuality, and the definition of "family." In some instances, fierce resistance to the concept of women's rights was met with an equally defiant resolve to defend those rights.[14] Some religious groups voiced concern that "reproductive health" could become an umbrella term encompassing and condoning abortion or the promotion of unsupervised sex education for adolescents. They also argued that certain terminology could legitimize family structures beyond legal heterosexual unions. Reproductive rights were considered threatening in terms of empowering individuals to make their reproductive decisions without respecting the obligations and responsibilities of married couples and the community, including the laws and cultures of diverse nation states.[15]

Many terminology problems that persisted during the negotiations were resolved through adroit rewording. Often, successful negotiations on thorny issues were achieved by inserting qualifying language that respected the laws, religions, and cultural values of diverse countries; for example, by highlighting the statement "in no case should abortion be promoted as a method of family planning"; by changing "fertility regulation"—which could be interpreted to include abortion—to "regulation of fertility"; and by more carefully defining "adolescent sexuality" and "household structure."[16]

The Holy See's controversial role at Cairo was widely reported in the media. There were, however, several aspects of WPOA the Holy See did support, such as the broader population and development approach, the emphasis on the role of the family, and language against coercive practices. After tough negotiations and the fine-tuned editing of certain passages, the Holy See joined the consensus on several chapters of the document. Women's health activists were satisfied that sexual and reproductive health had been brought into the public domain and were addressed in the final document. ICPD also succeeded in highlighting long-neglected issues of adolescent needs and the roles and responsibilities of men.

Achieving global population stabilization was a broadly endorsed goal, but the paucity of debate at Cairo over its importance was markedly unlike the attention during the 1980s that focused on the causative role of population growth in economic development.[17] During the three PrepCom meetings leading up to ICPD, disagreements were largely settled between so-called population alarmists who advocated vigorous targeted action to slow population growth and those who argued that human welfare, not demographic targets, should be the focus of population programs. The two sides of this debate found common ground in the emphasis at Cairo on high-quality reproductive health services. For the "alarmists," the emphasis on a broader reproductive health approach supported the importance of birth control. These advocates, while concerned that resources not be diverted from family planning, could hardly object to high-quality, broadly inclusive services. For the human welfare advocates, the emphasis on health, individual welfare, and individual rights to be stressed in population programs satisfied their concerns to avoid an excessive focus on demographics that might result in coercive or abusive practices or that might deemphasize critical social development issues. Furthermore, the late intervention by conservative Christian and extremist Islamic forces accelerated the "patching up" of unresolved differences in the interest of meeting greater threats.[18]

Several other population-related issues also received comparatively less attention. Environmental quality and resource conservation, the potential role of demographics on political stability and ethnic strife, and urbanization were hardly discussed. Although structural global economy issues linking population to North-South disparities in economic development—especially mass poverty and the consequences of structural adjustment policies—were raised by some Third World women's groups, they were omitted from the official exchange. It is unclear why these issues received so little attention; it might have been because more than 92 percent of the document had been agreed on prior to the conference, time was insufficient due to the abortion controversy, or delegates had decided to abandon disagreements to represent a unified opposition to religious attacks.

One contentious issue requiring lengthy negotiations to resolve was the reunification of families separated as a result of international migration, a matter covered in other international conventions such as the Convention on Child Rights. Some Southern countries advocated that such reunification should be recognized as a "right." Industrialized countries—notably "receiving countries," such as the United States, Canada, and many European countries—resisted the inclusion of language that implied such obligations in their domestic efforts to control immigration. Delegations from the receiving countries succeeded in changing the text of WPOA on this topic such that it urges countries "to promote" family reunification, without making it an absolute right.[19]

Women's NGOs: Influential and Vital

The most striking aspect of ICPD was the high level of participation by the nongovernmental community, building upon a process that characterized the Rio Earth Summit in 1992 and the Vienna Human Rights Conference in 1993.[20] ICPD attracted a diverse group of participants: population control advocates, feminist and health activists, academics and researchers, environmentalists, and religious groups. Some NGO participants were invited as members of their countries' official delegations. But unlike the official and NGO processes at the Rio conference, which were constrained by logistical barriers caused by the physical distance between the official conference site and the NGO Forum, these processes in Cairo were within walking distance of each other. This proximity allowed participants to move freely back and forth, promoting tremendous interconnectedness between the two processes.

The most visible and best organized in the NGO community were the women's health groups, which conducted a well-planned and well-coordinated advocacy effort.[21] A Women's Caucus coordinated women's NGO interactions, monitored the positions and negotiations of all delegations, advocated gender-sensitive positions, and proposed alternative language to bridge negotiation gaps.

Many women's groups worldwide had begun two years earlier to lay the groundwork for Cairo. In 1992, for example, women activists from the North and South developed and distributed the "Women's Declaration on Population Policies," which was reviewed, modified, and finalized by more than 100 women's organizations. It was additionally signed and endorsed by more than 2,500 individuals and organizations from over 110 countries.[22] In addition, networks of women from North and South as well as the Women's Caucus were widely acknowledged by the media, delegates, academics, and NGOs to have had a significant influence at PrepCom II in May 1993 by persuading government delegations to address reproductive health and rights, access to safe abortion, sexuality and gender equity, and broader socioeconomic development. In January 1994, about 200 women activists gathered in Rio de Janeiro to create a 21-point statement and to develop strategies for ensuring that women's perspectives and experiences were included at ICPD.[23] These strategies to influence and inform the public, official delegations, and the media were fine-tuned and applied throughout the process leading up to and at the Cairo conference itself. These efforts to harmonize diverse women's positions were vital for fostering North-South solidarity among women's NGOs during the Cairo deliberations.[24]

Conspicuously muted at the Cairo conference were tensions separating women's health groups from environmental scientists and activists—tensions that had been so prevalent at Rio. This relative quiescence can be partially attributed to concerted efforts by women's groups to seek common ground rather than to focus on their differences.[25]

Rather than targeting economic or political issues, the women's groups at Cairo focused on population issues. Until then, population had not been the highest priority issue of the global feminist movement, except perhaps in the United States, where the abortion debate is intense.[26] But over the past two decades, the international women's movement has increasingly recognized that sexuality and reproduction are critical issues in the struggle for gender equity. Without having control of their sexuality and reproduction, women argue, they will never achieve economic, social, and political equality.

Women have also felt compelled to become engaged in the population debate to redress the egregious abuses and failings they see in population policies and programs; among these are overemphasis on demographic goals at the expense of individual welfare, bureaucratic approaches to family planning that fail to address women's broader health needs, coercive practices, and no choice of contraceptive methods or an over-reliance on nonreversible methods.[27] Thus, even though the Social Summit in Copenhagen and the Women's Conference in Beijing are less than one year away, women's groups invested much time and effort to influence the Cairo proceedings and were tremendously successful.

Despite these successes, however, some NGOs, especially those from developing countries, claimed their views were overshadowed by those of well-organized lobbyists from the North.[28] Some groups argued that the burdens of limiting population growth still fall on the poor and that WPOA will result in conditions being attached to foreign aid for developing countries.

The "Unholy Alliance"

Even before the Cairo conference, the Holy See and some Islamic countries had criticized the draft WPOA—especially the sections dealing with abortion, and reproductive and sexual health and rights—in what the media coined as an "unholy alliance." That criticism escalated with conservative Islamic groups calling for a boycott of the conference and even threatening the safety of ICPD participants.

The Holy See's 15-member delegation fiercely attacked the draft document's recommendations on abortion, enlisting the support of several national delegations, including those from Argentina, Benin, Ecuador, the Philippines, Malta, Mauritius, Poland, and Slovakia. Nongovernmental "pro-life" activists and scientists, who advocated natural methods of birth control, also supported the Holy See's stance. Islamic criticism was articulated by Muslim scholars at Al-Ahzar University, the world's oldest Islamic university. In addition to abortion, their critique focused on the lack of attention given to poverty, the definition of family, and the treatment of extramarital sexuality, especially among adolescents.

To understand these religious concerns, it is necessary to understand how patriarchy and politics play into the issues involved. Some argue that the Holy See and fundamentalist Islam are basically patriarchal in their power structures and the norms they espouse. At Cairo, these religious groups resisted the strong women's agenda, underlined by feminist language, and succeeded in polarizing various participants over different social values and ethical norms. The divisions at Cairo were therefore more ideological and gender-based than religious or geographical.

The politics of the debate differed for the Holy See and Islamic groups. ICPD was held in Cairo, where longstanding anti-UN and anti-Western feelings in the Arab world had been exacerbated by the Gulf War and the incomplete peace accord between the Israelis and the Palestinians. Severe poverty and deeply entrenched social alienation characterize the development situation of the vast majority of the people living in countries governed by secular, Western-oriented elites. The Islamic arguments, therefore, could be interpreted as being about population issues more than as being a criticism of the Western modernization strategy of the secular governments in power in the region, such as that of Egypt. Recognizing the domestic implication of this threat, the Egyptian government went to great lengths to host a successful conference by ensuring the security of participants and enlisting the support of the Chief Cleric in Egypt (the Grand Mufti), who delivered a message of tolerance to Forum participants. Some argue that without the Egyptian government's strong political commitment, the conference could have been easily derailed.[29]

The Holy See's delegation at Cairo, after staking its position with moral absolutism and being overwhelmingly rejected by both official delegations and NGO groups, left the conference with its principles intact but its image seriously tarnished. By taking its dogmatic and unyielding stance, the Holy See has prompted queries about its status as a UN nonmember state and a permanent observer. For example, questions have been raised about why the Holy See has the privileged status that other religions do not.

Unlike after the Rio conference, the United States emerged from Cairo as a strong leader. Led by Undersecretary of State Timothy Wirth, the U.S. delegation played a constructive, consensus-building role during ICPD. Despite some criticism that the United States was leading "imperialist" attempts to impose homosexual and abortion rights on demand and to promote sexual promiscuity among teenagers, anti-U.S. rhetoric was generally absent. The U.S. delegation gained credibility with and the trust of both the NGO community and Southern delegates for several reasons. First, throughout the Cairo process, the U.S. delegation—more than half of which consisted of representatives from NGOs and the private sector, most of whom were women—sought the input and involvement of NGOs, especially women's health groups. Second, by bridging the North-South divide, the women's alliance, as described earlier, had created an environment that made it easier for delegates to encourage an exchange of ideas and to seek compromises among many Northern and Southern delegates. Third, the United States acknowledged politically sensitive issues, such as the "right to development" and the problem of overconsumption in industrialized countries. Finally, the U.S. government agreed to increases of financial commitments, albeit modest, for population programs overseas.

WPOA: An Unfinished Agenda

As a political statement and the product of global negotiations, WPOA is a lengthy and sometimes redundant document. Nonetheless, what emerges from its pages is a refreshingly new direction for population policies emphasizing reproductive health, the empowerment of women, and an endorsement of reproduction as part of universal human rights. As such, WPOA is a seminal document in dramatically shifting the central focus of population studies.

The challenge now is for governments, international organizations, and NGOs to translate WPOA's rhetoric into action. Implementation at the local and national levels will be difficult and cumbersome because it will require the support of the UN system, extensive bilateral and multilateral funding, committed follow-up by NGOs, and the engagement of academia, business, and the media. Many women's NGOs left Cairo with strategic plans to monitor the performance of donors and governments in implementing the WPOA—demanding transparency in donor and governmental activities—and to hold these accountable for the pledges they have made. How these follow-up activities will be executed is unclear. What seems likely, however, is that the next decade will witness considerable public evaluation and debate over the direction of all implementation efforts.

WPOA estimates that resource requirements to implement population activities will increase to $17 billion by the year 2000. Of this total, $10 billion is slated for family planning and $7 billion for reproductive health, sexually transmitted diseases and HIV/AIDS, and social investments such as the education of girls and women. WPOA proposes that developing countries meet about two-thirds of these required resources. Foreign aid, currently 25 percent of the total population expenditures, is suggested to increase to 33 percent of the total. This will require an increase of nearly six-fold in population aid by the year 2000: from about $1 billion to nearly $6 billion.[30]

In a larger context, the Cairo conference can be seen to have provided a model for the planetary management of complex global problems. Increasingly, the scale of some of these problems overwhelms the global community's institutional capacity to address them. Transnational problems are beyond the capacity of individual states to resolve. Thus, as demonstrated by Cairo, we are likely to witness a steady stream of piecemeal negotiations on certain global problems pressingly important at specific times, with the involvement of governments and concerned NGOs, presented to the public through the media. While these processes will be fluid and dynamic, with unpredictable outcomes, they can be viewed as part of a process of global democratization with the increasing participation of concerned global citizens through their governments and private associations. Cairo made a major contribution toward advancing international understanding and global problem solving.

Acknowledgments

We gratefully acknowledge the support of the John D. and Catherine T. MacArthur Foundation, the Global Stewardship Initiative of the Pew Charitable Trusts, and the Swedish International Development Authority. The views in this article are the authors' and not necessarily the views of their affiliated organizations.

Notes

1. The media coverage of ICPD was extensive. The Communications Consortium Media Center in Washington, D.C., has collected press clips on the coverage of ICPD and is preparing a videotape of network news coverage.
2. J. L. Finkle and B. B. Crane, "The Politics of Bucharest: Population, Development, and the New International Economic Order," *Population and Development Review* 1 (1975): 87–114.
3. Sharon L. Camp, "The Politics of U.S. Population Assistance," in Laurie Mazur, ed., *Beyond the Numbers* (Washington, D.C.: Island Press, 1994), 122–34.
4. G. Sen, "Development, Population, and the Environment: A Search for Balance," in G. Sen, A. Germain, and L. C. Chen, *Population Policies Reconsidered: Health, Empowerment, and Rights* (Cambridge, Mass.: Harvard University Press, 1994), 68.
5. See Camp, note 3 above.
6. "Cairo Conference Showcases Work of NGOs," *Christian Science Monitor*, 8 September 1994.
7. *POPULI 21*, no. 9 (October, 1994): 4.
8. United Nations, *World Programme of Action: International Conference on Population and Development* (New York: United Nations Population Fund, September 1994).
9. R. Boland, S. Rao, and G. Zeidenstein, "Honoring Human Rights in Population Policies: From Declaration to Action," in Sen, Germain, and Chen, note 4 above, pages 89–105.
10. The first PrepCom meeting, held in March 1991, defined the objectives and themes of ICPD; the second meeting in May 1993 was to agree on the form and substance of the final document to be adopted at Cairo; at the third meeting in April 1994, delegates attempted to agree on as much of the draft WPOA as possible.
11. R. Cassen, *Population and Development: Old Debates, New Conclusions*, Policy Perspectives no. 19 (Washington, D.C.: Overseas Development Council, 1994).
12. A. Germain, S. Nowrojee, and H. H. Pyne, "Setting a New Agenda: Sexual and Reproductive Health and Rights," in Sen, Germain, and Chen, note 4 above.
13. S. Corrêa and R. Petchesky, "Reproductive and Sexual Rights: A Feminist Perspective," in Sen, Germain, and Chen, note 4 above, page 107. See also S. Corrêa, *Population and Reproductive Rights, Feminist Perspectives from the South* (London: Zed Press, 1994).
14. Carmen Barroso, the John D. and Catherine T. MacArthur Foundation, personal communication with the authors, September 1994.
15. P. Chasek, *Earth Negotiations Bulletin* 6, no. 39 (Washington, D.C.: International Institute for Sustainable Development, 14 September 1994).
16. Ibid.
17. National Academy of Sciences, *Population Growth and Economic Development: Policy Questions* (Washington, D.C.: National Academy Press, 1986).
18. Susan Sechler, the Pew Global Stewardship Initiative, personal communication with the authors, September 1994.
19. The final text in Chapter X, "International Migration," states that "all Governments, particularly those of receiving countries, must recognize the vital importance of family reunification and promote its integration into their national legislation in order to ensure the protection of

the unity of the families of documented migrants." *World Programme of Action*, note 8 above.

20. Y. Kakabadse and S. Burns, "Movers and Shapers: NGOs in International Affairs," *International Perspectives on Sustainability* (Washington, D.C.: World Resources Institute, May 1994). See also *Environment*, October 1992, pages 6, 12 for further discussions of UNCED.
21. Many issues related to women's health—population policy, family planning, sexual and reproductive health, violence against women, and others—are part of the women's health movement, involving organizations ranging from national feminist NGOs to grassroots groups of varying sizes. Like all social movements, the women's health movement is shaped by many political, cultural, and socioeconomic factors. As such, the movement is by no means homogeneous, and there were indeed many debates at Cairo. For the purposes of this article, discussion of the role of the international women's movement at ICPD refers to the groups that did reach agreement. For further information on the women's health movement, see C. Garcia-Moreno and A. Claro, "Challenges from the Women's Health Movement: Women's Rights Versus Population Control," in Sen, Germain, Chen, note 4 above, pages 46–61. See also "Women's Groups Coalesce in Cairo," *Christian Science Monitor*, 8 September 1994.
22. In September 1992, women's health advocates representing women's networks in Asia, Africa, Latin America, the Caribbean, the United States, and Western Europe met to discuss how women's perspectives might best be included during the preparations for the 1994 ICPD and during the conference itself. The group suggested that a strong positive statement from women worldwide would make a unique contribution to reshaping the population agenda to better ensure reproductive health and rights. The declaration outlines fundamental ethical principles, minimum program requirements in the design and implementation of policies and programs, and the requisite conditions for women to control their own sexuality and reproductive health. See A. Germain, S. Nowrojee, and H. H. Pyne, "Setting a New Agenda: Sexual and Reproductive Health and Rights," in Sen, Germain, and Chen, note 4 above, pages 31–34.
23. International Women's Health Coalition and Citizenship, Studies, Information, Action, *Reproductive Health and Justice* report of the International Women's Health Conference (Rio de Janeiro, Brazil, 24–28 January 1994).
24. Garcia-Moreno and Claro, note 21 above.
25. Many environmentalists argue that population growth is a major cause of environmental degradation, while many women's health advocates argue that population growth, by itself, is not a major contributor to global environmental problems. In the past, some women's health groups feared that sounding a population "alarm" could legitimize a range of family planning abuses against women. See Sen, note 4 above, pages 63–73. See also S. A. Cohen, "The Road from Rio to Cairo: Toward a Common Agenda," *International Family Planning Perspectives* 19, no. 2 (Washington, D.C.: Alan Guttmacher Institute, June 1993).
26. Carmen Barroso, personal communication with the authors, November 1994.
27. Sen, note 4 above, pages 63–73. See also R. Dixon-Mueller, *Population Policy and Women's Rights: Transforming Reproductive Choice* (Westport, Conn.: Praeger Publishers, 1993), and B. Hartmann, *Reproductive Rights and Wrongs: The Global Politics of Population Control and Contraceptive Choice* (New York: Harper & Row, 1987).
28. See, for example, three articles from India: "Third World Feels Betrayed," *Telegraph*, 29 September 1994; "Foetal Distractions: Reductionisms by the West Are Threatening the Rights of Third World Women," *Telegraph*, 7 October 1994; "What Cairo Did Not Discuss," *Down to Earth*, 15 October 1994.
29. Barbara Ibrahim, Population Council, Cairo, Egypt, personal communication with the authors, September 1994.
30. *World Programme of Action*, note 8 above. ■

Questions

1. What does the term "reproductive rights" refer to?

2. What common ground did the "population alarmists" and the "human welfare advocates" find?

3. What two critical issues in the women's movement for gender equity are critical, and without which women will never achieve economic, social, and political equality?

Answers are at the back of the book.

9

Haiti is often called an ecological disaster. Barren mountains, eroded fields, arid countrysides, and eroded gorges exposing topsoil and rocks are just a few examples that characterize a nation that is over 90% deforested. The population of Haiti is increasing about 2% annually. Erosion has destroyed over half of the good cropland. Rural illiteracy is almost 90% and the average income is only a few hundred dollars per year. In order for Haitians to survive the collapse of their country's infrastructure, they have to understand that their problems are not only political but environmental as well. A long term solution would include political transformation followed by a decrease in illiteracy rates. A national health care system would need to be implemented with an emphasis in family planning. The International Planned Parenthood Federation has emphasized that women's literacy and empowerment are vital to the stabilization of a population.

Haiti on the Brink of Ecocide

Dwight Worker

Haitian boat people are environmental refugees, fleeing an island being devastated down to its very roots. Poverty and politics created this crisis, but the environment has paid the price.

As I rush down the street to avoid the beggars, a woman steps directly in front of me. She pushes her child into my arms. I grasp reflexively, and find myself holding a small naked boy. We look at each other. Dark, tired eyes. He begins to cry silently. He has skinny arms and legs, a distended stomach, blotchy skin, and patchy hair. He has kwashiokor, a disease of malnutrition— literally "the disease the first baby gets when the second one comes." Then the woman shouts and aggressively pushes her upheld hand into my stomach. Is she saying that I am fat? My friends wouldn't think so. Yet, compared to her, I am fat.

I have promised myself 10 times not to give more money to beggars. It doesn't do any good. Once you give to one, 50 more beggars will hound you in the streets. I hand the child back to the woman with a few "gourds," maybe 25 cents in the local currency. Then I walk away fast, before any other beggars can surround me. No, I am not in Africa.

If you want an "ecotour" to see an environmental disaster, you need not go to Africa or India. Save your money and look in our back yard at Haiti. It is our local "worst-possible-case" of eco-catastrophe.

I first came to Haiti in 1970, when "Papa Doc" Francois Duvalier had his "PRESIDENT FOR LIFE" sign over his White House. I arrived as the proverbially naive, optimistic college graduate, and I was staggered by the poverty and political oppression. The poverty was obvious—hungry children, beggars and the diseases of malnutrition. The oppression was more subtle, seen in the cautious looks on people's faces. At dinner, if you wanted to clear the table of Haitians, all you had to do was ask them for an opinion of Papa Doc Duvalier. In the streets, people were constantly pleading with me to take them "away from here" before they were killed by his private guards, the "Tonton Maucaute."

My ultimate trauma came at the post office in Port-au-Prince. I wanted to send some beautiful Haitian wood carvings to my parents. In front of the post, I was so besieged by a crowd of screaming desperate beggars that I literally could not move. Most of their bodies had gross physical injuries. Missing fingers, hands, arms and legs. Some had letters and even words scarred onto their faces. Moist pink flesh protruded where eyes used to be. Then I

Reprinted with permission from *E/The Environmental Magazine*. Subscription Department: P.O. Box 699, Mt. Morris, IL 61054 (815/734-1242). Subscriptions are $20/year.

felt something grip my ankle. One man with stumps for limbs had scooted up to me on a rolling cart and hooked his chin and shoulder around my ankle. He began pleading in Creole for anything. He didn't even have a hand for me to give to. I broke loose of them and, without mailing my package, ran back to my hotel, where I breathlessly told a German resident of my experience. "Oh, you have seen the 'mutilates,'" he said. "Mutilates?" "Yes. The ones Duvalier has disfigured so badly that they can no longer live normal lives. But Duvalier lets them live so that others can see what will happen if they too try to resist."

When I return to Haiti in 1993 I decide to go to the post office in Port-au-Prince. A Haitian friend has come with me this time "just in case." My camera is ready. But when we get there, I find no mob scene. Just some street vendors. I ask my friend where all the vendors are who used to sell the incredible wood carvings. "Not much of that anymore. No more good wood here for carving, and we cannot afford to import wood." "And where," I ask, "are all of the—mutilates—the victims of political oppression?" He shakes his head. "Oh, the army just kills us now. They don't bother with those details anymore. You see, we too have progress."

I have hired a driver to take me in his 4-wheel drive to Jean Rabel in Haiti's Northwest—the most deforested part of a deforested country. Our jeep creaks along the road at a walking pace as my driver dodges erosion gullies. He tells me that Haiti is celebrating its 150-year anniversary of non-stop kleptocracy—"government by the thieves." He says that there is no money for road repairs—or anything else. "The military has stolen all of it," he confides. "This road is returning to dirt. If the rains do come this year, they will close the road down indefinitely. Every year, everything gets worse. I know this is a fact. That is our history."

Columbus arrived in the Santa Maria on the north coast of the island of Hispanola (now shared by the Dominican Republic and Haiti) in 1492. An estimated 250,000 Arawak Indians lived here. Perhaps 100,000 lived in what is now Haiti. Deep forests covered over 75 percent of the land. Columbus called it "La Navidad." Because the Spaniards did not find much gold, they did not colonize Haiti. French buccaneers began preying upon the island in the 1600s. Settlers quickly followed, and they killed off almost all of the Arawak Indians with a combination of disease, massacre and overwork.

France gained complete control with the Treaty of Ryswick at the end of the 17th century, and Haiti soon became its richest colony. Their treatment of slaves was among the most brutal in the western hemisphere. Most slaves died within five years of importation, and over one million would die from work conditions and disease in the next century. In 1804, the slaves revolted, declared the country independent and renamed it "Haiti," the old Arawak name. But the new leaders quickly tried to restore the plantation system and recreate a "black-on-black" serfdom as brutal as the French system. The freed slaves deserted the plantations and fled into the mountains. The plantation system eventually failed, and since then the peasants have been dividing and subdividing their plots of farmland.

Today, land ownership in rural Haiti is among the most egalitarian in the Americas, unlike many other Latin American countries, where a tiny oligarchy controls much of the land. Yet, Haiti faces ecological disaster. The numbers are grim: Experts estimate Haiti to be from 90 to 97 percent deforested. (Wood exports from Haiti ended by 1900, so this deforestation can't be blamed on international exploitation, as it can be in many other parts of the world.) Haiti's population is now about seven million, growing at around two percent annually. Over half of the original good cropland of Haiti has been destroyed by erosion, and the farmers are destroying the remainder at about five percent a year. From 1950 to 1990, the per capita arable land dropped from about one acre to one-third of an acre per person.

The per capita income of the average Haitian in 1987 was $360 and declining. Rural illiteracy is close to 90 percent, among the highest in the world. And the average Haitian only gets 80 percent of their required calorie intake.

My driver pulls up to a river outside of the city of Port-au-Paix, on the north coast. This will be the only surface water that we will see for the next 50 miles. We descend a steep path to its edge. Hundreds of people in the river wash clothes, themselves and food. Children play and swim while mothers load their heads and backs with large containers of river water for drinking and cooking. My guide says that, during the drought, some of them must walk over 15 kilometers to get water here. In the rural villages, he adds, there used to be running water, some electricity, even phones. But no more. The wells are going dry, and the electricity and phones stopped working years ago. I look at everyone in the river and wonder if amoebic dysentery, intestinal parasites, cholera don't lie just around this river's bend...

We splash through the river and begin the climb up the mountains. More bone-jarring jolts as we wind through the passes. The countryside is arid, covered with cactus and parched bushes, with deeply eroded gorges exposing subsoil and rocks. In some places, the erosion gullies are 50 feet deep. The wind blows dust constantly.

I see several traces of smoke in the distance, so I ask the driver to stop. We hike through the erosion gorges until we come upon a man and his wife tending a charcoal pyre. His name is Jean Baptiste. He tells me he used to be a farmer, but no longer. It has not rained in five years and all of his fruit trees have died. He used to grow produce to sell in the city, but now he has cut up his dead fruit trees for charcoal, which he sells for money to buy food. If he didn't do it, he says, someone else would have come in the night and stolen his trees for wood. It never used to be like this, he says.

I offer Jean a few gourds of currency to photograph them making charcoal. He accepts gratefully. They take me to a pit that they have dug and lined with straw. There they stack wood until they have a pile shaped, ironically, like a large coffin. None of the wood is over three inches thick, so I ask them if the wood must be so small. "No. But there is nothing larger left." After the pile is about four feet high, they cover it with a thick coat of dirt. They punch a few holes in the dirt to allow the smoke to escape. Then they ignite the wood from below and let it smolder for a day. The heat drives off the volatile gases and leaves charcoal.

I look around and see no trees, not a one. I ask him where he got the wood for the pyre. He waves for me to follow. We cross a ravine and walk over a sun-parched field and stop at a hole with the remains of a tree stump and a machete. *They are digging out tree stumps with a machete to make charcoal.* He explains that they must search for tree stumps. But often, someone has already removed them. So they search for any roots that may be left.

The landscape is cratered with pits. There are almost no green plants remaining. Wherever a tree once stood, there is now a deep hole where someone took the stump and the roots. Roots that once held topsoil and moisture. It looks like something is deliberately removing all traces of life. And something is. I am looking at the ultimate environmental "scorched-Earth" policy. Except for strip mining, I can't think of a way to remove topsoil more quickly.

I ask Jean Baptiste if he thinks this is bad. "Terrible," he shakes his head. "But I have no choice. We have four children, and we must feed them today. It no longer rains here, so I cannot farm. What else is there to do?" Jean says that every three days he piles his bags of charcoal onto his three burros and takes them 10 miles to the nearest pickup point. There, a charcoal wholesaler buys it. With this money, he can buy enough beans and rice to feed his family for a few days. He used to take his burros all the way to Port-au-Paix and sell his charcoal retail for more money. But now, there is no water on the trail and so little grass that he would have to buy food for the burros for the trip. And he cannot afford that. The people who live along the trail now defend the meager brush and cactus on their land against anyone else's animals. As we prepare to leave, he points to the denuded hillside and says, "Yes, this is very bad. But when we are standing next to starvation, we have no choice."

We stop in the town of Jean Rabel for a few days. The market is full of loud women hawking food. The people appear well-fed. I do not see any hungry children or beggars. I ask my driver about the prosperity of Jean Rabel. "Yes," he agrees. "The stream still flows here, so the people can grow food. But downstream, it is dry. We pray that does not happen here."

My driver takes me to visit his uncle Michel. He is a neat, educated man, the principal at the local school. He graciously insists that I stay with him. He wants to see my camera. He tells me he is a photographer himself and invites me to visit his photo studio. As we walk, we are surrounded by enthusiastic children. I rub the boys' heads and they smile.

Michel shows me photographs at his studio, including some from when his grandfather was a boy that show Jean Rabel almost a century ago. The original marketplace, the first church, the river. I look at the river. It is wide and deep, big enough to hold 10 times the flow of the small stream that now flows through Jean Rabel. And on all the hillsides in these photos, I see trees. Trees and trees and trees. The hillsides are lush and forested.

I ask Michel where these hills are. "Right here," he answers. We step outside. He points to the hills on the other side of the stream. "There." I look at the brown rocky hills, then at the photos of the forested slopes. I must take his word that these are the same hills, for I can see no similarities whatsoever. In the old photos, I cannot see any rocks on the hillsides. And with my eyes right now, I cannot see one full tree on these hills.

I want to photograph some boat people preparing to leave Haiti. My driver suggests that we go west from Jean Rabel along the Atlantic coast. We find some men in a small boat along the shoreline. Their boat is filled with small branches of wood. My guide asks them where we can photograph some boat people getting ready to leave. Stony silence. I offer them money for information. One of them volunteers that no one who was planning on sailing to the U.S. would tell any stranger about it. Then they go back to their boat. We watch them sail up to some small mangroves growing in the shallow bay. They hack away at the inch-thick stalks and toss them into their boat. *They are cutting the mangroves beneath the water level to make charcoal.* There go the coastal estuaries.

For the next five days, wherever we travel, we see more of the same: completely barren hills, stripped of all topsoil, with smoke around us from the ubiquitous charcoal pits. Along the coastline, we find men in small boats hacking away at the mangroves. I feel that I am making two journeys here. The physical one is obvious: a journey into northwest Haiti, the most deforested and poorest area in the western hemisphere.

But my second journey is back in time. I am seeing the complete collapse of what the West calls "infrastructure." The roads, electricity, running water, phone service, even mail. I am journeying back into a vast, unplanned "demodernization" of a region. No one ever decided to do it. But it's happening with incredible rapidity. I wonder if this is to be the future of much of the world. Of the Somalias, Angolas, Bangladeshes, Indias, and yes, the Bosnias and El Salvadors too. Is this getting ever closer to home?

On my return, I stop at a CARE office in Gonaives. I meet an assistant director, whom I shall call Jacque. I comment that this office appears to spend a lot of its budget on new vehicles. He agrees and says that nowadays, they are mostly in the food delivery business. The roads are so bad and there is so much looting along the way that they must deliver it themselves. Sometimes they must give food to the military. They used to spend more time training people to farm, but now with the continuous drought and all the erosion, it is becoming futile to teach agriculture in a desert.

Jacque says that, in the northwest, people get over half their calories from imported food from the U.S. "Miami rice," he calls it. "How do they pay for it?" I ask. "They don't. It is aid. If it ever stops, hundreds of thousands will starve—maybe millions. And my family lives there, so I worry for them daily." Then he asks me, "Do you think the U.S. will stop sending food aid?" I tell him that I have no idea. "But what do you personally think U.S. policy to Haiti will be?" he probes.

I remain silent for a moment. I think the U.S. policy planners wish that the "Haiti problem" would just go away. The U.S. has no pressing economic needs in Haiti, not even its dirt-cheap labor pool. Clinton has broken his campaign promises to offer Haitian boat people asylum and better treatment, but African-Americans and others won't oppose him because of it. The "Haiti problem" may be embarrassing and awkward for the U.S., but is it strategic in the post-cold war era? Not at all.

Fidel Castro said recently that if the U.S. truly opened its borders, one billion people would immediately come running, sailing, flying, crawling, and swimming into the U.S. He's probably right. With open borders, perhaps half of all Haitians would opt to come here. But we don't want them. The U.S. wants to contain the problem in Haiti. And since

Haiti is an island with easy-to-patrol coastal borders, we can expect continued interception and "repatriation" of the Haitian boat people. Emigration is not going to solve Haiti's crisis.

So what about the embargo that the United States has imposed to force the Haitian military to return the reins of power to President Jean-Bertrand Aristide, now in exile in Washington, DC? The Haitian military says that it hurts the poor more than anyone else. Everything I have seen in Haiti supports that contention. You can be assured that when Haiti is down to its last barrel of oil or bag of rice, the Haitian military will have it.

If Aristide returns to Haiti without armed escort, the Haitian military will execute him. For Aristide to remain in power, he must have the support of an international military force, i.e. French, U.S., or U.N. troops to protect him. He will then need his own elite guard to protect him from the next coup. The small but influential middle class does fear Aristide, and the military surely has plans right now on how to dispose of him. The Haitians by themselves will not "restore" the democracy that they have never had.

Jacque brings me a cup of good coffee. "You have just been in the countryside. You have seen it. Tell me, what do you personally think should be done? What do you really think?"

What do I really think? I think that it could literally take centuries to heal the land and rebuild the topsoil. Geological time. I think that we are doing everything wrong in Haiti that we can possibly do wrong. I think that without the continuous food assistance programs, Haiti could experience a massive die-off. But as I think of Jacque's family in the northwest, I dare not tell him what I really think. It's too cold-blooded.

So what would we have to do to make me feel that Haitians have any long-term chance? First the Haitians would have to understand that their most serious problems are not just political, but environmental. Yes, the brutal kleptrocracy must end. But it will not help in the long-term if Aristide (a pronatalist Catholic priest, no less) or any other "democrat" is restored to power, unless they recognize the gravity of Haiti's eco-catastrophe.

After a political change in Haiti, the Haitians would have to end most illiteracy within the next 15 years. The World Bank Development Report on Haiti has called for universal access to primary education for all six-year-olds by the year 2000. They would also have to incorporate a national health system with family planning as an integral part of services. Paul Ehrlich, author of such books as *The Population Explosion*, points out that population control can work in poor countries. Take Sri Lanka and China, for example, which have lowered birth rates by improving women's rights, focusing on maternal health and infant survival and using peer pressure as a motivator. The International Planned Parenthood Federation has stressed that women's literacy and empowerment are essential to stabilizing populations.

Teaching and implementing sustainable agricultural techniques are essential as well. CARE and other groups in Haiti are training local field agents to teach others proper terracing, soil retention and irrigation techniques. These programs need to continue and grow.

Haiti needs to develop and rigidly enforce a national program to restore the topsoil and reforest the mountains. US AID, our foreign aid program, has funded the Agroforestry Outreach Project in Haiti to replant the hillsides with new species of fast-growing trees. They are teaching the farmers and charcoal makers to take an economic interest in allowing trees to grow for several years.

All goats and sheep must be banned. This may not be popular in Haitian culture, but it must be tried. Plant life won't reappear and replenish itself on areas already overgrazed by the "horned locusts."

Finally, Haiti must develop a long-term bridge with international agencies and private aid groups to finance these programs, and run them honestly—for once.

Will it happen? Not without international support. Already, according to the World Bank, two-thirds of all agricultural programs and three-fourths of all education are financed by international donor programs. The World Bank states that many of these programs are administered inefficiently, if not corruptly. But even so, these donor programs are the difference between life and death for many Haitians. Any reductions in them would be devastating.

Should the U.S. intervene? Given the level of environmental devastation in Haiti, there are no short-term solutions. True land restoration in Haiti could take centuries. Can the U.S. demonstrate this kind of commitment? Just look at the U.S. withdrawal from Somalia. Then keep in mind that Haiti is an international tar baby. Intervention will be costly, unrewarding and politically unpopular within the U.S.

But what is the alternative to intervention? The alternative is right in front of me, wherever I look. The bare mountains, the eroded fields, the hungry beggars waiting for me at the gate. Haiti is what happens when we have no long-term sustainable policies, and when we believe that problems are basically just political in nature, not environmental. The Haitians risk exposure, drowning and interdiction on the high seas because they feel that the risk of death by remaining in Haiti is even greater. They are truly environmental refugees. Without intervention, we can expect continued military brutality and increased prospects of mass starvation.

I gather up my things at the table and put my hand on Jacque's. "I cannot imagine the United States not sending food to the Haitians," I tell him. "I think the U.S. will not permit massive starvation so near to its borders. I think the U.S. will continue sending food to Haiti—for as long as it can.

As I get up to leave, he squeezes my hand. Then I walk quickly out the gate—before any beggars can intercept me. ■

Questions

1. In northwest Haiti, how do most of the people get over half their calories?

2. Name two categories in which most of Haiti's problems fall.

3. How long could true land restoration take in Haiti?

Answers are at the back of the book.

10

Environmental degradation, depletion of natural resources, and the shrinkage of yield-raising technologies have contributed to the reduction of food production in recent decades. With seafood, there is a limit enacted by nature as fishing exceeds capacity. Rangelands are being overgrazed at or beyond capacity worldwide. Water is also being depleted at an alarming rate. In the U.S. alone, more than one-fourth of cropland is irrigated, however, aquifer depletion will eventually put an end to irrigation. In 1992, the U.S. National Academy of Sciences and the Royal Society of London issued a report that expressed concern over continued population growth. Earth can only support a limited number of people and the food supply will be the dominant factor. Recognizing food scarcity as the primary threat to the future is an important step in working towards a solution. Governments can introduce family planning programs, which would decrease fertility rates, reduce illiteracy, and diminish poverty. This would also help to conserve the environment and the natural resources by protecting the soil and water, which in turn would help stabilize agriculture and help to protect and sustain the food supply.

Earth is Running Out of Room

Lester R. Brown

Food scarcity, not military aggression, is the principal threat to the planet's future.

The world is entering a new era, one in which it is far more difficult to expand food output. Many knew that this time would come eventually; that, at some point, the limits of the Earth's natural systems, cumulative effects of environmental degradation on cropland productivity, and shrinking backlog of yield-raising technologies would slow the record increase in food production of recent decades. Because no one knew exactly when or how this would happen, food prospects were debated widely. Now, several constraints are emerging simultaneously to slow that growth.

After nearly four decades of unprecedented expansion in both land-based and oceanic food supplies, the world is experiencing a massive loss of momentum. Between 1950 and 1984, grain production expanded 2.6-fold, outstripping population growth by a wide margin and raising the grain harvested per person by 40%. Growth in the fish catch was even more spectacular—a 4.6-fold increase between 1950 and 1989, thereby doubling seafood consumption per person. Together, these developments reduced hunger and malnutrition throughout the world, offering hope that these biblical scourges would be eliminated one day.

In recent years, these trends suddenly have been reversed. After expanding at three percent a year form 1950 to 1984, the growth in grain production has slowed abruptly, rising at scarcely one percent annually from 1984 until 1993. As a result, grain production per person fell 12% during this time.

With fish catch, it is not merely a slowing of growth, but a limit imposed by nature. From a high of 100,000,000 tons, believed to be close to the maximum oceanic fisheries can sustain, the catch has fluctuated between 96,000,000 and 98,000,000 tons. As a result, the 1993 per capita seafood catch was nine percent below that of 1988. Marine biologists at the United Nations Food and Agriculture Organization report that the 17 major oceanic fisheries are being fished at or beyond capacity and that nine are in a state of decline.

Rangelands, a major source of animal protein, also are under excessive pressure, being grazed at or beyond capacity on every continent. This means that rangeland production of beef and mutton may not increase much, if at all, in the future. Here, too, availability per person will decline indefinitely as population expands.

With both fisheries and rangelands being pressed to the limits of their carrying capacity, future growth in food demand can be satisfied only by expanding output from croplands. The increase in demand for food that was satisfied by three food systems must now be satisfied by one.

Until recently, grain output projections for the most part were simple extrapolations of trends. The past was a reliable guide to the future. However, in a world of limits, this is changing. In projecting food supply trends now, at least six new constraints must be taken into account:

- The backlog of unused agricultural technology is shrinking, leaving the more progressive farmers fewer agronomic options for expanding food output.
- Growing human demands are pressing against the limits of fisheries to supply seafood and rangelands to supply beef, mutton, and milk.
- Demands for water are nearing limits of the hydrological cycle to supply irrigation water in key food-growing regions.
- In many countries, the use of additional fertilizer on currently available crop varieties has little or no effect on yields.
- Nations that already are densely populated risk losing cropland when they begin to industrialize at a rate that exceeds the rise in land productivity, initiating a long-term decline in food production.
- Social disintegration by rapid population growth and environmental degradation often is undermining many national governments and their efforts to expand food production.

New Technologies are Not Enough

In terms of agricultural technology, the contrast between the middle of the 20th Century and today could not be more striking. When the 1950s began, a great deal of technology was waiting to be used. Except for irrigations, which goes back several thousand years, all the basic advances were made between 1840 and 1940. Chemist Justus von Liebig discovered in 1847 that all nutrients taken from the soil by crops could be replaced in mineral form. Biologist Gregor Mendel's work establishing the basic principles of heredity, which laid the groundwork for future crop breeding advances, was done in the 1860s. Hybrid corn varieties were commercialized in the U.S. during the 1920s, and dwarfing of wheat and rice plants in Japan to boost fertilizer responsiveness dates back a century.

These long-standing technologies have been enhanced and modified for wide use through agricultural research and exploited by farmers during the last four decades. Although new developments continue to appear, none promise to lead to quantum leaps in world food output. The relatively easy gains have been made. Moreover, public funding for international agricultural research has begun to decline. As a result, the more progressive farmers are looking over the shoulders of agricultural scientists seeking new yield-raising technologies, but discovering that they have less and less to offer. The pipeline has not run dry, but the flow has slowed to a trickle.

In Asia, rice crops on maximum-yield experimental plots have not increased for more than two decades. Some countries appear to be "hitting the wall" as their yields approach those on the research plots. Japan reached this point with a rice yield in 1984 at 4.7 tons per hectare (2.47 acres), a level it has been unable to top in nine harvests since then. South Korea, with similar growing conditions, may have run into the same barrier in 1988, when its rice yield stopped rising. Indonesia, with a crop that has increased little since 1988, may be the first tropical rice-growing nation to see its yield rise lose momentum. Other countries could hit the wall before the end of the century.

Farmers and policymakers search in vain for new advances, perhaps from biotechnology, that will lift food output quickly to a new level. However, biotechnology has not produced any yield-raising technologies that will lead to quantum jumps in output, nor do many researchers expect it to. Donald Duvick, for many years the director of research at Iowa-based Pioneer Hi-Bred International, one of the world's largest seed suppliers, makes this point all too clearly: "No breakthroughs are in sight. Biotechnology, while essential to progress, will not produce any sharp upward swings in yield potential except for isolated crops in certain situations."

The productivity of oceanic fisheries and rangelands, both natural systems, is determined by nature. It can be reduced by overfishing and overgrazing or other forms of mismanagement, but once sustainable yield limits are reached, the contribution of these systems to world food supply can not be expanded. The decline in fisheries is not limited to developing countries. By early 1994, the U.S. was experiencing precipitous drops in fishery stocks off the coast of New England, off the West Coast, and in the Gulf of Mexico.

With water—the third constraint—the overpumping that is so widespread eventually will be curbed to bring it into balance with aquifer recharge. This reduction, combined with growing diversion of irrigation water to residential and industrial uses, limits the amount of water available to produce food. Where farmers depend on fossil aquifers for their irrigation water—in the southern U.S. Great Plains, for instance, or the wheat fields of Saudi Arabia—aquifer depletion means an end to irrigated agriculture. In the U.S., where more than one-fourth of irrigated cropland is watered by drawing down underground water tables, the downward adjustment in irrigation pumping would be substantial. Major food-producing regions where overpumping is commonplace include the southern Great Plains, India's Punjab, and the North China Plain. For many farmers, the best hope for more water is from gains in efficiency.

Perhaps the most worrisome emerging constraint on food production is the limited capacity of grain varieties to respond to the use of additional fertilizer. In the U.S., Western Europe, and Japan, usage has increased little if at all during the last decade. Utilizing additional amounts on existing crop varieties has little or no effect on yield in these nations. After a tenfold increase in world fertilizer use from 1950 to 1989—from 14,000,000 to 146,000,000 tons—use declined to the following four years.

A little-recognized threat to the future world food balance is the heavy loss of cropland that occurs when countries that already are densely populated begin to industrialize. The experience in Japan, South Korea, and Taiwan gives a sense of what to expect. The conversion of grainland to nonfarm uses and to high-value specialty crops has cost Japan 52% of its grainland; South Korea, 42%; and Taiwan, 35%.

As the loss of land proceeded, it began to override the rise in land productivity, leading to declines in production. From its peak, Japan's grain output has dropped 33%; South Korea's, 31%; and Taiwan's, 74%.

Asia's densely populated giants, China and India, are going through the same stages that led to the extraordinarily heavy dependence on imported grain in the three smaller countries that industrialized earlier. In both, the shrinkage in grainland has begun. It is one thing for Japan, a country of 120,000,000 people, to import 77% of its grain, but quite another if China, with 1,200,000,000, moves in this direction.

Further complicating efforts to achieve an acceptable balance between food and people is social disintegration. In an article in the February 1994 Atlantic entitled, "The Coming Anarchy," writer and political analyst Robert Kaplan observed that unprecedented population growth and environmental degradation were driving people from the countryside into cities and across national borders at a record rate. This, in turn, was leading to social disintegration and political fragmentation. In parts of Africa, he argues, nation-states no longer exist in any meaningful sense. In their place are fragmented tribal and ethnic groups.

The sequence of events that leads to environmental degradation is all to familiar to environmentalists. It begins when the firewood demands of a growing population exceed the sustainable yield of local forests, leading to deforestation. As firewood become scarce, cow dung and crop residues are burned for fuel, depriving the land of nutrients and organic matter. Livestock numbers expand more or less apace with the human population, eventually exceeding grazing capacity. The combination of deforestation and overgrazing increases rainfall runoff and soil erosion, simultaneously reducing aquifer recharge and soil fertility. No longer able to feed themselves, people become refugees, heading for the nearest city of food relief center.

Crop reports for Africa now regularly cite weather and civil disorder as the key variables affecting harvest prospects. Not only is agricultural progress difficult, even providing food aid can be a challenge under these circumstances. In Somalia, getting food to the starving in late 1992 required a UN peacekeeping force and military expenditures that probably cost 10 times as much as what was distributed.

As political fragmentation and instability spread, national governments no longer can provide the physical and economic infrastructure for development. Countries in this category include Afghanistan, Haiti, Liberia, Sierra Leone, and Somalia. To the extent that nation-states become dysfunctional, the prospects for humanely slowing population growth, reversing environmental degradation, and systematically expanding food production are diminished.

Other negative influences exist, but they have emerged more gradually. Among those that affect food production more directly are soil erosion, the waterlogging and salting of irrigated land, and air pollution. For example, a substantial share of the world's cropland is losing topsoil at a rate that exceeds natural soil formation. On newly cleared land that is sloping steeply, soil losses can lead to cropland abandonment in a matter of years. In other situations, the loss is slow and has a measurable effect on land productivity only over many decades.

Growing Pessimism

Until recently, concerns about the Earth's capacity to feed ever-growing numbers of people adequately was confined largely to the environmental and population communities and a few scientists. During the 1990s, however, these issues are arousing the concerns of the mainstream scientific community. In early 1992, the U.S. National Academy of Sciences and the Royal Society of London issued a report that began: "If current predictions of population growth prove accurate and patterns of human activity on the planet remain unchanged, science and technology may not be able to prevent either irreversible degradation of the environment or continued poverty for much of the world."

It was a remarkable statement, an admission that science and technology no longer can ensure a better future unless population growth slows quickly and the economy is restructured. This abandonment of the technological optimism that has permeated so much of the 20th century by two of the world's leading scientific bodies represents a major shift, though perhaps not a surprising one, given the deteriorating state of the planet. That they chose to issue a joint statement, their first ever, reflects the deepening concern about the future within the scientific establishment.

Later in 1992, the Union of Concerned Scientists issued a "World Scientists' Warning to Humanity," signed by some 1,600 of the planet's leading scientists, including 102 Nobel Prize winners. It observes that the continuation of destructive human activities "may so alter the living world that it will be unable to sustain life in the manner that we know." The scientists indicated that "A great change in our stewardship of the earth and the life on it is required, if vast human misery is to be avoided and our global home on this planet is not to be irretrievably mutilated."

In November, 1993, representatives of 56 national science academies convened in New Delhi, India, to discuss population. At the end of their conference, they issued a statement in which they urged zero population growth during the lifetimes of their children.

Between 1950 and 1990, the world added 2,800,000,000 people, and an average of 70,000,000 a year. Between 1990 and 2030, it is projected to add 3,600,000,000 or 90,000,000 a year. Even more troubling, nearly all this increase is projected for the developing countries, where life-support systems already are deteriorating. Such population growth in a finite ecosystem raises questions about the Earth's carrying capacity. Will the planet's natural support systems sustain such growth indefinitely? How many people can the Earth support at a given level of consumption?

Underlying this assessment of population carrying capacity is the assumption that the food supply will be the most immediate constraint on population growth. Water scarcity could limit population growth in some locations, but it is unlikely to do so for the world as a whole in the foreseeable future. A buildup of environmental pollutants could interfere with human reproduction, much at DDT reduced the reproductive capacity of bald eagles, peregrine falcons, and other birds at the top of the food chain. In the extreme, accumulating pollutants in the environment could boost death rates to the point where they would exceed birth rates, leading to a gradual decline in human numbers, but this does not seem likely. For now, it appears that the food supply will be the most immediate, and therefore the controlling, determinant of how many people the Earth can support.

Grain supply and demand projections for the 13 most populous countries—accounting for two-thirds of world population and food production—show much slower growth in output than the official projections by the Food and Agriculture Organization and the World Bank. If those projections of relative abundance and an continuing decline of food prices materialize, governments can get by with business as usual. If, on the other hand, the constraints discussed above continue, the world needs to reorder priorities.

The population-driven environmental deterioration/political disintegration scenario described by Robert Kaplan not only is possible, it is likely in a business-as-usual world. However, it is not inevitable. This future can be averted if security is redefined, recognizing that food scarcity, not military aggression, is the principle threat to the future. Government must give immediate attention to filling the family planning gap; attacking the underlying causes of high fertility, such as illiteracy and poverty; protecting soil and water resources; and raising investment in agriculture. ■

Questions

1. During what time frame were most of the basic technological advances made?

2. What two food systems are determined by nature?

3. What do crop reports for Africa cite as determining factors affecting harvest possibilities?

Answers are at the back of the book.

11

Approximately 1,600 years ago, the Polynesians stepped ashore on Easter Island and found a paradise. The famous gigantic stone statues they left behind indicate a complex, social society made possible by abundant natural resources. Archeologists estimate that the 64 mile island had a population of 7,000 and may have supported up to 20,000. This would not have been unreasonable considering the fertility of the island. But records show that the Polynesians were destroying their forests by the year 800. Through time, statuettes with sunken cheeks and visible ribs show that the people were starving. Within a few short centuries, the forests were gone, the animals had disappeared, and so had the people. All that remains today is an impoverished and barren grassland. Could similar events occur on a global or regional scale? Some people think so.

Easter's End

Jared Diamond

In just a few centuries, the people of Easter Island wiped out their forest, drove their plants and animals to extinction, and saw their complex society spiral into chaos and cannibalism. Are we about to follow their lead?

Among the most riveting mysteries of human history are those posed by vanished civilizations. Everyone who has seen the abandoned buildings of the Khmer, the Maya, or the Anasazi is immediately moved to ask the same questions: Why did the societies that erected those structures disappear?

Their vanishing touches us as the disappearance of other animals, even the dinosaurs, never can. No matter how exotic those lost civilizations seem, their framers were humans like us. Who is to say we won't succumb to the same fate? Perhaps someday New York"s skyscrapers will stand derelict and overgrown with vegetation, like the temples at Angkor Wat and Tikal.

Among all such vanished civilizations, that of the former Polynesian society on Easter Island remains unsurpassed in mystery and isolation. The mystery stems especially from the island's gigantic stone statues and its impoverished landscape, but it is enhanced by our associations with the specific people involved: Polynesians represent for us the ultimate in exotic romance, the background for many a child's, and an adult's, vision of paradise. My own interest in Easter was kindled over 30 years ago when I read Thor Heyerdahl's fabulous accounts of his *Kon-Tiki* voyage.

But my interest has been revived recently by a much more exciting account, one not of heroic voyages but of painstaking research and analysis. My friend David Steadman, a paleontologist, has been working with a number of other researchers who are carrying out the first systematic excavations on Easter intended to identify the animals and plants that once lived there. Their work is contributing to a new interpretation of the island's history that makes it a tale not only of wonder but of warning as well.

Easter Island, with an area of only 64 square miles, is the worlds most isolated scrap of habitable land. It lies in the Pacific Ocean more than 2,000 miles west of the nearest continent (South America), 1,400 miles from even the nearest habitable island (Pitcairn). Its subtropical location and latitude—at 27 degrees south, it is approximately as far below the equator as Houston is north of it—help give it a rather mild climate, while its volcanic origins make its soil fertile. In theory, this combination of blessings should have made Easter a miniature paradise, remote from problems that beset the rest of the world.

The island derives its name from its "discovery" by the Dutch explorer Jacob Roggeveen, on Easter (April 5) in 1722. Roggeveen's first impression was not of a paradise but of a wasteland: "We originally, from a further distance, have considered the said Easter Island as sandy; the reason for that is this, that we counted as sand the withered grass, hay, or other scorched and burnt vegetation, because its wasted appearance could give no other impression than of a singular poverty and barrenness."

The island Roggeveen saw was a grassland without a single tree or bush over ten feet high. Modern botanists have identified only 47 species of higher plants native to Easter, most of them grasses, sedges, and ferns. The list includes just two species of small trees and two of woody shrubs. With such flora, the islanders Roggeveen encountered had no source of real firewood to warm themselves during Easter's cool, wet, windy winters. Their native animals included nothing larger than insects, not even a single species of native bat, land bird, land snail, or lizard. For domestic animals, they had only chickens.

European visitors throughout the eighteenth and early nineteenth centuries estimated Easter's human population at about 2,000, a modest number considering the island's fertility. As Captain James Cook recognized during his brief visit in 1774, the islanders were Polynesians (a Tahitian man accompanying Cook was able to converse with them). Yet despite the Polynesians' well-deserved fame as a great seafaring people, the Easter Islanders who came out to Roggeveen's and Cook's ships did so by swimming or paddling canoes that Roggeveen described as "bad and frail." Their craft, he wrote, were "put together with manifold small planks and light inner timbers, which they cleverly stitched together with very fine twisted threads.... But as they lack the knowledge and particularly the materials for caulking and making tight the great number of seams of the canoes, these are accordingly very leaky, for which reason they are compelled to spend half the time in bailing." The canoes, only ten feet long, held at most two people, and only three or four canoes were observed on the entire island.

With such flimsy craft, Polynesians could never have colonized Easter from even the nearest island, nor could they have traveled far offshore to fish. The islanders Roggeveen met were totally isolated, unaware that other people existed. Investigators in all the years since his visit have discovered no trace of the islanders' having any outside contacts: not a single Easter Island rock or product has turned up elsewhere, nor has anything been found on the island that could have been brought by anyone other than the original settlers or the Europeans. Yet the people living on Easter claimed memories of visiting the uninhabited Sala y Gomez reef 260 miles away, far beyond the range of the leaky canoes seen by Roggeveen. How did the islanders' ancestors reach that reef from Easter, or reach Easter from anywhere else?

Easter Island's most famous feature is its huge stone statues, more than 200 of which once stood on massive stone platforms lining the coast. At least 700 more, in all stages of completion, were abandoned in quarries or on ancient roads between the quarries and the coast, as if the carvers and moving crews had thrown down their tools and walked off the job. Most of the erected statues were carved in a single quarry and then somehow transported as far as six miles—despite heights as great as 33 feet and weights up to 82 tons. The abandoned statues, meanwhile, were as much as 65 feet tall and weighed up to 270 tons. The stone platforms were equally gigantic: up to 500 feet long and 10 feet high, with facing slabs weighing up to 10 tons.

Roggeveen himself quickly recognized the problem the statues posed: "The stone images at first caused us to be struck with astonishment," he wrote, "because we could not comprehend how it was possible that these people, who are devoid of heavy thick timber for making any machines, as well as strong ropes, nevertheless had been able to erect such images." Roggeveen might have added that the islanders had no wheels, no draft animals, and no source of power except their own muscles. How did they transport the giant statues for miles, even before erecting them? To deepen the mystery, the statues were still standing in 1770, but by 1864 all of them had been pulled down, by the islanders themselves. Why then did they carve them in the first place? And why did they stop?

The statues imply a society very different from the one Roggeveen saw in 1722. Their sheer number and size suggest a population much larger than 2,000 people. What became of everyone? Furthermore, that society must have been highly organized. Easter's resources were scattered across the island: the best stone for the statues was quarried at Rano Raraku near Easter's northeast end; red stone, used for large crowns adorning some of the statues, was quarried at Puna Pau, inland in the southwest; stone carving tools came mostly from Aroi in the northwest. Meanwhile, the best farmland lay in the south and east, and the best fishing grounds on the north and west coasts. Extracting and redistributing all those goods required complex political organization. What happened to that organization, and how could it ever have arisen in such a barren landscape?

Easter Island's mysteries have spawned volumes of speculation for more than two and a half centuries. Many Europeans were incredulous that Polynesians—commonly characterized as "mere savages"—could have created the statues or the beautifully constructed stone platforms. In the 1950s, Heyerdahl argued that Polynesia must have been settled by advanced societies of American Indians, who in turn must have received civilization across the Atlantic from more advanced societies of the Old World. Heyerdahl's raft voyages aimed to prove the feasibility of such prehistoric transoceanic contacts. In the 1960s the Swiss writer Erich von Däniken, an ardent believer in Earth visits by extraterrestrial astronauts, went further, claiming that Easter's statues were the work of intelligent beings who owned ultramodern tools, became stranded on Easter, and were finally rescued.

Heyerdahl and Von Däniken both brushed aside overwhelming evidence that the Easter Islanders were typical Polynesians derived from Asia rather than from the Americas and that their culture (including their statues) grew out of Polynesian culture. Their language was Polynesian, as Cook had already concluded. Specifically, they spoke an eastern Polynesian dialect related to Hawaiian and Marquesan, a dialect isolated since about A.D. 400, as estimated from slight differences in vocabulary. Their fishhooks and stone adzes resembled early Marquesan models. Last year DNA extracted from 12 Easter Island skeletons was also shown to be Polynesian. The islanders grew bananas, taro, sweet potatoes, sugarcane, and paper mulberry—typical Polynesian crops, mostly of Southeast Asian origin. Their sole domestic animal, the chicken, was also typically Polynesian and ultimately Asian, as were the rats that arrived as stowaways in the canoes of the first settlers.

What happened to those settlers? The fanciful theories of the past must give way to evidence gathered by hardworking practitioners in three fields: archeology, pollen analysis, and paleontology.

Modern archeological excavations on Easter have continued since Heyerdahl's 1955 expedition. The earliest radiocarbon dates associated with human activities are around A.D. 400 to 700, in reasonable agreement with the approximate settlement date of 400 estimated by linguists. The period of statue construction peaked around 1200 to 1500, with few if any statues erected thereafter. Densities of archeological sites suggest a large population; an estimate of 7,000 people is widely quoted by archeologists, but other estimates range up to 20,000, which does not seem implausible for an island of Easter's area and fertility.

Archeologists have also enlisted surviving islanders in experiments aimed at figuring out how the statues might have been carved and erected. Twenty people, using only stone chisels, could have carved even the largest completed statue within a year. Given enough timber and fiber for making ropes, teams of at most a few hundred people could have loaded the statues onto wooden sleds, dragged them over lubricated wooden tracks or rollers, and used logs as levers to maneuver them into a standing position. Rope could have been made from the fiber of a small native tree, related to the linden, called the hauhau. However, that tree is now extremely scarce on Easter, and hauling one statue would have required hundreds of yards of rope. Did Easter's now barren landscape once support the necessary trees?

That question can be answered by the technique of pollen analysis, which involves boring out a column of sediment from a swamp or pond, with the most recent deposits at the top and relatively more ancient deposits at the bottom. The absolute age of each layer can be dated by radiocarbon methods. Then begins the hard work: examining tens of thousands of pollen grains under a microscope counting them, and identifying the plant species that produced each one by comparing the grains with modern pollen from known plant species. For Easter Island, the bleary-eyed scientists who performed that task were John Flenley, now at Massey University in New Zealand, and Sarah King of the University of Hull in England.

Flenley and King's heroic efforts were rewarded by the striking new picture that emerged of Easter's prehistoric landscape. For at least 30,000 years before human arrival and during the early years of Polynesian settlement, Easter was not a wasteland at all. Instead, a subtropical forest of trees and woody bushes towered over a ground layer of shrubs, herbs, ferns, and grasses. In the forest grew tree daisies, the rope-yielding hauhau tree, and the toromiro tree, which furnishes a dense, mesquite-like firewood. The most common tree in the forest was a species of palm now absent on Easter but formerly so abundant that the bottom strata of the sediment column were packed with its pollen. The Easter Island palm was closely related to the still-surviving Chilean wine palm, which grows up to 82 feet tall and 6 feet in diameter. The tall, unbranched trunks of the Easter Island palm would have been ideal for transporting and erecting statues and constructing large canoes. The palm would also have been a valuable food source, since its Chilean relative yields edible nuts as well as sap from which Chileans make sugar, syrup, honey, and wine.

What did the first settlers of Easter Island eat when they were not glutting themselves on the local equivalent of maple syrup? Recent excavations by David Steadman, of the New York State Museum at Albany, have yielded a picture of Easter's original animal world as surprising as Flenley and King's picture of its plant world. Steadman's expectations for Easter were conditioned by his experiences elsewhere in Polynesia, where fish are overwhelmingly the main food at archeological sites, typically accounting for more than 90 percent of the bones in ancient Polynesian garbage heaps. Easter, though, is too cool for the coral reefs beloved by fish, and its cliff-girded coastline permits shallow-water fishing in only a few places. Less than a quarter of the bones in its early garbage heaps (from the period 900 to 1300) belonged to fish; instead, nearly one-third of all bones came from porpoises.

Nowhere else in Polynesia do porpoises account for even 1 percent of discarded food bones. But most other Polynesian islands offered animal food in the form of birds and mammals, such as New Zealand's now extinct giant moas and Hawaii's now extinct flightless geese. Most other islanders also had domestic pigs and dogs. On Easter, porpoises would have been the largest animal available—other than humans. The porpoise species identified at Easter, the common dolphin, weighs up to 165 pounds. It generally lives out at sea, so it could not have been hunted by line fishing or spearfishing from shore. Instead, it must have been harpooned far offshore, in big seaworthy canoes built from the extinct palm tree.

In addition to porpoise meat, Steadman found, the early Polynesian settlers were feasting on seabirds. For those birds, Easter's remoteness and lack of predators made it an ideal haven as a breeding site, at least until humans arrived. Among the prodigious numbers of seabirds that bred on Easter were albatross, boobies, frigate birds, fulmars, petrels, prions, shearwaters, storm petrels, terns, and tropic birds. With at least 25 nesting species, Easter was the richest seabird breeding site in Polynesia and probably in the whole Pacific.

Land birds as well went into early Easter Island cooking pots. Steadman identified bones of at least six species, including barn owls, herons, parrots, and rail. Bird stew would have been seasoned with meat from large numbers of rats, which the Polynesian colonists inadvertently brought with them; Easter Island is the sole known Polynesian island where rat bones outnumber fish bones at archeological sites. (In case you're squeamish and consider rats inedible, I still recall recipes for creamed laboratory rat that my British biologist friends used to supplement their diet during their years of wartime food rationing.)

Porpoises, seabirds, land birds, and rats did not complete the list of meat sources formerly available on Easter. A few bones hint at the possibility of breeding seal colonies as well. All these delicacies were cooked in ovens fired by wood from the island's forests.

Such evidence lets us imagine the island onto which Easter's first Polynesian colonists stepped ashore some 1,600 years ago, after a long canoe voyage from eastern Polynesia. They found themselves in a pristine paradise. What then happened to it? The pollen grains and the bones yield a grim answer.

Pollen records show that destruction of Easter's forests was well under way by the year 800, just a few centuries after the start of human settlement. Then charcoal from wood fires came to fill the sediment cores, while pollen of palms and other trees and woody shrubs decreased or disappeared, and pollen of the grasses that replaced the forest became more abundant. Not long after 1400 the palm finally became extinct, not only as a result of being chopped down but also because the now ubiquitous rats prevented its regeneration: of the dozens of preserved palm nuts discovered in caves on Easter, all had been chewed by rats and could no longer germinate. While the hauhau tree did not become extinct in Polynesian times, its numbers declined drastically until there weren't enough left to make ropes from. By the time Heyerdahl visited Easter, only a single, nearly dead toromiro tree remained on the island, and even that lone survivor has now disappeared. (Fortunately, the toromiro still grows in botanical gardens elsewhere.)

The fifteenth century marked the end not only for Easter's palm but for the forest itself. Its doom had been approaching as people cleared land to plant gardens; as they felled trees to build canoes, to transport and erect statues, and to burn; as rats devoured seeds; and probably as the native birds died out that had pollinated the trees' flowers and dispersed their fruit. The overall picture is among the most extreme examples of forest destruction anywhere in the world: the whole forest gone, and most of its tree species extinct.

The destruction of the island's animals was as extreme as that of the forest: without exception, every species of native land bird became extinct. Even shellfish were overexploited, until people had to settle for small sea snails instead of larger cowries. Porpoise bones disappeared abruptly from garbage heaps around 1500; no one could harpoon porpoises anymore, since the trees used for constructing the big seagoing canoes no longer existed. The colonies of more than half of the seabird species breeding on Easter or on its offshore islets were wiped out.

In place of these meat supplies, the Easter Islanders intensified their production of chickens, which had been only an occasional food item. They also turned to the largest remaining meat source available: humans, whose bones became common in late Easter Island garbage heaps. Oral traditions of the islanders are rife with cannibalism; the most inflammatory taunt that could be snarled at an enemy was "The flesh of your mother sticks between my teeth." With no wood available to cook these new goodies, the islanders resorted to sugarcane scraps, grass, and sedges to fuel their fires.

All these strands of evidence can be wound into a coherent narrative of a society's decline and fall. The first Polynesian colonists found themselves on an island with fertile soil, abundant food, bountiful building materials, ample lebensraum, and all the prerequisites for comfortable living. They prospered and multiplied.

After a few centuries, they began erecting stone statues on platforms, like the ones their Polynesian forebears had carved. With passing years, the statues and platforms became larger and larger, and the statues began sporting ten-ton red crowns—probably in an escalating spiral of one-upmanship, as rival clans tried to surpass each other with shows of wealth and power. (In the same way, successive Egyptian pharaohs built ever-larger pyramids. Today Hollywood movie moguls near my home in Los Angeles are displaying their wealth and power by building ever more ostentatious mansions. Tycoon Marvin Davis topped previous moguls with plans for a 50,000-square-foot house, so now Aaron Spelling has topped Davis with a 56,000-square-foot house. All that those buildings lack to make the message explicit are ten-ton red crowns.) On Easter, as in modern America, society was held together by a complex political system to redistribute locally available resources and to integrate the economies of different areas.

Eventually Easter's growing population was cutting the forest more rapidly than the forest was regenerating. The people used the land for gardens and the wood for fuel, canoes, and houses—and, of course, for lugging statues. As forest disappeared, the islanders ran out of timber and rope to transport and erect their statues. Life became more uncomfortable—springs and streams dried up, and wood was no longer available for fires.

People also found it harder to fill their stomachs, as land birds, large sea snails, and many seabirds disappeared. Because timber for building seagoing canoes vanished, fish catches declined and porpoises disappeared from the table. Crop yields also declined, since deforestation allowed the soil to be eroded by rain and wind, dried by the sun, and its nutrients to be leeched from it. Intensified chicken production and cannibalism replaced only part of all those lost foods. Preserved statuettes with sunken cheeks and visible ribs suggest that people were starving.

With the disappearance of food surpluses, Easter Island could no longer feed the chiefs, bureaucrats, and priests who had kept a complex society running. Surviving islanders described to early European visitors how local chaos replaced centralized government and a warrior class took over from the hereditary chiefs. The stone points of spears and daggers, made by the warriors during their heyday in the 1600s and 1700s, still litter the ground of Easter today. By around 1700, the population began to crash toward between one-quarter and one-tenth of its former number. People took to living in caves for protection against their enemies. Around 1770 rival clans started to topple each other's statues, breaking the heads off. By 1864 the last statue had been thrown down and desecrated.

As we try to imagine the decline of Easter's civilization, we ask ourselves, "Why didn't they look around, realize what they were doing, and stop before it was too late? What were they thinking when they cut down the last palm tree?"

I suspect, though, that the disaster happened not with a bang but with a whimper. After all, there are those hundreds of abandoned statues to consider. The forest the islanders depended on for rollers and rope didn't simply disappear one day—it vanished slowly, over decades. Perhaps war interrupted the moving teams; perhaps by the time the carvers had finished their work, the last rope snapped. In the meantime, any islander who tried to warn about the dangers of progressive deforestation would have been overriden by vested interests of carvers, bureaucrats, and chiefs, whose jobs depended on continued deforestation. Our Pacific Northwest loggers are only the latest in a long line of loggers to cry, "Jobs over trees!" The changes in forest cover from year to year would have been hard to detect: yes, this year we cleared those woods over there, but trees are starting to grow back again on this abandoned garden site here. Only older people, recollecting their childhoods decades earlier, could have recognized a difference. Their children could no more have comprehended their parents' tales than my eight-year-old sons today can comprehend my wife's and my tales of what Los Angeles was like 30 years ago.

Gradually trees became fewer, smaller, and less important. By the time the last fruit-bearing adult palm tree was cut, palms had long since ceased to be of economic significance. That left only smaller and smaller palm saplings to clear each year, along with other bushes and treelets. No one would have noticed the felling of the last small palm.

By now the meaning of Easter Island for us should be chillingly obvious. Easter Island is Earth writ small. Today, again, a rising population confronts shrinking resources. We too have no emigration valve, because all human societies are linked by international transport, and we can no more escape into space than the Easter Islanders could flee into the ocean. If we continue to follow our present course, we shall have exhausted the world's major fisheries, tropical rain forests, fossil fuels, and much of our soil by the time my sons reach my current age.

Every day newspapers report details of famished countries—Afghanistan, Liberia, Rwanda, Sierra Leone, Somalia, the former Yugoslavia, Zaire—where soldiers have appropriated the wealth or where central government is yielding to local gangs of thugs. With the risk of nuclear war receding, the threat of our ending with a bang no longer has a chance of galvanizing us to halt our course. Our risk now is of winding down, slowly, in a whimper. Corrective action is blocked by vested interests, by well-intentioned political and business leaders, and by their electorates, all of whom are perfectly correct in not noticing big changes from year to year. Instead, each year there are just somewhat more people, and somewhat fewer resources, on Earth.

It would be easy to close our eyes or to give up in despair. If mere thousands of Easter Islanders with only stone tools and their own muscle power sufficed to destroy their society, how can billions of people with metal tools and machine power fail to do worse? But there is one crucial difference. The Easter Islanders had no books and no histories of other doomed societies. Unlike the Easter Islanders, we have histories of the past—information that can save us. My main hope for my sons' generation is that we may now choose to learn from the fates of societies like Easter's. ■

Questions

1. What was the most common tree in the forest? And what was it used for?

2. What were some of the meat sources on Easter?

3. What replaced the meat sources?

Answers are at the back of the book.

Section Two

Problems of Resource Scarcity

12

In 1989, estimates showed that reserves of oil and gas on land and in coastal waters would be depleted by 2020. But this assessment has been disputed, partly because categories were not included in the report, the supporting data was not complete, and not surprisingly, the assessment methods used were questioned. A 1995 assessment report is now available on a CD-ROM containing more than 10,000 pages of documentation. The results from the report show that instead of 78 billion barrels of oil in 1989, we have 112 billion barrels. Gas resources, which were 504 trillion cubic feet (TCF) more than doubled to 1074 TCF in the 1995 report. A broader field of resources was used in the 1995 study than had been previously estimated. The team that produced the report also developed new techniques for estimating reserve growth and the volume of oil or gas buildup. These factors contributed greatly to the reported increased resources that can be utilized.

USGS Oil & Gas CD Hits the Charts

R.C. Burruss

What a difference a day—or in this case six years—makes. In 1989 it appeared that our nation's estimated reserves of oil and gas on land and in coastal waters might be gone by 2020. The U.S. Geological Survey's 1995 assessment of national oil and gas resources changes that picture dramatically. The USGS periodically assesses these resources onshore and in state waters, while the Minerals Management Service considers oil and gas resources beneath federal offshore waters. The assessments provide the basis for national evaluations of land use, energy and economic policy and strategic plans, and development of environmental policy.

The USGS released its latest assessment—the results of a five-year study—in February at the 10th McKelvey Forum, held in Washington, D.C. The new resource numbers are, in project chief Don Gautier's words, "astonishing" (*Oil and Gas Journal*, March 6,1995, p.84-88). The total amount of technically recoverable oil and gas remaining beneath onshore lands and state waters is significantly greater than the 1989 study indicated. And, the 1995 assessment is the first to appear in a digital medium. It is available in an interactive, CD-ROM format (USGS Digital Data Series 30). A printed summary (USGS Circular 1118) is also available.

Why a CD-ROM?

The Survey's 1989 assessment of oil and gas resources was surprisingly controversial—for two main reasons. First, the volume of assessed resources, especially undiscovered natural gas, decreased significantly from the assessment published in 1981.

Second and perhaps more importantly, certain categories of resources were not reported, and documentation of methodology and supporting data was incomplete. People who disagreed with the USGS estimates, therefore, doubted our assessment methods. As a result, the American Association of State Geologists, the National Academy of Sciences, and the American Petroleum Institute reviewed the study extensively. The USGS received a clear message for future assessment projects: Publish the supporting information at the time the numbers are released. That task is daunting.

The 1995 assessment used a geologic play-based methodology to assess 570 confirmed or hypothetical petroleum plays. Therefore, we needed to publish 570 maps of the individual plays and all associated text, statistical charts, and graphs, including text on methodology, bibliographies, and related studies. At up to 30 pages per play, we faced a challenge of producing more than 10,000 pages of documentation. Using normal USGS paper publication procedures, this task would be impossible. A digital medium seemed the only viable publication option, and a CD-ROM was the logical choice.

Having committed to a CD-ROM, the format became a critical issue. The sheer volume of information required a display mechanism that was easy to use. The CD-ROM development team, led by Ken Takahashi, saw a clear need for both interactivity and the ability to use the same CD in both Macintosh and IBM compatible (Windows) systems. After examining commercial and public-domain software, Takahashi chose Macromedia's Macromind Director.™ This choice simplified publication. Because Director™ supports runtime versions for both Macintosh and Windows that use the same files, we could release a single CD for both machines. By using a common set of menus, buttons, and icons, the user can view the oil and gas information as a linear sequence of text and graphics—essentially as a digital book called up through a table of contents—or by accessing national and regional maps. The user can, thereby, view maps of individual plays, texts, statistics, and graphics in any order that is meaningful.

Astonishing Results

The 1995 assessment reports a resource of 112 billion barrels of oil compared to 78 billion barrels in 1989. The number for gas resources, 1074 trillion cubic feet (TCF), is more than double the 1989 number of 504 TCF. These large increases in resource volume in the 1995 assessment are due to several critical factors. The 1995 assessment is broader in scope than previous assessments. This time, we considered three types of resources:

- conventional, undiscovered, but technically recoverable oil and gas;
- additions to reserves in known fields (reserve growth); and
- oil and gas in continuous-type accumulations which are generally equivalent to other analysts' categories of "unconventional" resources.

The greater significance of reserve growth became clear from production data accumulated during the domestic drilling boom of the late 1970s and early 1980s. These data were not available during preparation of the 1989 assessment, which used drilling data from 1979 and earlier.

Finally, the assessment team developed new methods for estimating future reserve growth and the volume of technically recoverable oil or gas in "continuous-type" accumulations.

Reserve growth is the largest category of oil resources in the 1995 assessment. These resources add to reserves through extensions of known fields, additions of new pools in previously discovered fields, and revisions of reserve estimates. This class of resources is affected by the development of new technology, including enhanced recovery and improved subsurface imaging techniques. For example, 3-D seismic allows better definition of the size of oil and gas accumulations. USGS methods to predict future reserve growth are based on models of historical field growth through 1992, developed by David Root, Emil Attanasi, and Richard Mast.

Continuous-type accumulations are the largest natural-gas resource, although reserve growth in conventional fields is of similar magnitude. Continuous-type resources are largely equivalent to definitions in the 1989 study of unconventional resources, including tight-gas sandstones, coal-bed methane, and gas in fractured shales. USGS defines these resources as accumulations that are geographically extensive and generally lack distinct oil-water and gas-water contacts. By using this geological definition, we can distinguish continuous-type resources from conventional, discrete accumulations, and eliminate our reliance on regulatory or engineering criteria. Other published estimates of this resource vary widely, with the high end at several thousand trillion cubic feet of gas. In this assessment, we limited our estimates to only those resources judged to be technically recoverable.

The Right Product?

The assessment results have been favorably received by state geologists, the American Association of Petroleum Geologists Evaluation Committee, and congressional staffs. Exploration geologists and geophysicists, financial analysts, and land-use managers are similarly supportive, offering comments such as "this is the best product in many years" or "it's the right product at the right time." At the AAPG annual meeting in March in Houston, we ran a hands-on demonstration and distributed almost 1,000 copies of the CD-ROM in two-and-a-half days. "I can use that in my company," commented many visitors at the demonstration booth. We find these responses heartening and hope that they reflect progress toward the Survey's goal of meeting the needs of its customers.

So where do we go from here? The five-year implementation plan for the USGS Energy Resource Surveys Program shows oil and gas assessment activities peaking in 1995. Later this year, USGS will publish two archival CD-ROMs which will contain all the maps of oil and gas plays in a geographic information-system format. Moreover, 1995 marks the beginning of a new assessment activity—the National Coal Resource Assessment project, led by Hal Gluskoter Branch of Coal Geology, Reston. Our petroleum assessment activities will decrease as we shift resources to produce the coal assessment by 1999. We are building on the lessons learned from the petroleum assessments, and by 1999, USGS hopes that a few more of its CD-ROMs will "hit the charts." ■

Questions

1. Why was the 1989 assessment of oil and gas resources controversial?

2. Why was a CD-ROM used?

3. For the 1995 assessment, what three types of resources were considered?

Answers are at the back of the book.

13

Natural gas is on the rise and could shape the future of energy. It is a relatively clean and versatile hydrocarbon and has the potential to replace substantial amounts of coal and oil. Because of its lower carbon content, natural gas produces 30% less carbon dioxide per unit of energy than oil, and 43% less than coal. It is relatively easier to process and transport compared to coal. It is also more adaptable than coal or oil and can be utilized in more than 90% of energy applications. Natural gas is the most popular heating fuel in North America and now natural gas is becoming common in other energy markets, including electricity generation. Low costs and low emissions permitted natural gas to dominate the market for new power plants in the United States and the United Kingdom in the early 1990s. As a result of this dominance, natural gas could also become the primary fuel for new power plants in many countries. Natural gas is a shift towards a more efficient and clean energy system.

The Unexpected Rise of Natural Gas

Christopher Flavin and Nicholas Lenssen

With growing advantages to its use, natural gas may usurp oil as the world's energy resource of choice.

When the U.S. Senate called a hearing in 1984 to assess the prospects for natural gas, almost everyone expected a gloomy session. At the time, gas production in the United States had been falling for 12 years and prices had tripled in a decade. It seemed a textbook example of a rapidly depleting resource.

Few were surprised when Charles B. Wheeler, senior vice president at Exxon—the world's largest oil company—told the Senate that natural gas was essentially finished as a major energy source. "We project a shortfall of economically available gas from any source," said Wheeler.

Only one voice interrupted the gloom that pervaded the hearing room—that of Robert Hefner, an iconoclastic geologist who headed a small Oklahoma gas-exploration company and grandson of one of the earliest oil wildcatters. Hefner told those in attendance, "My lifetime work requires that I respectfully have to disagree with everything Exxon says on the natural-gas resource base."

A decade later, legions of government and industry analysts have had to eat their words, while Hefner has turned his contrarian views on natural gas into a comfortable fortune. Natural-gas prices in the United States fell sharply after 1986, and production climbed. By 1993, the nation was producing 15% more gas. For the world as a whole, gas production has risen 30% since the mid-1980s, with increases recorded in nearly every major country.

The world now appears to be in the early stages of a natural gas boom that could profoundly shape our energy future. If natural-gas production can be doubled or tripled in the next few decades (as Hefner and a growing number of geologists believe), this relatively clean and versatile hydrocarbon could replace large amounts of coal and oil. Because it is easy to transport and use—even in small, decentralized technologies—natural gas could help accelerate the trend toward a more-efficient energy system and, over the long run, the transition to renewable sources of energy.

Advantages of Natural-Gas Use

The environmental advantages of natural gas over other fossil fuels were a strong selling point from the start. Methane is the simplest of hydrocarbons—a carbon atom surrounded by four hydrogen atoms—

with a higher ratio of hydrogen to carbon than other fossil fuels. Natural gas helped reduce the dangerous levels of sulfur and particles in London's air during the 1950s. In fact, these two contaminants are largely absent from natural gas by the time it goes through a separation plant and reaches customers. Natural-gas combustion also produces no ash and smaller quantities of volatile hydrocarbons, carbon monoxide, and nitrogen oxides than oil or coal do. And, unlike coal, gas has no heavy metals.

As a gaseous fuel, methane tends to be combusted more thoroughly than solids or liquids are. Due to its lower carbon content, natural gas produces 30% less carbon dioxide per unit of energy than oil does and 43% less than coal, thus reducing its impact on the atmosphere. It is also relatively easy to process compared with oil and less expensive to transport (via pipeline) than coal, which generally moves by rail.

To be fair, methane gas is not entirely benign. When not properly handled, it can explode. And as a powerful greenhouse gas in its own right, it can contribute to the warming of the atmosphere. But with careful handling, both of these problems can be reduced dramatically.

Gas as a Power Generator

Natural gas is far more versatile than either coal or oil, and with a little effort can be used in more than 90% of energy applications. Yet, until recently, its use has been largely restricted to household and industrial markets, in which it has thrived. In North America, for example, natural gas is far and away the most popular heating fuel. By the early 1990s, nearly two-thirds of the single-family homes and apartment buildings built in the United States has such heating systems.

In recent years, new technologies such as gas-powered cooling systems and heat pumps have even allowed this energy source to challenge electricity's dominance of additional residential and commercial applications. More significantly, natural gas has begun to find its way into energy markets from which it was excluded in the past, including electricity generation.

Gas has always been an attractive fuel for electric power generation, but high prices and legal strictures deterred its use by utilities during the 1970s and 1980s. Most of the plants built then were fueled by coal or nuclear power. By 1990, gas constituted only 8% of the fuel used in electricity generation in North America and only 7% in Europe.

Until recently, most power plants used a simple Rankine cycle steam turbine. The heat that was generated by burning a fuel produced steam, which spun a turbine connected to an electricity generator. Although this technology had progressed steadily for decades, by the 1960s its efficiency in turning the chemical energy of fossil fuels into electricity had leveled off at about 33%, meaning that nearly two-thirds of the energy was still dissipated as waste heat. The inefficiency of this process made it desirable to use as cheap a fuel as possible. Until recent years, natural gas did not fit the bill.

This situation changed, however, as natural-gas prices fell and turbine technologies improved during the 1980s. Much of the recent gas turbine renaissance is focused on the combined-cycle plant—an arrangement in which the excess heat from a gas turbine is used to power a steam turbine, thus boosting efficiency. Combined-cycle plants reached efficiencies of more than 40% during the late 1980s, with the figure climbing to 45% for a General Electric plant opened in South Korea in 1993. At about the same time, Asea Brown Boveri announced plans for a combined-cycle plant with an efficiency of 53%.

These generators are inexpensive to build (roughly $700 per kilowatt, or a little more than half as much as the average coal plant) and can be constructed rapidly. The huge 1,875 megawatt Teeside station completed in the United Kingdom in 1992 took only two and a half years to complete.

Natural gas powered turbines and engines are also helping to drive the growing use of combined heat and power systems, in which the waste heat from power generation is used in factories, district heating systems, or even individual buildings. Small-scale "cogeneration" has already become popular in Denmark and other parts of northern Europe.

Gas turbines plants also have major environmental advantages over conventional oil or coal plants, including no emissions of sulfur and negligible emissions of particulates. Nitrogen-oxide emissions can be cut by 90% and carbon dioxide by 60%. Indeed, the combination of low cost and low emissions has allowed natural gas to dominate the market for new power plants in the United States and the United Kingdom during the early 1990s. Even larger markets are unfolding in southern Asia, the Far East, and Latin America.

In the future, this technology could spur utilities to convert hundreds of aging coal plants into gas-burning combined-cycle plants—for as little as $300 per kilowatt. Worldwide, some 400,000 megawatts worth of gas turbine plants could be built by 2005, according to forecasts by General Electric. Units are already up and running in countries as diverse as Austria, Egypt, Japan, and Nigeria. A secondary result of this boom could be the emergence of natural gas as the dominant fuel for new power plants in many countries.

On the Road to Natural Gas

Interest in natural gas as a vehicle fuel blossomed in the early 1990s as cities such as São Paulo and Mexico City struggled to cope with intractable air pollution. In the United States, many state and local governments began to promote natural-gas vehicles in public and private fleets, while car manufacturers built gas-powered versions of some of their auto and light truck models, and gas-distribution companies converted gasoline-powered cars to the use of natural gas. In many regions, natural gas has eclipsed both ethanol and methanol, the two new automotive fuels that commanded most of the attention in the 1980s. An industry study estimates that as many as 4 million natural-gas vehicles could be on U.S. roads by 2005.

Natural gas is beginning to break oil's stranglehold on the transportation market. Compared with gasoline and diesel fuel, natural gas has both economic and environmental advantages. In the United States, for instance, its wholesale price was less than half that of gasoline in 1993, a disparity caused in part by the cost of refining gasoline. As in other applications, the chemical simplicity of methane is a major advantage, reducing emissions and allowing for less engine maintenance. Until recently, compress-gas vehicles were confined to just a few countries—nearly 300,000 on Italy's roads, and more than 100,000 on New Zealand's.

The main challenge in using natural gas in motor vehicles lies in storing the fuel in the car—usually in cylindrical, pressurized tanks. While early tanks were bulky and heavy, manufacturers are now producing lightweight cylinders made of composite materials that will make it possible to build virtually any kind of natural-gas vehicle with a range similar to a gasoline-powered one. Engineers believe they can design a tank into the smallest passenger car without even sacrificing trunk space.

Switching to natural gas will be even easier for buses, trucks, and locomotives, as their size means that finding room for the tanks is not an issue. Many local bus systems are already switching over, in order to avoid the cancer-causing particulates and other pollutants that flow from current diesel-powered engines. Operators of local delivery services are moving in the same direction. The United Parcel Service in the United States, for example, is testing natural gas in its vehicles. The idea of switching train locomotives from the currently dominant diesel-electric systems to gas-electric ones is just beginning to be studied. In the United States, Union Pacific and Burlington Northern are both testing the use of liquefied natural gas in their engines. Preliminary data indicates favorable economics and excellent environmental performance.

Converting service stations so that they can provide natural gas is also straightforward, and several oil companies have begun to do this. In Europe and North America, virtually all cities and many rural areas have gas pipes running under almost every street, and they simply need to provide service stations with compressors for putting the gas into pressurized tanks. And it may well be possible for residential buildings, millions of which are already hooked up to gas lines, to be fitted with compressors, meaning fewer trips to a service station. As of early 1994, about 900 U.S. service stations were selling natural gas, with four of five more joining their ranks each week.

How Long Will It Last?

Geologists disagree vehemently about how much natural gas remains to be found, but the trend is clear—as knowledge grows, the estimated size of the resource base expands with it. The U.S. experience provides insights, since it has the most-extensive gas industry and its resources are the most heavily exploited. The sharp increase in U.S. gas production since the mid-1980s has been accompanied by a reevaluation of the resource base. A 1991 National Research Council study of official estimates made by the U.S. Geological Survey (USGS) found that, "after a detailed examination of the [USGS's] databases, geological methods, and statistical methods, the committee judged that there may have been a systematic bias toward overly conservative estimates."

As with virtually all other energy technologies, the techniques for locating and developing new gas fields are advancing rapidly. Part of this is due to the advent of computer software that makes it possible to generate three-dimensional seismic images of the subsurface geology and to determine how much gas may be there. As a result, the real cost of finding and extracting gas has declined markedly.

U.S. gas resources are only a tiny fraction of the world total, and discoveries are now proceeding more rapidly in other regions. During the past two decades, enormous amounts of natural gas have been discovered in Argentina, Indonesia, Mexico, North Africa, and the North Sea, among other areas. Each either is or could become a major exporter of natural gas. In addition, some of the former Soviet republics in central Asia have extensive gas resources, which are relatively inaccessible but are being studied by major Western oil and gas companies.

Russia, the former seat of Soviet power, is one of the keys to the global gas outlook. It is the largest producer and has the most identified reserves. While oil production has declined catastrophically with the collapse of the communist economic system, the flow of gas has fallen only slightly. Western experts have reassessed the Russian data and decided that the gas fields are even richer than previously believed. According to estimates by USGS scientists, the total Russian resource is close to 5,000 exajoules—enough to meet current world demand for 60 years. Because it is located in Siberia and other remote areas, much of this gas must be moved long distances. But it is still within reach of more than half of the world's energy consumers, including the 1.8 billion who live in China, Japan, and Europe. And at least 50 additional countries have natural-gas reserves that are minor on a global scale but sufficient to fuel their economies for decades.

Even as reliance on natural gas grows during the next few decades, one of its most important features will become apparent: It is the logical bridge to what some scientists believe will become our ultimate energy carrier—gaseous hydrogen produced from solar energy and other renewable resources. Because these two fuels are so similar in their chemical composition—hydrogen can be thought of as methane without the carbon—and in the infrastructure they require, the transition could be a relatively smooth one. Just as the world shifted early in this century from solid fuels to liquid ones, so might a shift from liquids to gases be under way today—thereby increasing the efficiency and cleanliness of the overall energy system. ■

Questions

1. Why do gas turbine plants have environmental advantages over conventional oil and gas plants.

2. What is the main problem in using natural gas in motor vehicles?

3. What country could feasibly meet current world demands for natural gas for the next 60 years and why?

Answers are at the back of the book.

14

Due to technological advances of the last ten years, wind power has become competitive with fossil fuels and nuclear energy. The capability of wind power is immense. Just one percent of Earth's winds could feasibly meet the world's energy requirements. Not only is wind power abundant, it is environmentally advantageous. Wind turbines are recyclable and do not contribute to air pollution, acid rain, global warming, or to ozone destruction. They do not produce hazardous waste or siphon water from rivers. Even though wind turbines produce less than one percent of the country's electricity, their popularity is growing. The U.S. Department of Energy predicts a sixfold increase in the use of wind power in the next 15 years. By the year 2050, the American Wind Energy Association, anticipates that wind power will supply at least 10 percent of the nation's energy requirements.

The Forecast for Windpower

Dawn Stover

In its mad rush toward the ocean, the Columbia River long ago carved a deep gorge through the Cascade mountain range of the Pacific Northwest. The gorge is famous for its winds, which in the summer blow hundreds of brightly colored sailboards back and forth across the river. During the warm season, desert air in eastern Oregon and Washington rises as it heats, sucking cooler air from the western forests and ocean through the passageway formed by the river. In the winter, the winds reverse as high-pressure systems in the east force air toward the ocean.

Strong, predictable winds make the Columbia River Gorge as popular with turbine builders as it is with windsurfers. High on the bluffs overlooking the water and its glinting bits of sail, developers plan to construct three large wind power plants. One site will be dotted with big, heavy machines modeled after trusty farm equipment. Another will have small, lightweight turbines designed with an emphasis on aerodynamics. And the third will have variable-speed wind machines that rely on sophisticated electronics.

Not only are these three designs much more reliable than the turbines of a decade ago, they are also so efficient that they are making wind power economically competitive with fossil fuels and nuclear energy. "Wind power today is a lot cheaper than most people realize," says Kenneth C. Karas, president of Zond Systems, a wind-turbine maker in Tehachapi, Calif.

Wind turbines are not a recent invention. Persians devised crude wind machines to grind grain as early as the seventh century. By the seventeenth century, Don Quixote was tilting at them on the plains of La Mancha. In the United States, farmers and ranchers used more than 500,000 windmills to pump water before the advent of rural electrification. By the 1970's people living "off the gird" had begun using wind turbines to produce power for homes. But only since 1981 have wind machines been used on a large scale to generate electricity for the utilities.

The potential of wind power is enormous. Just one percent of Earth's wind could theoretically meet the entire world's energy needs. Within the United States, wind could generate all of the electricity used today, even when land restrictions are taken into account. North Dakota alone could supply more than a third of the country's electricity needs.

Not only is wind power plentiful, but it is popular with the public because of its environmental benefits. Wind turbines are recyclable. They do not contribute to air pollution, acid rain, global warming, or ozone destruction. They do not create hazardous waste or unsightly mines. They do not siphon water from rivers. They do not chop fish into little pieces. They do, however, sometimes kill birds.

Today more than 17,000 turbines spin power into the U.S. grid, mostly in California. Wind machines currently produce less than 1 percent of the nation's electricity, but their contribution is growing quickly. The U.S. Department of Energy forecasts a six fold increase in the nation's wind-energy use during the next 15 years. The American Wind Energy Association, the industry's trade group, predicts that wind will provide at least 10 percent of the nation's energy needs by the year 2050.

The Columbia River Gorge is just one of many windy areas targeted for development. Within the next year or so, large wind farms are also planned for Wyoming, Iowa, Minnesota, Texas, Vermont, and Maine. These farms are expected to add about 400 megawatts in the United States. Existing wind turbines already supply enough power for close to a half-million homes.

Wind energy is growing internationally too. Europe already has almost 2,000 megawatts of generating capacity and aims to double that by the year 2000. One European country, Denmark, has set a goal of producing 10 percent of its electricity from the wind by turn of the century. Major wind projects are also underway in Canada, India, Argentina, Ukraine, China, Costa Rica, and New Zealand.

Advanced turbine technology is the biggest reason for wind power's declining cost. "No longer will wind energy be seen as the domain of a disheveled miller with corn flour in his hair, furling the cloth sails on his wooden windmill," writes Paul Gipe in his new book, Wind Energy *Comes of Age*. "This archaic image has given way to that of trained professionals tending their sleek aeroelectric generators by computer." Larger turbines, new blade designs, advanced materials, smarter electronics, flexible hub structures, and aerodynamic controls are all contributing to improvements in efficiency.

"The wind industry today is in many ways analogous to the airplane industry of the 1930's," says Steven P. Steinhour, lands and permits director of Kenetech Windpower, the world's largest wind company. Indeed, some of the most interesting new technologies rely on ideas borrowed from aircraft designers. For example, many of the prototype turbines sprouting at wind farms have computer-designed airfoils. A few even have aircraft-type control surfaces like ailerons.

Wind blowing past a turbine doesn't "push" its blades around like the arms of a pinwheel. Rather, the blades work somewhat like the wings of an airplane. Air passing over a blade's upper surface travels farther than air crossing the underside, resulting in a pressure difference that creates lift. As lift drives the blades forward, they turn a drive shaft connected to a generator.

Not long ago, engineers made turbine blades from helicopter rotors and sailplane wings. But as it turns out, these components are not ideal for wind machines. Aircraft wings are designed to maximize lift and prevent stalls, even if that means increased drag and the resulting reduced efficiency. But turbine blades don't carry precious human cargo, so stalling isn't so worrisome. In fact, some turbine designers intentionally create blades that will stall, or slow, in high winds to prevent turbine damage. Stalling occurs when the blade's angle of attack becomes so steep that the airflow around the blade is too turbulent to produce lift.

The National Renewable Energy Laboratory has helped a number of U.S. wind companies develop airfoils that are designed specifically for wind turbines. When a conventional turbine was retrofitted with these airfoils, its energy capture increased by about 30 percent. In a sign of the changing times, the lab recently constructed a wind technology center at Rocky Flats, a former nuclear-weapons plant outside Denver.

The shape of the new airfoils also makes them less susceptible than aircraft wings to "roughening" from collisions with insects. This isn't a problem for aircraft, which fly high enough to avoid insects most of the time. But roughening of a turbine blade's leading edge can substantially reduce lift and increase drag. The new shapes developed at the Colorado lab minimize the energy losses for bug-spattered blades.

The Renewable Energy Lab is also funding the

development of new turbine designs. The goal of the lab's Next-Generation Turbine program is to develop machines by the year 2000 that will generate electricity at a cost of four cents per kilowatt-hour at sites with only moderate wind speeds—an average of 13 mph at a height of 30 feet.

One way to capture more energy at lower wind speeds is to use longer blades and taller towers. As a turbine's blades increase in length, its power output rises exponentially.

"Over the last decade, you can trace the increasing size of wind machines," says Randall Swisher, executive director of the American Wind Energy Association. The machines installed in the early 1980s had capacities of about 40 to 50 kilowatts, not much more powerful than a farm windmill. Today, wind turbines typically generate 300 to 500 kilowatts apiece.

The largest U.S. machine is the 500-kilowatt Z-40 (the number stands for the rotor diameter in meters) made by Zond. Modeled after sturdy farm equipment used in Europe, a single Z-40 can produce enough electricity to power 150 to 200 homes. Zond has tested two prototypes of its Z-40 in Tehachapi, Calif., which could well be called the wind capital of the world. There are more than 5,000 turbines arrayed along the ridges of the Tehachapi Pass, where cool air rushes through the mountain to replace hot air rising from the Mojave desert. Standing atop one of these ridges, you can see more than a dozen different types of wind turbines whirling in the breeze.

Cost considerations will probably keep wind turbines from growing too much larger than the Zond machine. "At some point, the bearings and other components become so big that non-standard parts are required," says Karas. Jumbo turbines also mean oversize cranes and shipping containers.

Zond's proposal for the Columbia River Gorge would put 50 Z-40 turbines on Oregon's Sevenmile Hill in 1996. Across the river, in Washington, a consortium of small public utility districts plans to equip its wind power plant with small, lightweight turbines designed by Seattle-based Advanced Wind Turbines. Representing a different school of turbine thought, the designers of the AWT-26 opted for brains over brawn. While the Z-40 is built like a tractor, in keeping with the European tradition, the AWT-26 embodies an American predilection for aerodynamic engineering.

FloWind Corp. of San Rafael, Calif., which sells the AWT-26, is also working on a new turbine that looks like a giant eggbeater. The advantage of this vertical-axis design is that the extruded aluminum blades can capture wind from any direction. Also, heavy components like the gearbox and generator can be placed on the ground rather than atop a tower. The disadvantage is that the winds near the ground are usually weaker than the winds at the top of a tower.

The only company making such vertical-axis wind turbines, FloWind has installed more than 500 in California. The company stubbornly refuses to give up on the design, and many engineers say it is not inherently inferior to horizontal-axis turbines. "I think it's going to end up being a site-specific thing," says wind energy program manager Sue Hock of the National Renewable Energy Laboratory.

FloWind is currently developing a new vertical-axis turbine called the EHD, short for Extended-Height-to-Diameter ratio. In essence, the company is "stretching" its turbine to reach the winds at higher altitudes. The new design also allows denser turbine spacing.

Vertical-axis turbines aren't the only wind machines that don't look like traditional windmills. Over the years, dozens of alternative designs have been proposed, and some are still being built. Among the most bizarre ideas is the installation of small turbines in highway median barriers to take advantage of the wind created by passing vehicles. Another proposal would mount large turbines at the base of a 3,300-foot-tall cylindrical tower erected in a desert near the sea: Saltwater sprayed into the tower would cool as it evaporated, creating a downdraft to drive the turbines.

Even among conventional windmill-style turbines, there are variations in size, number of blades, and orientation to the wind. The rotors on modern turbines are typically located upwind of the tower. Sensors tell the machine which direction the breeze is coming from, and a motor automatically rotates the turbine to face the wind.

A few machines like the AWT-26 have a downwind orientation. These machines are simpler and less expensive, because they eliminate the sensors and motor needed for point a turbine into the wind. But the blades of the AWT-26 are more vulnerable to fatigue, because they must pass through the tower's "wind shadow" on every rotation. Engineers have attempted to minimize this problem by mounting turbines on narrow, flexible towers: one version uses a slender tube anchored with guy lines. Another has a tubular tower wrapped in a Slinky-like spiral structure that helps break up the wind shadow.

But even these measures can't eliminate the noise produced as each blade passes through the tower's wind shadow. It sounds like a giant bullwhip swishing through the air.

The design for the Northwind 250 turbine eliminates this problem. Created by New World Power Technology Co. of Moretown, VT, the Northwind will be a 250-kilowatt turbine with a motor that yaws the rotor into the wind. "We thought that the penalties for the tower shadow and noise were greater than the cost penalties for the yaw drive," explains company president Clint (Jito) Coleman. Like the AWT-26, the Northwind turbine takes advantage of aerodynamic principles for more efficient operation and uses a two-blade teetered design.

Ultimately, the success of lightweight turbines like the AWT and the Northwind will depend on how reliable they prove to be in the field. "The question is whether more pounds in the air, or more sophistication, will win," says Coleman. "It's a lot easier to make something heavier and stronger than to work out all the details for a teetered machine."

While New World's Northwind and Advanced Wind Turbine AWT use sophisticated aerodynamics for cost savings, Kenetech Windpower of San Francisco looks to sophisticated electronics. Kenetech's turbine, the most efficient on the market, relies heavily on electronic controls. Called the KVS-33, it's a variable-speed machine with a rotor diameter of 33 meters, or about 108 feet. Variable-speed machines have been around since the mid-1980s, but Kenetech's is the first to be widely accepted by the utilities.

The KVS-33 is more efficient than comparable constant-speed machines because its rotor speeds up or slows down to match shifts in wind velocity. The twin generators produce alternating current with a range of frequencies, rather than the precise 60Hz of the U.S. power grid. A power-control system rectifies the fluctuating alternating current and then inverts is back into 60Hz alternating current.

Kenetech also uses advanced electronics to operate the KVS-33. A controller monitors wind gusts and other conditions and automatically adjusts each turbine. Individual turbine controllers report to a central computer that keeps track of utility load requirements. For example, a station located at Altamont Pass in Northern California controls not only nearby turbines but also other in Southern California and Minnesota.

The KVS-33 went into production three years ago, and 577 of the machines had been manufactured by the end of 1994. Kenetech expects to build another 900 this year. The company hopes to erect as many as 345 KVS-33s in the Columbia River Gorge during the next few years.

The Gorge could soon become a battleground for the three design approaches represented by the Zond, AWT, and Kenetech turbines. But already, the technologies in these different designs are converging as engineers begin devising turbines for the 21st century. European machines have begun losing weight, American machines are putting on pounds, and virtually every wind company is working on a variable speed control system for its turbines. The German company Enercon has also developed a 500-kilowatt turbine with a direct-drive generator—instead of a gearbox, it has a huge ring generator coupled directly to the rotor's drive shaft.

Wind experts at federal laboratories predict that the turbines entering the market in five years will be larger and lighter than those used today. They will have 200 foot tall towers, advanced airfoils, variable-speed operation, and direct-drive generators. The experts see room for at least three different configurations of "dream machines" in the marketplace of the year 2000.

As turbine size and efficiency increase, the cost of wind power is expected to continue dropping. Since the early 1980s, the price of electricity from wind has plummeted by more than 80 percent. The average cost is now between six and nine cents per kilowatt-hour, but the wind plants that will go online this year will sell electricity for as little as 3.9 cents per kilowatt-hour. That's competitive with prices for oil, coal, natural gas, and nuclear energy, and nearly as low as prices for hydroelectric power. And, experts say, wind development creates more jobs per dollar invested than these other energy sources.

If the cost of conventional energy sources, such as oil, included surcharges for air pollution, wind power would look like quite a bargain. But even without these considerations, wind has quietly gained ground on its nonrenewable rivals.

Wind power has advanced to this point even as tax incentives for its development have dwindled. Substantial tax credits, especially in California, gave the industry a jump-start in the 1970s and '80s. But all that's left today is a federal tax credit of 1.5 cents per kilowatt-hour of electricity produced during the first ten years of a turbine's lifetime.

Even without hefty tax credits, a few wind companies have thrived. And in recent years, big companies like Westinghouse and Siemens have shows renewed interest in wind. Utilities are also going with the wind. A few years ago, the American Wind Energy Association had only five utility members; today there are 35.

There is trouble ahead, however. Natural gas prices are now so low that some utilities are scrapping plans to expand wind power and are focusing exclusively on gas-fired generating plants. Wind turbines look simple, but they are expensive. About 80 percent of the cost of wind power goes toward building turbines, with only 20 percent for operation and maintenance. For gas-fired plants, fuel is the primary cost, and it is spread out over many years. Even with wind costs at an all-time low, gas is still slightly cheaper. "We're battling over tenths of a cent," says New World Power's Coleman. "That's the game we're playing."

Some utilities are investigating the idea of "green marketing"—selling electricity produced from renewable sources to environmentally aware customers who are willing to pay a slightly higher price. But even wind power is not an environmental panacea. The plants typically require at least 15 acres of land for each megawatt of capacity, although much of the land can simultaneously be used for activities such as farming or grazing. Some people also find the sound and appearance of wind turbines offensive. And worst of all, turbines at some sites have taken a heavy toll on birds, especially raptors.

Although wind power has made big strides in the last few years, its progress has bogged down in states where there is little political support for renewable energy. The nation's strongest winds are in the Great Plains. "North Dakota has a lot of wind, but not a lot of people live there, and there aren't a lot of transmission lines," says Kenetech's Clarence Grebey. California has strongly supported wind power, but at least 14 other states have better wind resources.

Even in places where the wind blows, it doesn't usually blow at a constant speed throughout the year. But in areas where energy requirements soar during one particular season, wind power can sometimes be matched to peak demand. For example, the wind in Maine's boundary mountains blows strongest during the winter, and that is when down-easters most need electricity to heat their homes.

As wind power moves beyond California, turbine engineers are challenged to design machines that perform reliably in cold weather. For example, Kenetech is experimenting with black blades that would absorb heat to prevent icing. And many companies are building tubular towers that offer maintenance workers more protection form frostbite.

New World Power is even designing a turbine to generate electricity for scientific and communications equipment in Antarctica. NASA is interested in the project, because preliminary calculations indicate that there might be enough wind on the chilly plains of Mars to generate power there. ■

Questions

1. What are the contributing factors that are improving the efficiency of wind turbines?

2. Why is wind power popular with the public?

3. How much power can a single Z-40 generate?

Answers are at the back of the book.

15

Can sea power provide cheap energy? Sea power is expressed in waves, currents, and tides. A square mile of surface water is the equivalent of over 7,000 barrels of oil. Ocean Thermal Energy Conversion (OTEC) theorizes that it can extract stored heat energy from the sea directly. OTEC generates electricity by taking the temperature difference of the water between the surface of tropical waters and the frigid depths below. OTEC plants could supply enough power and water to the tropical areas, therefore making them independent of expensive fuel imports. An advantage for OTEC is that as long as the sun continues heating the ocean the "fuel" is free. However, OTEC has a very low thermal efficiency rating. In order for OTEC to run efficiently, it has to develop a system that produces more energy than is needed to run the plant. During the 1970s the U.S. government allotted $260 million to OTEC research, but since 1980 federal support has foundered. OTEC has immense potential, but it also has enormous engineering and cost issues. Once these issues have been overcome, some futurists see OTEC as an essential part of a global exchange from petroleum to hydrogen fuels.

Sea Power

Mariette DiChristina

The world's largest solar collector absorbs an awesome amount of the sun's energy: equal to 37 trillion kilowatts annually—or 4,000 times the amount of electricity used by all humans on the planet. A typical square mile of that collector—otherwise known as the surface waters of Earth's vast oceans—contains more energy than 7,000 barrels of oil.

From the earliest water wheels, humans have sought to tap sea power, expressed in waves, currents, and tides. But a more promising idea extracts that stored heat energy directly: Ocean Thermal Energy Conversion, or OTEC, generates electricity by using the temperature differences between tropical waters drawn from the sun-warmed surface, and those from the chilly 2,500-foot depths below. Near lush Kailua-Kona, on an old black-lava bed on Hawaii's west coast, a test plant produces up to 100 kilowatts net. Rather than creating air pollutants or spent radioactive fuel, OTEC's by-product is not only harmless, it's downright useful: 7,000 gallons per day of desalinated ocean water with a crisp taste that rivals the best bottled offerings.

Using largely conventional components, OTEC plants built on costs or moored offshore could provide enough power and water to make tropical areas, including the Hawaiian islands, independent of costly fuel imports, say proponents. On drawing boards are plans by Sea Solar Power of York, Pa., for a 100-megawatt floating OTEC plant off the Indian state of Tamil Nadu. Other proposals include smaller plants in the Marshall and Virgin Islands. Some 98 tropical nations and territories could benefit from the technology, according to one study.

OTEC has advantages over other ocean-energy schemes. The largest wave-powered devices have produced only a few kilowatts, for example. Waves and currents have low energy potential—that is, they are not consistently vigorous enough to provide much power to run generators. Tides have greater power potential, but the technology to tap them is costly and limited to a few coastal spots where the tide regularly rises and falls at least 16 feet and can be harnessed. One, built across an estuary in Brittany, can generate 240 megawatts. The only North American demonstration project, on Nova Scotia's Annapolis River, can produce 50 megawatts.

OTEC isn't affected by capricious tides and waves, however. The solar energy stored in the seas is always available. Better yet, "that 'fuel' is free as long as the sun hits the ocean," adds Luis Vega, the shorts-clad director of the Kailua-Kona demonstration project.

That turns out to be a necessary bit of good fortune for OTEC. Tropical-ocean surface waters are typically some 80ºF, while those far below hover several degrees above freezing. That temperature gradient gives OTEC a typical energy conversion of 3 or 4 percent. As any engineer knows, the greater the temperature difference between a heat source (in this case, the warm water), the greater the efficiency of an energy-conversion system. In comparison, conventional oil- or coal-fired steam plants, which may have temperature differentials of 500ºF, have thermal efficiencies around 30 to 35 percent.

To compensate for its low thermal efficiency, OTEC has to move a lot of water. That means OTEC-generated electricity has a glut of work to do at the plant before any of it can be made available to the community power grid. Some 20 to 40 percent of the power, in fact, goes to pump the water through intake pipes in and around an OTEC system. While it takes roughly 150 kilowatts of juice to run the Kailua-Kona test plant, larger commercial plants would use a lower percentage of the total energy produced, says Vega.

That's why, a century after the idea was first conceived, OTEC researchers are still striving to develop plants that consistently produce more energy than is needed to run the pumps, and that operate well enough in the corrosive marine climate to justify the development and construction. "It's a beautiful process," says Vega. "But it needs large, costly components." During the 1970s the U.S. government invested $260 million in OTEC research. After the 1980 election, federal support fizzled.

One thing is not in doubt: The theory works. Georges Claude, a frenchman who also invented the neon sign, proved it. In 1930, Claude designed and tested an OTEC plant on Cuba's north coast. His patented invention, a version of OTEC called open cycle, generated 22 kilowatts of power—but consumed more than that in operating, partly because of the poor site choice. Claude's next attempt, a floating plant off Brazil, was thwarted by storm damage to an intake pipe; the luckless inventor died virtually bankrupt from his OTEC efforts.

It's been smoother sailing for the Kailua-Kona plant, operated by the Pacific International Center for High Technology Research of Honolulu. Las September, the Kailua-Kona project took Claude's open-cycle concept to an OTEC world record, generating 255 kilowatts gross of electricity, and 104 net. Operated in a $12-million, five-year project, the plant's power is used by neighboring enterprises at the Natural Energy Laboratory, a Hawaiian facility devoted to developing solar and ocean resources.

Imagine a boiling hurricane. That's essentially what you see through the circular viewing portal when the Kailua-Kona OTEC plant is running. Inside a chamber, air froths from ocean water, forming whitecaps on the turbulent surface. More seawater—9,000 gallons a minute—pours in from 13 upright white plastic pipes. As the pressure inside drops to that of the atmosphere at 70,000 feet, the water abruptly goes ballistic, and 72°F steam shoots about. "That steam is cool enough to touch," a technician advises, "but in that vacuum your hand would blow apart."

After the resulting steam rushes through a turbine-generator, it's condensed back to liquid—desalinated water—by frigid deep-ocean water pulled from other pipes. Less than 0.5 percent of the incoming ocean water becomes steam. So large amounts of water must be pumped through the plant to create enough steam to run the large, low-pressure turbine. That limits an open-cycle system to no more than three megawatts of gross power; the bearing/support system needed for larger, heavier turbines may not be practical. Vega has a solution for this problem, however. "I was influenced by the movie *The Graduate*," he says, referring to the promising career path suggested to a young college graduate. "You know—plastics." Designed with new kinds of lighter-weight plastic or composite turbines, a series of open-cycle-system modules might together create ten-megawatt-size plants, he says. That's still not impressive as conventional power plants go. A large nuclear reactor, for example, can produce 1,000 megawatts.

Another type of OTEC system, called closed cycle, can more easily be scaled up to a larger industrial size; it can theoretically reach 100 megawatts. In 1881, French engineer Jacques Arsene d'Arsonval (who was later to become Claude's teacher and friend) originally conceived this version, although he never tested it.

In closed-cycle OTEC, warm surface water vaporizes pressurized ammonia via a heat exchanger. The ammonia vapor then drives a turbine-generator. The cold deep-ocean water condenses the ammonia back to liquid at another heat exchanger. Closed cycle's high-water mark to date was a floating test plant called Mini-OTEC that produced 18 kilowatts of net power in 1979.

Turbines are already commercially available for use with a pressurized-ammonia system, which gives closed cycle an advantage for installations that would require large amounts of electricity. The technology nonetheless requires large, expensive heat exchangers. New heat exchangers will begin testing next January at a 50-kilowatt (gross) closed-cycle experimental plant that is soon to be constructed at the Kailua-Kona site. The heat-exchanger will employ roll-bonded aluminum, which is less costly than the titanium previously used in OTEC experimental plants.

Researchers there will also monitor the aquaculture tanks located downstream. They want to determine the effects on marine life from any ammonia that might escape from the plant, as well as from the small amounts of chlorine added to the ocean water to prevent equipment fouling from algae and other varieties of marine creatures.

The Kailua-Kona test plants will also help reveal the answer to one of the biggest OTEC unknowns: the eventual life cycle of components, which are continuously besieged by the ocean's corrosive salt spray and biofouling. "We're discovering how to deal with rust," says Vega.

Because open cycle doesn't scale up easily, and closed cycle produces no drinking water, "the jury's outon which way to go—open or closed," says Vega.

Combining the two systems may yield the best of both: A hybrid OTEC could first produce electricity by closed cycle. Then, the hybrid system could desalinate the resulting warm and cold seawater effluents using the open-cycle process. Adding such a second stage to an open-cycle plant could also double water production.

Ultimately, OTEC has great potential—along with a generous share of remaining engineering and cost issues. Futurists see OTEC as an essential part of a worldwide switch from petroleum to hydrogen fuels; ocean-based OTEC plantships could electrolyze water for hydrogen. "OTEC is environmentally benign and could provide all of humanity's energy needs," declares Tom Daniel, scientific/technical program manager of the Natural Energy Laboratory.

Funding for the Kailua-Kona open-cycle plant runs out after 1995. The next step, as Vega sees it, must be to construct a scaled commercial plant of about five megawatts, and to operate it for one to two years. Such a plant could cost about $100 million over a five-year construction and development period—a stiff price, perhaps, for the tropical locales that would most benefit from OTEC's eventual use. "We need to go through a money-losing proposition to prove the money-making one," emphasizes Daniel. "That's where I believe the government should come in."

Like other forms of renewable energy, OTEC won't play well if that government considers only the immediate bottom line. Large OTEC plants could become cost-competitive if oil doubles from its current $18 or so a barrel, says Vega. Oil prices don't include what Vega and others call "externalities," such as money spent coping with the polluting effects of burning hydrocarbons or military defense of oil fields. Factoring in oil defense alone would make oil's "true" cost $100 a barrel, says energy guru Amory Lovins. (Among the closed-cycle test plant's funders is the Department of Defense Advanced Research Projects Agency, which consider the development of new fuel sources to be of importance to the nation's defense.)

Like every other method of generating power, OTEC is not entirely innocent of environmental consequences. The flow of water from a 100-megawatt OTEC plant would equal that of the Colorado River. And that water would also be some 6°F above or below the temperature it was when it was originally drawn into the plant. The resulting changes in salinity and temperature could have unforeseen consequences for the local ecology.

Can the tide turn for OTEC? A self-acknowledged dreamer, Vega professes to have no illusions. "Some people think I'm conservative, and some think I'm crazy," he sighs. The truth is somewhere in between." ■

Questions

1. How does OTEC compensate for its low thermal efficiency?

2. Briefly explain OTEC's "open cycle."

3. Briefly explain OTEC's "closed cycle."

Answers are at the back of the book.

16 *The large expanse of land surrounding the Chernobyl nuclear power station, Reactor 4, which exploded nine years ago, is inhabited—not by people, but by genetically altered plants, insects, and field mice. Wild boar, roe deer, herons, and swans are in plentiful evidence. Ironically, the worst nuclear catastrophe in history is providing scientists with a rare opportunity to study how wild life adapts to severe adversity. The multitude of evolutionary deviations some of the animal species are going through since the accident is greater than would typically occur in 10 million years. The genetic impact of this disaster, for the animals and for the people who once lived close to Chernobyl, is far from clear.*

The Truly Wild Life Around Chernobyl

Karen F. Schmidt

Many animals are in evolutionary overdrive.

At first glance, the glistening marshes of Glebokye Lake in Ukraine appear to be paradise. Wild boars stomp by, roe deer leap through the waist-high grass and myriad herons and swans feed in the shallows. Look on the horizon, though, and the skyline of Pripyat, a ghost city abandoned by 45,000 people, is visible. So is the red-and-white-striped tower of the Chernobyl nuclear power station's infamous Reactor 4, which exploded nine years ago. Paradise? Not even close. Glebokye is one of the most radioactive lakes in the world.

There is no disputing that the worst nuclear disaster in history was a human tragedy. But paradoxically, it is providing scientists with a unique opportunity to study how life adapts to extreme adversity. For more than two years, researchers affiliated with the University of Georgia's Savannah River Ecology Laboratory—armed with respirators, dosimeters and protective clothing—have gone to Ukraine to study Chernobyl's flourishing wildlife. On a recent trip, they allowed a *U.S. News* reporter unprecedented access to observe their work. "All your life you're told of the dangers of radiation, and here are all these organisms living with it," says team leader Ronald Chesser. "How are they managing to survive?"

One clue is being revealed this week at a meeting of the Society for the Study of Evolution in Montreal. Robert Baker of Texas Tech University in Lubbock, and a member of Chesser's group, is presenting startling evidence that Chernobyl field mice are undergoing an extremely rapid rate of evolution. Indeed, he says the amount of evolutionary change in some animal species since the accident is greater than would normally occur in 10 million years.

Life Altering

Investigating how small creatures such as Chernobyl field mice are adapting and evolving is part of an emerging scientific discipline called evolutionary toxicology—the study of how radioactive and chemical pollutants alter the life course of species. "Man is deflecting the path of evolution; 200 years from now, we may be living with organisms that are genetically quite different from today's," says John Bickham of Texas A&M University at College Station.

Few predicted a swift comeback of wildlife around the Chernobyl plant. Flaws in the reactor's design combined with judgment errors by the operators caused an explosion on April 26,1986, which belched into the atmosphere at least 10 times the amount of radiation released by the atomic bomb dropped on Hiroshima in 1945. More than 5 million acres of prime farmland were contaminated, and more than 160,000 people were forced to abandon their homes. The accident also caused nearly 1,500 acres of the surrounding forest to die almost immediately. (Pine trees have large chromosomes that are particularly sensitive to radiation.) Soviet researchers noted steep declines in wild animals and die-offs of small insects and worms living in the forest litter. Cattails with two and three "heads" became common.

Today, much of the original radioactivity has disappeared. Contaminants with short half-lives, such as radioiodine, have completely decayed and longer-lived ones, such as plutonium, radiocesium and radiostrontium, have settled deep into soils, says Richard Wilson, a physicist at Harvard University and a Chernobyl expert. However, high surface radioactivity remains in patches, and contaminants still circulate within the food chain. Mushrooms and berries set Geiger counters screaming. And boars, deer and mice captured in the zone have taken up radiocesium in their muscles and radiostrontium in their bones.

The abundance of wildlife is a puzzle, given what is known about the biological effects of radiation. Indeed, high-tech probing of the cells of animals from the region challenges the conventional wisdom that animals cannot tolerate a high rate of genetic change. Here's why: Extensive studies from Hiro-shima and cold war laboratories have shown that radiation breaks chromosomes and the strands of the DNA double helix, which contain the blueprints for making the body's proteins. Most of the time, genetic damage signals a cell to die a programmed death, or else it enlists repair enzymes to restore the genetic code. Problems arise when genetic mistakes aren't fixed or are repaired incorrectly, and persist as mutations. In the body, such genetic errors can lead to birth defects in offspring and cancer. More subtly, mutations can cause cells to produce faulty proteins, such as those important for immunity.

While Chernobyl mice don't look like "mutants," on closer inspection, they have many breaks in their DNA strands and a phenomenally high mutation rate. Baker and Ron Van Den Bussche at Texas Tech analyzed DNA from five voles—a type of field mouse—captured within the contaminated zone and compared it with DNA from voles living outside the zone. To search for signs of genetic mutation, they read the code of a gene called cytochrome b that, because it is passed down directly from mother to offspring and changes slowly, is used by scientists as a sort of "genetic clock" for estimating genetic relatedness. As expected, the voles from the area outside the zone had essentially the same cytochrome b gene. But among the Chernobyl voles, the gene sequences as well as the proteins from all five animals were different. Indeed, the differences in the genes between two Chernobyl voles were greater than those normally found between mice and rats, two species that diverged about 15 million years ago.

How Adaptable?

During mammalian evolution, the rate of spontaneous mutation of one letter in the genetic code has been estimated at 1 in a billion per generation, says David Hillis, who studies molecular evolution at the University of Texas at Austin. But at Chernobyl, the mutation rate in the cytochrome b gene is 1 in 10,000. The important question now is whether rapid mutation observed in the cytochrome b gene is also occurring in other genes. "If that kind of high mutation rate can be tolerated across the entire genome," says Hillis, "it would indicate that mammals in particular are a whole lot more resilient than anyone ever guessed."

Still, the full impact on the mice of this rapid rate of mutation is far from clear. While populations seem to thrive in the Chernobyl environment, individual mice may be living on the edge of their ability to adapt to stress or paying a price with shorter life spans, says Chesser. His lab is now investigating whether Chernobyl mice have had to maximize their resistance to cancer by keeping a well-known tumor suppressor gene called p53 turned on full throttle.

People from the region are clearly paying a heavy price for the accident as cancer and other disorders related to genetic mutations are rising in those who received the heaviest doses. Last month, scientists reported in the journal *Nature* that rates of thyroid cancer in Ukrainian children have climbed fivefold overall and 30-fold in those who lived nearest to Chernobyl. Higher rates of spontaneous abortion and birth defects have been documented in Belarus, but it's not clear that radiation exposure is the primary cause, says Martin Cherniack, an associate professor of medicine and international health at Yale University. Researchers also expect at least a small increase in leukemias and other cancers to show up over the years, but that, too, may be hard to trace to the Chernobyl accident.

The human toll triggers a conflict of emotions for Chesser. "The research potential is very exciting, but I also feel the sadness of all the people who were betrayed," he says. The picturesque landscape near Chernobyl will be repopulated in the coming years, but more by genetically adapted plants, insects and field mice than by people. ■

Questions

1. Define evolutionary toxicology.

2. If rapid mutation observed in the cytochrome b gene is occurring in other genes, what would this indicate?

3. What factors contributed to the Chernobyl explosion and how much radiation was released compared to that of the atomic bomb dropped on Hiroshima?

Answers are at the back of the book.

17

Pollution, drainage, and changes in water flow over a long period of time have contributed to the decline of the Florida Everglades. Approximately 55 endangered species are threatened. Vertebrate populations have been reduced 75 to 90 percent, and the wading-bird population has decreased 90 percent. The restoration of the Everglades is the most ambitious effort in North America to restore an ecosystem. Over the next 15 to 20 years, the U.S. Army Corps of Engineers, and other federal agencies will collaborate with local citizens to achieve this. Wetlands managers from around the world are keeping a close watch on the restoration. If it works, wetlands from Australia to Brazil can be restored and the Everglades will become a worldwide model.

Bringing Back the Everglades

Elizabeth Culotta

Amid great scientific and political uncertainty, ecosystem managers in Florida are pushing ahead with the boldest—and most expensive—restoration plan in history.

When steamships plied central Florida's Kissimmee River early in this century, passengers on ships traveling in opposite directions would spot each other across the marshes in the morning, then traverse the serpentine waterway for a full day before meeting. But in the 1960s, the U.S. Army Corps of Engineers straightened out the Kissimmee. In the name of efficiency and flood control, they dug 56 miles of straight canal to replace 103 miles of meanders—and destroyed at least 1.2 million square meters of wetlands in the process. The river was once home to flocks of white ibis; today it boasts the cattle egret, accompanying herds of cows grazing on the canal's linear banks.

But at one spot on the central Kissimmee, boats must again follow the twists and turns of the old river channel. The Corps is slowly putting the kinks back into the Kissimmee. By working with the state of Florida to restore the wetlands, they hope to bring back the invertebrates, fish, and, eventually, the wading birds that once nested here. With an estimated price tag of $370 million, this is the most ambitious river restoration in U.S. history.

It is, however, a mere drop in the watershed compared to plans for the rest of south Florida. Over the next 15 to 20 years, at a cost of roughly $2 billion, the Corps and state and other federal agencies plan to replumb the entire Florida Everglades ecosystem, including 14,000 square kilometers of wetlands and engineered waterways. It's an urgent task, planners say. For after decades of drainage, altered water flow, and pollution, the Everglades is dying, and as they go, so goes the region.

If wetlands that once replenished underground aquifers stay dry, cities may face future water shortages. Anoxic conditions threaten fish in Florida Bay, saltwater intrudes into marshes and drinking wells, and wildlife—including 55 endangered or threatened species—is at risk. "This is not rescuing an ecosystem at the last minute. This is restoring something that has gone over the edge," says George Frampton, assistant secretary of the Department of the Interior and chair of the federal interagency South Florida Ecosystem Restoration Task Force.

More than the ecology of southern Florida is at stake. Wetland managers from Australia to Brazil are keeping a close eye on the project as they search for ways to restore their own ravaged regions. If planners can pull it off, the Everglades restoration will become a world model, says wetlands expert Joy Zedler of San Diego State University, who notes

Reprinted with permission from *Science*, June 23, 1995, Vol. 268, pp. 1688–1690.

that most restorations "are the size of a postage stamp compared to the Everglades." James Webb, Florida regional director of the Wilderness Society and a member of the Governor's Commission for a Sustainable South Florida, puts it another way: "If we can't do it in the Everglades, we can't do it anywhere."

The overall goal of the restoration is to take engineered swampland riddled with canals and levees and transform it into natural wetlands that flood and drain in rhythm with rainfall. Planners hope the entire ecosystem—plants and animals—will blossom as a result. "Wet it and they will come" is the unofficial motto. But because no one understands all the complex ecology involved, planners must accept a hefty dose of scientific uncertainty. "We really don't know what we're going to get out there," says biologist John Ogden of Everglades National Park. And the Corps and the South Florida Water Management District (SFWMD), cosponser of the restoration, still haven't come up with a final blueprint for the replumbing.

The other big unknown in the Everglades is political. Would-be rescuers represent a surprisingly broad coalition of interests and money, from federal and state agencies to environmentalists and urban developers, who want a steady water supply. But holding such a diverse coalition together over the planned life of the project will be tricky. Moreover, the steep price tag—of which one third is supposed to come from the federal budget—and extensive federal involvement run counter to Washington's current budget-cutting mood. Indeed, some of the agencies now contributing expertise and money, such as the National Oceanic and Atmospheric Administration, are high on the list of candidates for political extinction (*Science*, 19 May, p. 964). "We have the technical knowledge to do the restoration," says Ogden. "But I worry about sustaining the political will."

A River of Grass

In the late 1800s, when fewer than 1000 people lived in what are now Dade, Broward, and Palm Beach counties, water spilled over the banks of Lake Okeechobee in the wet season and flowed lazily southward to Florida Bay. This was the "River of Grass," a swath of saw grass and algae-covered water 50 miles wide and only a foot or two deep. People found the vast swamp inhospitable—too wet and too many bugs—but its mosaic of wetland habitats supported a stunning diversity of animals and plants, including huge colonies of wading birds.

Then the human migration to Florida began. In order to make the River of Grass and adjacent marshlands suitable for cities and agriculture, about half of the Everglades was drained in successive waves of development starting early this century. The mammoth flood-control project, built by the Corps at the behest of the state of Florida, transformed the hydrology of both public and private lands. Today, water is channeled swiftly through 1600 kilometers of canals and 1600 kilometers of levees, stored in parks called "water conservation areas," and partitioned by countless water-control structures. The River of Grass is interrupted by the world's largest zoned farming area, the Everglades Agricultural Area (EAA), south of Lake Okeechobee. To prevent flooding, "extra water" is diverted east and west to the Gulf of Mexico and the Atlantic Ocean. The whole system is completely artificial, says Lewis Hornung, a Corps engineer responsible for undoing much of the work of his predecessors on the Kissimmee.

The old Corps engineers recognized that their work would alter the natural world, says Hornung. But no one predicted the devastating effects. For example, hundreds of thousands of birds once nested around the headwaters of the Shark River in the southern Everglades. But as water was drained away further north, the marshes dried out more often, salinities rose—and the birds left. Throughout the Everglades, wading-bird populations are down by 90%. All other vertebrates, from deer to turtles, are down from 75% to 95%, says Ogden. "What we have out there is not the Everglades," he says. "It's a big wet area with spectacular sunsets, but functionally it's not working at all. The animal life in many places is no better than you'd see in roadside ditches in Florida in the summer."

The Best Laid Plans

The good news is that hydrological damage may be reversible, explains ecologist Lance Gunderson of the University of Florida. "It's all there except the water. ... If we redo the hydrology, it will explode," says Richard Ring, superintendent of Everglades National Park.

But will the flora and fauna come back? Anecdotal reports from marshes in the northern part of the park suggest that the wetlands do indeed revive when fresh water returns, says Steve Davis, senior ecologist at SFWMD. "And it's sort of common sense," adds Robert Johnson, chief hydrologist at the national park. "Wetlands need to be wet." Still, to date scientists can't cite the results of any large-scale reflooding study to prove this point. Says Ogden: "Hydrological restoration doesn't equal ecological restoration. This is a big uncertainty, and we need to design flexible plans to deal with it."

Plans are already shifting. In late 1994, the Corps released a preliminary study that outlined six alternatives for revamping the hydrology, although they didn't endorse any specific option. Planners now say none of the six is likely to be the solution, admits Stuart Appelbaum, who directs the Corps' Everglades planning process. There's simply no consensus yet on exactly how to increase water storage and flow while guarding against floods. Nor have restorers made tough decisions about which lands to acquire from private owners for water management. The Corps has gone back to its planning; Appelbaum says a coordinated restoration blueprint is due in 6 years.

Frampton and others want a plan sooner. But in the meantime, restorers point to three smaller, independent hydrological efforts that are already entering construction. One is the Kissimmee. A second project will funnel more water to Shark Slough in the northeastern part of Everglades National Park, and a third will create a buffer strip between wetlands and drained crop fields along the park's eastern border. "At least there are three projects you can point to that are more than just words or paper, where things are actually happening," says Colonel Terrence "Rock" Salt, executive director of the federal task force.

These planners are using an approach they call "adaptive management," which basically means learning by doing. For example, as part of reflooding Shark Slough, Hornung's crew needs to move water from one water-conservation area to another. To do so, he could either build a canal—which models say is more efficient—or simply tear down the levee between the areas. He's experimenting by degrading part of the levee and watching what happens.

To researchers, such experiments are nothing out of the ordinary, but admitting that the outcome is unknown is a new idea for engineers accustomed to having a plan and sticking to it, says Salt. "In our legal system (and there have been many lawsuits over the Everglades already), uncertainty is an admission," says Davis of the SFWMD. "And now here we are starting off up front admitting and defining it."

Quality Control

One thing ecologists do know is that water quality, as well as quantity, will be a crucial part of any restoration. Reflood the swamps with polluted water, and the historic system in unlikely to return. Says biologist Douglas Morrison of the National Audubon Society in Miami: "You can say, 'Wet it and it will grow'—but then the next question is what will grow?" In the Everglades, the answer is often cattails. These tall plants were once only a small part of Everglades vegetation, cropping up around high-nutrient areas like alligator holes. But today in some places, cattails nearly 4 meters tall completely blanket the wetlands, says ecologist Ronald Jones of Florida International University. "It's a massive conversion at the landscape lever," agrees biologist Wiley Kitchens of the National Biological Service in Gainesville, Florida.

The culprit: phosphorus. The historic Everglades had extremely low concentrations of this nutrient, says Jones. Today, extra phosphorus enters the system from the EAA, where water used to irrigate fertilized sugarcane fields picks up a load of phosphorus, then is swiftly channeled to the water-conservation areas. There is spurs nutrient-loving vegetation like cattails and blue-green algae. Jones argues that to be true to the historic system, the Everglades needs very low levels of phosphorus—perhaps as low as 10 parts per billion. "The sugar growers say we want it cleaner than Perrier—and that's true, for phosphorus. That's just the character of the Everglades," he says.

Not surprisingly, the sugar growers are

unconvinced. "Ten parts per billion—what's the basis for that? Parts of the Chesapeake Bay watershed are at around 400 ppb," says Peter Rosendahl, vice president of environmental communications at Flo Sun, one of the major sugar companies. He points out that no one really knows how much phosphorus the Everglades can handle; studies are under way now. "There's no real reason to believe that extra nutrients are the cause of the decline in the Everglades," he says.

A partial solution to the problem, one mandated by an act passed by the state legislature last year, calls for a ring of artificial marshes around the EAA to filter phosphorus form the water. A test marsh full of cattails is already up and running.

There are other thorny water quality issues, however. Chief among them is mercury, which is mysteriously contaminating fish and wildlife in the heart of the remote Everglades, to the point that fishers are advised not to eat their catch. So far, no one knows where the mercury is coming from or just how much damage it's causing, says Dan Scheidt, south Florida coordinator at the Environmental Protection Agency. But whether the issue is phosphorus or mercury, it's increasingly clear that specific goals for water quality will have to be addressed in the coordinated restoration plan. "We have some movement on the hydrology," says Salt. "But we haven't yet looked at water-quality issues holistically—and we need to.

Supporting the Swamp

The depth of the political backing for the plan also concerns planners. In the current political climate, it's hard to count on ongoing federal commitments. Webb of the Wilderness Society worries that popular support is "like the River of Grass itself—miles wide and only a few inches deep." There's also the small matter of aligning dozens of government agencies and interest groups, from sugar-cane growers to Indian tribes. For example, sugar-cane researcher Barry Glasz of the Department of Agriculture says he doesn't even like the work "restore," because to him it suggests turning the clock back to a time before agriculture. Indeed, many environmental groups would like nothing better than to reduce the sugar industry's presence in South Florida. "The EAA has about half a million acres of sugar. We'd like to see maybe one third of that taken out of production and become wetland or water-retention areas," says Ron Tipton of the World Wildlife Fund.

On the other hand, surveys have shown strong public support for saving the Everglades, says Davis of the SFWMD. And urban planners and utility officials—who want to guard the water supply—agree with environmentalists that some hydrological restoration is needed. In the historic system, wetlands cached rainfall for months and so recharged the ground water of the Biscayne Aquifer, which supplies the thirsty cities of Florida's southeast coast, explains Tom Teets, water supply plannner for the SFWMD. Now much of the rainfall is shuttled out to sea long before it seeps into the ground. Water supplies are adequate for the 4.1 million people who lived in Florida's urban southeast coast in 1990, but Teets and others worry about the 6 million expected to live there by 2010. "We get 60 inches of rainfall, but we can't retain it because the water has been managed so poorly," says Jorge Rodriquez, deputy director of the Miami-Dade Water and Sewer Department. "So we feel everyone can benefit from restoration."

Adjacent to the test fill in the central Kissimmee, water is once again flowing through the ancient oxbow turns. The area affected is too small to see a large influx of wildlife, says Louis Toth of the SFWMD, the Kissimmee's resident biology expert. But, vegetation is slowly colonizing the filled-in canal, and game fish are spawning in the newly restored flood plain. Whether uncertain science and precarious political support can engineer a similar recovery for the whole Everglades, however, is still too far downstream to see clearly. ■

Questions

1. What is the estimated cost for this ambitious restoration project.

2. Wading bird populations in the Everglades have been reduced by how much?

3. What will restoration do to the 56 miles of straight canal?

Answers are at the back of the book.

18

Biologists today face the challenge of educating the public to the immediate seriousness of biological diversity, climate change and human population growth issues. By creating more effective means of interaction with social scientists, biologists can contribute significantly to sustainable development by integrating the biological processes model with the economics model. Population growth and economic degradation have continued unimpeded despite promising advances such as biological survey, ecosystem management and adaptive management. Limited human resources in the environmental sciences is the most difficult obstacle in managing global environmental problems. A basic knowledge of biology and its societal effects is necessary in order for the public to support policy initiatives. Unless biologists recognize that now is biology's moment in history and act to educate the public, the end result may be social chaos and conflicts over scarce resources.

Will Expectedly the Top Blow Off?

Environmental Trends and the Need for Critical Decision Making.

Thomas E. Lovejoy

The real challenge is how we as biologists can create a sense of urgency about biological diversity, climate change, and human population growth.

Six years ago—when the population was 5.0 billion compared with today's 5.5 billion—I spoke at the AIBS annual meeting about environmental trends and the imperative to make critical decisions in the next decade (Lovejoy 1988). Today provides an opportunity to review how successful we have or have not been. On the face of the matter, it does not look very good. Trends in population growth, atmospheric levels of carbon dioxide, and deforestation continue largely unaltered. To these trends one can now add the lugubrious state of all major fisheries; as Hardy Eshbaugh of Miami University of Oxford, Ohio, expresses it, we have clear-cut the seas.

Let us begin, however, by looking at the population issue which can overwhelm all others. First, at the plus side of the ledger, in September 1994, sovereign states are to meet in Cairo for the International Conference on Population and Development. The population policy of the United States has been revised into a humane, proactive effort to bring human numbers under control. The president of the United States—for what I believe to be the first time in history—has made a strong policy statement even though it has been ignored by the media. The role and empowerment of women are recognized as integral to any successful progress. We have learned that there are ways to make progress through education, particularly women's education, and through the availability of contraception. That is good news, because we cannot afford the increase in population that would follow were we to wait for the effects of increases in standards of living and declines in infant mortality. As with many aspects of the environmental challenge, to control the population we need to work on several fronts simultaneously. Arrayed against these encouraging steps are ideological forces that somehow manage to ignore the basic verity that abortion represents the failure of family planning.

In another positive step, the United States has reversed its awkward stance on the Biodiversity Convention. We have signed the convention despite

Thomas E. Lovejoy, Will Expectedly the Top Blow Off? *BioScience* from Supplement 1995, pp. S–3 to S–6.

its imperfections, and the ratification is now before the entire Senate due to the foreign affairs committee vote of 16–3 in its favor.[1]

In March 1994 the United States sent an entirely scientific delegation to the science meeting under the convention, signaling the constructive outlook of the new policy. The United States has not waited for ratification or formal international action before improving its national policy and actions with respect to biological diversity. In 1993 the Secretary of the Interior created the National Biological Survey[2] to consolidate into a single agency the field biology work of the US Department of the Interior. This agency has a large agenda and precious little new funding, and it has yet to receive appropriate statutory authority. Nonetheless, on Capitol Hill, its constructive scientific purpose is now better than it was initially understood. A first-class ecologist, Ron Pulliam, has been recruited as its first director. The new agency has indicated from the outset that it can only hope to succeed in its survey function through broad collaborative efforts within and without government.

If anything is now clear, it is that we in the United States—in fact, human society generally—can no longer approach environmental problems in unrelated increments and fragmented jurisdictions. Indeed, institutional fragmentation is as serious an environmental problem as habitat fragmentation. Nowhere is this situation clearer than in south Florida, where the accumulation of decades of decisions, each of which appeared reasonable in its own time and context but each made by institutions and interests largely in isolation from one another, has produced ecosystem degradation that is visible from space. Scarcely a drop of water of the Everglade's famous River of Grass flows naturally anymore, with ill consequences for south Florida, Florida Bay, and the reef system off the Keys.

The only possible way to address the problem is through collaborative planning and decision making, which is hardly easy once matters have gone so far. But there is no other solution to what we now call ecosystem management, which when successful maintains ecosystem function and characteristic biodiversity. The concept of ecosystem management, although grown from multiple roots, has as its essence just good common sense: If one approaches management of a large enough unit of landscape early, multiple options provide more flexibility for human aspirations to be met than if one considers small landscape fragments only after damage has been done. Ecosystem management recognizes that we must move from thinking of nature as something that can be set aside discretely and thus protected within a human-dominated landscape to thinking of human populations and activities as taking place within a natural landscape.

Equally profound, and perhaps of more interest to us as scientists, is the emergence of the concept of adaptive management, where management plans are designed as actual experiments. The results, both successes and failures, can thus be evaluated scientifically. In a way, this notion is inherent in the Biosphere Reserve concept of the Man and the Biosphere (MAB) program. The notion of a core area of undisturbed natural community against which one can compare the effects of manipulating a surrounding area makes good scientific sense. Indeed, an adequate network of biosphere reserves can become a national set of ecological standards. In this context, the importance of wilderness areas far transcends the recreational experience that a limited few enjoy within them, because these areas also provide the ultimate context for science and society to judge how the biology of the planet is being managed.

Biological survey, ecosystem management, and adaptive management all presuppose better, more effective, more coordinated, and more open science than US government programs have previously provided. That is not to say that there have not been some superb government science programs. But as the Committee for the National Institute for the Environment, in Washington, DC, has noted (CENR 1995), the programs have been far too fragmented, uneven in quality, and impervious to outside evaluation.

[1]The Senate failed to vote on ratification. It is clearly in the interest of US science and industry to do so, and we need to make this case to the new Congress.

[2]Note added in proof: In January 1995, the name was formally changed to the National Biological Service.

These are problems that the Committee on Environment and Natural Resources (CENR), operating under the National Science and Technology Council chaired by President Clinton, is designed to address. A national forum was held at the National Academy of Sciences (NAS) in late March 1994 to assist in the development of a government-wide strategy in these areas of science, and most recently the CENR Subcommittee on Biodiversity and Ecosystem Dynamics shared working drafts of the implementation plan with the NAS Commission on Life Sciences. The CENR subcommittee has been particularly successful, no doubt due largely to the advances in thinking within the scientific community. Representative of these advances are the Sustainable Biosphere Initiative produced by the Ecological Society of America and the Systematics Agenda 2000, produced by a consortium of scientific societies. As someone who sometimes thinks of life and work in the nation's capital as a gigantic tableau of social primate behavior, I have in my participation in these developments never experienced less of a sense of territoriality between departments, agencies, and subcommittees. The ultimate test, of course, is the future meshing of the conceptual achievements of these documents with the reality of the budget process.

Promising as these advances may be, they nonetheless appear diminutive when compared to the unabated trends in population growth and environmental degradation and the glacially slow progress of the international multilateral environmental agenda. The North/South positions are too ritualized and the rhetoric too generalized and ideological to serve as meaningful solutions to the actual problems. Particularly disturbing is a trend, exemplified by the science meetings under the biodiversity convention, to subvert science with politics. It is essential for the scientific community to remain vigilant and vocal about the need for scientific assessment to proceed independently.

Too frequently opportunities to make progress on environmental problems run aground on the shoals of North/South posturing. Fingers are pointed at northern consumption patterns (of which those of the United States are amongst the highest), and suspicions are raised that the environmental concern emanating from industrialized nations is really a stalking horse to prevent the developing nations from attaining their God-given right to development and higher living standards. Why, it is asked, do you (the northern nations) point fingers at us (the southern nations) about population growth when you are the ones consuming so much of the world's resources ? There is something to those accusations of course, but as clear as it is that 5.5 billion people cannot live US lifestyles, it is equally clear that 5.5 billion people cannot live as hunter-gatherers. Both population growth and consumption patterns are problems. We simply have to recognize that consumption does not equate one-on-one with quality of life and that the consumption patterns that might be labeled as *Yankee* occur in at least some segments of most countries in the world. The real point is that we urgently need to get on with solving these problems rather than engaging in deadlocking rhetoric. Because there is an ethical imperative for each individual to have some minimal level of quality of life, it is fundamentally easier to deal with the consumption issue than with the issue of additional population.

As biologists we have something to contribute to this discussion and this agenda. First, generally speaking, we are more aware of the state of the environment than anyone else. Biological diversity is, after all, the most sensitive indicator of environmental change. Furthermore, it is in our direct interest as scientists to be engaged in the discussion, because the biotic impoverishment of the planet automatically impoverishes the potential growth of the life sciences. Imagine the howls from astrophysicists if someone were to propose eliminating a number—a large number, somewhat at random but including some of the most interesting—of celestial bodies. Similarly, biologists need to stand up and be counted.

Biologists also have an extraordinary amount to contribute to the main solution to the environmental crisis, namely sustainable development. While some consider sustainable development an oxymoron, and while in fact no development will be sustainable if current patterns and bents continue, I believe it is abundantly clear that an important segment of sustainable development inevitably will be biologically based—in fact, derived from biological diversity.

I have made much in the last three years of the multibillions of dollars of economic activity that have derived from the enzyme from the Yellowstone hot spring bacterium *Thermus aquaticus*, which was described by Thomas Brock of the University of Wisconsin in Madison and which makes the polymerase chain reaction (PCR) possible. This reaction—so central to diagnostic medicine and forensic medicine (even making news in the murder trial of football star O. J. Simpson)—has already fed back into strengthened systematic science and population biology. The technique, and thus the enzyme, also is essential to the human genome project and all the incalculable potential that project holds for human society. Little wonder that Kary Mullis shared the 1993 Nobel Prize for chemistry for conceiving of this reaction.

Let us not forget that all this activity based on PCR is possible because of science concerned with biological diversity and because of biological collections (in this case the American Type Culture Collection). These scientific activities were coupled with the lucky accident that the scenic beauty captured in Thomas Moran's watercolors, rather than biological diversity, inspired the US Congress to set aside Yellowstone as the world's first national park in 1872. *Thermus aquaticus*, in fact, thus becomes an argument in itself for biological survey and ecosystem management.

Pursuing this chain of coincidence even further, it is important to bear in mind that the molecular scissors—the endonucleases—that the genetic engineers, biotechnologists, and molecular biologists employ for society's benefit also derive from a biological-diversity toolbox. In the end, molecular biology and the ability to generate wealth at the level of the molecule derive in significant degree from biological diversity.

Exciting new science and practical applications are resulting from the study of microorganisms with weird metabolisms and weird appetites. Bacteria that can break down aromatic compounds and chlorofluorocarbons (CFCs) have been discovered in nature and—together with similar oddities—are part of bioremediation using biological processes for environmental clean-up. Some observers believe that bioremediation is likely to have a short flush of success and then become largely unnecessary when industries reduce pollution at the source. I believe that, to the contrary, bioremediation will be used to reduce pollution at the source. In addition, as industrial ecology grows in sophistication and in practice, bioremediation in the factory is likely to be used to make the waste stream of one industry acceptable feedstock for another.

Organisms and their enzymes are already being used in bioindustry to produce chemicals such as acrylamides. Biological processes for chemical manufacture eliminate the need for toxic catalysts and in some instances high-pressure processes. The biological processes are cheaper and cleaner, and when a more effective enzyme or organism is identified, there is no need to rebuild the factory to accommodate the substitute process.

As biologists, we have to find more effective ways to interact with social scientists. Biologists see the biosphere as ultimately run by biological processes largely driven by solar energy. Economists view the world as largely driven by economics, money, supply, and demand. I believe we should work together to integrate these two models.

There are interesting questions. How should the American oyster population of the Chesapeake Bay be valued? Is its value what it brings to market as seafood annually? Or is the value that the current population filters a volume of water equal to the entire bay once a year, and its value before degradation of the bay that it filtered that same enormous volume once a week? Our economies are riddled with such beneficial subsidies from nature, for which there is no current accounting. Similarly, our economies are riddled with subsidies and incentives that lead to environmental degradation. There is something akin to a Gordian knot here that can only be unraveled by biological and social scientists together.

In the midst of the environmental crisis, organismal biologists in particular are suddenly finding themselves moving from the shadow of the laboratory sciences to the spotlight of world issues. The trick is to accept that responsibility and to be willing to bridge the gap (often a false one) between basic science and its application to societal problems. We particularly have to avoid what often seems like selfish yammering for money for research. While research is a true need, we are far more likely to attain a positive response from society if we are seen as wanting to develop the information necessary to produce good public policy rather than as wanting only to pursue our private, esoteric intellectual pleasures.

At the same time we are probably nanoseconds away—in terms of graduate training time—from recognizing that the toughest limiting factor in addressing global environmental problems is the limited human resources in the environmental sciences, particularly systematics and ecology. Now is the time to be bold and increase graduate training in these fields, even before the specific jobs are in sight.

We also need to seek ways to use experts' time more effectively. One way is to pursue the paramedic model, as Costa Rica's INBio has done with parataxonomists. Another way is to push the frontier of interactive electronic media, as has Australia's CSIRO. They have produced a CD-ROM, for example, that in essence permits anyone to determine the family or subfamily of any beetle larva. While creating these electronic products requires a large specialist contribution, as with the example of the expensive energy-efficient light bulb, the ultimate savings in specialists' time is staggering. These products are essentially redefining the boundary between the amateur and the specialist: empowering the amateur and parascientist while reserving the time of the specialist for those tasks for which that person is uniquely suited.

There also is a tremendous challenge before us with respect to education. Part of this challenge derives from the failure of our much-vaunted system of higher education to provide a minimal modicum of understanding about biology and how it relates to our existence. It is nothing short of scandalous that one can still graduate from most of our universities and colleges without that rudimentary knowledge. A basic knowledge of biology and its implications for society is requisite to responsible citizenship. Even if it were possible to quickly rectify this failing, it would not help the present citizenry make responsible decisions in the home or voting booth.

Every one of us has a particular responsibility to help with public education. There are some encouraging signs, such as the preliminary results of a study by the National Environment Education and Training Foundation showing that environment was a concern even for disadvantaged urban youth. But the discouraging reality is that probably 95% of US citizens do not understand even something as simple as exponential increase.

If that is the case, how can we expect the public to understand the threat of biotic impoverishment and global climate change and to support policy initiatives to address the threat? Those of us in the scientific community have a special responsibility to explain these issues to the public. These issues include explaining that:

- Biological diversity and ecosystems are important to science, to society, and to sustainable development;
- Artificially elevated levels of carbon dioxide are likely to cause ripples through the structure and function of biological communities, because there is no reason to expect every plant species to respond in the same way and degree; and
- Biological diversity, which is largely surviving in landscapes as isolated natural areas, is highly vulnerable to even natural climate change, because species will be unable to disperse and track their requisite climatic conditions.

Most important, we need to explain that even though uncertainty tends to be measured more effectively in science than in other forms of human endeavor, uncertainty is part of almost every kind of decision society makes. Rather than using uncertainty as an excuse for a blasé, lethargic approach to energy policy and greenhouse-gas emissions, we need to explain that the real policy issue is whether we oppose, or favor by default, total-planet experiments that bet the biosphere, if there is even a small chance we may regret the result. After all, there is not even an experimental control planet to colonize if we lose at biosphere roulette.

There is an important lesson to be learned from multilateral negotiations on the environment. One negotiation stands out as particularly successful in producing prompt action: the Montreal Protocol dealing with CFCs and the ozone layer. One can argue that the problem was relatively simple, and the solution was clear and inescapable. Those familiar with international obligations assert that a real sense of urgency was the more important factor in making the protocol work. I would assert that until there is such a sense of urgency about other environmental issues, international negotiation is likely to be dominated by short-sighted self service rather than long-term societal benefit.

The real challenge is how we as biologists can create that sense of urgency about biological diversity, climate change, and human population growth. These are problems that grow by increments and that may not seem of particularly great consequence, but which in aggregate are disastrous. No group is in a better position than are biologists to make this case and make it eloquently. It is likely to be hard, and maybe even impossible, to make significant progress unless we biologists enter the fray with greater energy and passion than we have so far. How can we possibly do otherwise with impending extinction rates projected at 10,000 times the past rates (May et al. in press)?

My speech six years ago was entitled "Will unexpectedly the top blow off?" borrowing from an Archibald MacLeish poem about a circus crowd so entranced by the show that nobody notices a problem until the entire big top of the circus tent blows off. My thesis in part was that environmental problems may grow so large, social chaos and quarreling over dwindling resources are likely to ensue, thwarting any possibility of remedial action—a notion that was given some flesh in R. D. Kaplan's *Atlantic Monthly* article (Kaplan 1994).

Today it is appropriate to pose the question differently: "Will expectedly the top blow off?" The answer, I believe, is yes, unless it is recognized that now is biology's moment in history. We biologists must recognize it first. We need to act now.

Now.

References Cited

Committee on Environment and Natural Resources (CENR). 1995. CENR Integrated Strategy Document. The White House, Washington, DC.

Lovejoy, T. E. 1988. Will unexpectedly the top blow off? *BioScience* 38: 722–726.

Kaplan, R. D. 1994. The coming anarchy. *The Atlantic Monthly* 273(2): 44–63.

May, R. M., J. H. Lawton, and N. E. Stork. In press. Assessing extinction rates. In J. H. Lawton and R. M. Mays, ed. *Extinction Rates*. Oxford University Press, London, UK. ■

Questions

1. What is adaptive management?

2. What three programs provide more effective, more coordinated, and more open science than U.S. government programs?

3. What is the most sensitive indicator of environmental change?

Answers are at the back of the book.

19

Although forests once covered more than 40% of the land surface, they have since been decreased by one-third. Indeed, tropical forests have lost half of their initial area in the past 50 years. Forests shelter many different species, protect soils, affect hydrological cycles, and are crucial to the energy budget and reflectivity of Earth. Most species extinctions today occur from deforestation. The causes of forest decline lie in the lack of understanding of forest values and lack of economic ability to measure their outputs. When we understand more about the intrinsic value of forests, then instituting policy measures will not be as inconvenient as living without the forests.

The World's Forests: *Need for a Policy Appraisal*

Norman Myers

There is need for a fresh policy approach toward forests. An organization is soon to established for this purpose, the World Commission on Forests and Sustainable Development. It is hoped that the commission will move us beyond the negative clamor about forest destruction, and toward a constructive appraisal of how forests can best confer their manifold benefits on society, now and in the future.

Forests once covered more than 40% of Earth's land surface, but their expanse has been reduced by one-third. The most rapid decline has occurred since 1950—tropical forests have lost half their original expanse in the past 50 years, the fastest vegetation change of this magnitude in human history. Temperate forests are in steady state for the most part, but certain boreal forests have started to undergo extensive depletion. In the absence of greatly expanded policy responses, many of the world's forests appear set to decline at ever-more rapid rates, especially as global warming overtakes them.

Forests can supply such an exceptional array of goods and services that they should be reckoned among our most valuable natural resources. Only a few products are generally harvested, however, but with degradation of the forests' many other potential outputs. Thus, forests are overexploited and underutilized.

The consequences of forest loss are far from being recognized in their full scope, especially by political leaders and policy-makers. Forests protect soils. They play a major role in hydrological cycles. They exert a gyroscopic effect in atmospheric processes and other factors of global climate, with an influence second only to that of the oceans. They are critical to the energy budget and the albedo (reflectivity) of Earth. And they harbor a majority of species on land.[1] Thus, there is a vital linkage between forests and the two recent conventions on climate and biodiversity, although the latter are of limited effectiveness without a parallel initiative for forests.[2]

A policy appraisal of forests should address both the scope of changes necessary for forests to undergo sustainable development, and the scope required for forests to contribute fully to sustainable development in the countries concerned and in the world at large. Both prospects can be facilitated by the new commission through an authoritative assertion of all forests' values to society. Forestry has so far been dominated by private interests,

Reprinted with permission from *Science,* May 12, 1995, Vol. 268, pp. 823–824.

commercial for the most part. Certain of these interests could well have an expanded role in the future, but public interests deserve to be better represented in the policy arena, especially the fast-growing interests at a global level.[2]

In light of their exceptional potential to support humanity, why are forests allowed to decline? Well over half of all tropical deforestation is due to slash-and-burn agriculture by displaced landless peasants, sometimes known as "shifted cultivators" (by contrast with shifting cultivators of tradition, who cause no long-term injury to forest ecosystems).[3] Comprising several hundred million of the world's 1.3 billion people living in absolute poverty, these communities should have their plight relieved on humanitarian grounds, let alone to reduce deforestation. They are driven to migrate into the forests by poverty, population pressures, and land hunger, among other reasons.[4] Thus, the source of most tropical deforestation lies in an amalgam of factors that are usually far removed from the forests—and lie outside the purview of traditional forestry measures.

Boreal forests in Siberia are newly declining, primarily through clear-cut logging and fires.[5] The annual loss of these forests encompasses an area twice as large as deforestation in Brazilian Amazonia.[6] Boreal forests in northeastern North America and northern and central Europe are experiencing acid precipitation, with commercial losses of $30 billion a year in Europe alone.[7]

The ultimate source of forest decline lies both in our lack of scientific understanding of forests' overall values and our lack of economic capacity to evaluate many of their outputs. Instead of enjoying their proper place in the mainstream of development, forests tend to be relegated to the sidelines in the councils of power.[8] The Food and Agriculture Organization, the leading forestry agency in the United Nations, has reduced its budget allocation to forestry from a mere 5% in 1975 to 3% today. As a result of its "Cinderella status," forestry's case often falls through a plethora of institutional cracks.

The principal challenge for the commission will be to formulate a policy vision for forests, especially with regard to their role in the biosphere and the world. Here, I provide a selection of possible policy options.

First, the encouragement of sustainable development. Through the myriad goods and services they provide, forests should be enabled to support development sectors as diverse as energy, agriculture, fisheries, water, health, biodiversity, and climate. They can generally do this through their simple existence, and hence do it sustainably. In the spirit of this newly expansive approach with its emphasis on development both within and beyond forests, the maintenance of watershed functions should be seen as a form of "development" that ranks alongside timber harvesting. A national park is as legitimate a form of land use as a paper pulp plantation. Genetic reservoirs count together with agroforestry. Certain forest tracts can serve as extractive reserves. All forests constitute carbon sinks. In a few localities, development can even entail outright preservation of forest ecosystems, some of the most productive and diverse on Earth, for scientific research. Many of these functions can be served simultaneously as well as sustainably.

Second, enhancing forests' institutional status. When forests are treated as the poor relation by those in the corridors of power, forest policy is effectively set by departments of economic planning, agriculture, employment, human settlements, trade, and other entrenched bureaucracies. These agencies decide what forms of government investment, and hence of land use, will predominate, to the detriment of forests.[9] Although it is generally not recognized, basic forest policy is seldom formulated by foresters.

In order to dispel the Cinderella syndrome, policy planners need to appreciate forest outputs in their full scope, both actual and potential. A major reason why this is not done is that forest benefits often accrue to widely dispersed communities in the country concerned or to those in other countries, as in the case of watershed functions, biodiversity, and climate. Over half of the environmental and other eternality benefits of sustainable forest management in Costa Rica accrues to the global community.[10] A rational response would be for the global community to compensate forest countries that supply worldwide benefits, through a mechanism such as the Global Environment Facility. This organization already disburses $700 million per annum to make up the gap between what a country gains through environmental activities and what it loses in benefits to the global community.

Third, the removal of "perverse" subsidies. Much deforestation is fostered by government subsidies. In the United States, subsidies for below-cost timber sales alone amounted to $323 million in 1993, including $35 million for the Tongass National Forest,[11] a rainforest depleted through overlogging more rapidly than most rainforests in Amazonia or Borneo. Covert subsidies in the Philippines, in the form of the government's undervaluation of forest resources, led to revenue losses of $250 million in 1987. Much the same has applied in Indonesia, Malaysia, and the Ivory Coast, among other leading tropical timber countries. Subsidies for cattle ranching in Brazilian Amazonia caused commercial timber losses of $2.5 billion annually during the mid-1980s.[12] These perverse subsidies persist in part because certain governments remain unaware of the all-round and enduring value of their forests, and hence they view the forests as capital to be liquidated.

Fourth, calculating the costs of inaction. It is generally easy to calculate the costs of a specific action—for example, the budget for a fuelwood plantation—by using any of a number of marketplace indicators. It is less easy to calculate the concealed costs of inaction. Thus there is an asymmetry of evaluation. Nevertheless, it is possible to provide surrogate estimates of such costs. For instance, the opportunity costs of those who trek far afield to find fuelwood and thus utilize time that could otherwise have been spent on farm activities amounts to at least $50 billion per year.[13] This contrasts with the costs of tree planting to meet fuelwood needs—$12 billion per year— costs that, in the absence of a comparative evaluation, are viewed as "too high."

A similar reasoning applies to the costs of saving tropical forest biotas, in the absence of figures for the covert costs of losing them. Pharmaceuticals from tropical forest plants have a commercial value of $25 billion a year and an economic value at least twice as large,[14] but this reflects only a small part of the much greater biotic impoverishment that would ensue from grand-scale deforestation.[15] What price tag should we attach to the decline of watershed services in numerous deforested catchments? In India, annual flood damage attributable to deforested catchments amounted to $1 billion to $2 billion in the early 1980s.[16] What value will be lost if we reduce forests' stabilization of the global climate system? Tropical forests with the largest carbon stocks are theoretically worth $1000 to $3000 per hectare per year in terms of global warming injuries prevented[17]—yielding a far higher rate of return than any alternative form of current land use in the forests.

These cost estimates are preliminary and exploratory. They urgently need to be firmed up, as do the many other benefits inherent in forests and amenable to creative economic analysis. Only then will we be in a position to give "real world" regard to the immediate costs of saving forests.

An alternative approach to tackling the asymmetry of evaluation is to shift the burden of proof as it concerns forest exploitation. The once-and-for-all exploiter can generally go ahead with little hindrance. This leaves the conservationist to argue the case for sustainable forms of forest use—a challenge that, in light of the many incommensurable and intangible values at stake, can be taxing indeed. What about requiring an exploiter to demonstrate that his form of forest use will generate economic returns of a sustainable sort exceeding those of any other option?

Fifth, the promotion of forests as global commons resources. By virtue of their many outputs that indivisibly benefit not just forest nations but the world community as well, forests constitute a type of global commons resource. This raises the issue of national rights and international responsibilities on the part of forest nations. Forests lie within the sovereign jurisdiction of individual nations and are subject to the policy discretion of individual governments. At the same time, the environmental services of forests extend far beyond national boundaries by virtue of their watershed basins, atmospheric processes, and climate systems ("the winds carry no passports").

We need to reconcile national prerogatives with international interests, and in a manner that recognizes the environmental interdependencies of the planetary ecosystem. The new commission should foster a coalition of interests as a basis for an eventual international instrument or set of instruments. The more the commission can establish a consensus about the world's forests and their value for all, the greater the chance that individual governments will engage in enlightened forest policies as an authoritative expectation of the community of nations. Instituting many of these policy measures will be difficult—but not as difficult as living in a world that has lost many of its forests.

References and Notes

1. N. Myers, *The Primary Source: Tropical Forests and Our Future* (Norton, New York,1992); G. M. Woodwell, in *World Forests for the Future*, K. Ramakrishna and G. M. Woodwell, Eds. (Yale Univ. Press, New Haven, CT, 1993), pp.1–20.
2. K. Ramakrishna and G. M. Woodwell, Eds., *World Forests for the Future* (Yale Univ. Press, New Haven, CT, 1993). The policy purposes can also be promoted by the new Intergovernmental Panel on Forests under the United Nations.
3. N. Myers, in *The Causes of Tropical Deforestation*, K. Brown and D. W. Pearce, Eds. (University College London Press, London, 1994), pp.27–40. See also Food and Agriculture Organization, *Forest Resources Assessment 1990* (Food and Agriculture Organization, Rome, 1993).
4. R. Bilsborrow and D. Hogan, Eds., *Population and Tropical Deforestation* (Oxford Univ. Press, New York, in press); W. J. Peters and L. F. Neuenschwander, *Slash and Burn Farming in Third World Forests*. (Univ. of Idaho Press, Moscow, ID, 1988).
5. V. Alexeyev, *Human and Natural Impacts on the Health of Russian Forests* (Institute of Forest and Timber Research, Moscow, 1991); A. Shvidenko and S. Nilsson, *Ambio* **23** (no. 7), 396 (1994).
6. A. Rosencrantz and A. Scott, *Nature* **355**, 29 (1992).
7. S. Nilsson, Ed., *European Forest Decline: The Effects of Air Pollutants and Suggested Remedial Policies* (Royal Swedish Academy of Agriculture and Forestry, Stockholm, 1991).
8. N. Myers, *The Primary Source: Tropical Forests and Our Future* (Norton, New York, 1992), pp. 263–265.
9. J. MacNeill, in *Tropical Forests and Climate*, N. Myers, Ed. (Kluwer, Dordrecht, Netherlands, 1992).
10. World Bank, *Costa Rica Forestry Sector Review* (World Bank, Washington, DC, 1992).
11. W. Devall Ed., Clear *Cut: The Tragedy* of *Industrial. Forestry* (Island Press, Washington, DC, 1994).
12. R. Repetto, *Sci Am.* **262**, 36 (April 1990).
13. N. Myers, in *Scarcity or Abundance: A Debate on the Environment*, N. Myers and J. Simon (Norton, New York, 1994), p. 174.
14. P. P. Principe, in *Tropical Forest Medical Resources and the Conservation of Biodiversity*, M. J. Balick *et al.*, Eds. (Columbia Univ. Press, New York, 1993).
15. G. M. Woodwell, Ed., *The Earth in Transition: Patterns and Processes of Biotic Impoverishment* (Cambridge Univ. Press, New York, 1990).
16. Centre for Science and Environment, *State of India's Environment 1982* (Centre for Science and Environment, New Delhi, 1982).
17. D. W. Pearce, *Global Environmental Value and the Tropical Forests* (University College London, London,1994). See also T. Panayotou and P. S. Ashton, *Not by Timber Alone* (Island Press, Washington, DC, 1992).

18. For their constructive comments on early drafts, I thank J. Kent, J. MacNeill, J. Maini, R. Schmidt, Spears, O. Ullsten, and G. Woodwell. ■

Questions

1. Give reasons why forests should be protected?

2. Why are forests in decline?

3. Name five possible policy options needed in order to protect forests.

Answers are at the back of the book.

20

The Endangered Species Act is in jeopardy. Opponents accuse the ESA of violating private property rights and wasting tax dollars. A reform proposal in the Senate would give the secretary of interior the control to authorize a species to become extinct. Occasionally, when resources are scarce, even people who support species protection do not favor rescue efforts. A new law which focuses on protecting areas that support an abundance of life would be practical. Appreciation for natural ecosystems has increased with the understanding that some functions of ecosystems include, managing climate, purifying water, pollinating crops, and building and replenishing soils. The creation of a law, focusing on the protection of areas with a broad spectrum of species by allocating habitat protection throughout an entire region, would make economic and biological sense.

Is He Worth Saving?

Betsy Carpenter

The potent new campaign to overturn the Endangered Species Act

The Endangered Species Act, a law passed with virtually no legislative opposition 22 years ago, is now nearly as imperiled as the creatures it was designed to protect. Last week's Supreme Court ruling upholding the government's power to protect plants and animals on private land has only emboldened critics in Congress who charge that the law violates private-property rights, ravages local economies and wastes precious tax dollars on useless creatures that were on the verge of extinction anyway.

Opponents are now pushing for a major overhaul of the law. Already, legislators have effectively barred new endangered species listings for the remainder of the fiscal year. A reform proposal in the Senate would remove nearly mandatory protections for endangered creatures, giving the secretary of interior the power to permit a listed species to go extinct. Ironically, many ardent advocates of species protection also embrace the idea that in a time of scarce resources not all life forms will merit intensive, expensive rescue efforts. But the national debate over the act has become so polarized that an emerging strategy favored by many scientists and policy makers to protect biologically rich landscapes instead of individual species may well founder amid the partisan fighting.

Unmitigated Unsuccess

No one is happy with the current system. According to an analysis by Charles Mann and Mark Plummer, authors of the recent book *Noah's Choice: The Future of Endangered Species,* the species protection program has produced few success stories. By the end of 1994, only 21 species had been struck from the list, and of those 21, only six were delisted because they had gained enough ground to warrant removal. The others were crossed off when they were declared extinct or found not to have been endangered in the first place. Further, even the six species whose status improved did not always owe their recovery to the ESA. The arctic peregrine falcon, for instance, which was delisted last October, owes its revival primarily to bans on pesticides in the 1960s and 1970s.

As the act's advocates point out, there are more ways of measuring success than tallying up delistings. Many species that linger on the list, such as the whooping crane and the red wolf, are in much better shape than they were 20 years ago, says the Environmental Defense Fund's Michael Bean. Many others are so beleaguered by the time they are listed that recovery is naturally a lengthy process. Further, funding has always been inadequate, contends Bean..

But while the debate over whether the act is saving species continues, there is no question that the law has angered many Americans. "It creates enemies of wildlife," says Ike Sugg of the Competitive Enterprise Institute in Washington, D.C. Once an endangered species is listed, private development and public projects, such as roads and dams, are often halted, at least until federal officials work out a recovery plan. In some cases, that has taken years. Rather than face land-use restrictions, some landowners destroy endangered wildlife and wildlife habitat.

All Species Are Not Equal

At the same time landowners' disenchantment with the law has grown, scientists have increasingly questioned the act's central premise that all life forms are of equal value. "We have to admit that there's a state of triage, and we have to choose where to focus our energies," says Melanie Stiassny of the American Museum of Natural History in New York.

This quiet revolution reflects a fundamental change in scientists' understanding of ecosystems and biological diversity. Twenty years ago, most thought of ecosystems like fragile webs—pluck out one insect or plant and the whole structure might collapse. But a number of new studies looking at a range of ecosystems, from tropical forests to prairie grasslands, have suggested that Mother Nature builds many spare parts into natural communities. "There are species that could disappear without really impacting the ecosystem," says Daniel Simberloff of Florida State University in Tallahassee.

Still, while many scientists now see some species as expendable, their appreciation for the value of functioning, natural ecosystems has soared. "It's only recently we've considered the extraordinary life support services nature provides humans for free," says Harold Mooney of Stanford University. Left to their own devices, ecosystems regulate climate, purify water, pollinate crops, maintain the gaseous composition of the atmosphere, and build and replenish soils, among other vital services. "The [first] act was about protecting big, fuzzy creatures. But now we need an approach that protects whole suites of species," Mooney says.

According to many endangered species experts, a new law centered on protecting landscapes that support a wealth of species would make economic as well as biological sense. Such plans would distribute the burden of habitat protection throughout an entire region, rather than drop it on the shoulders of a luckless few, say proponents. Public lands often could make up the core of a regional plan. In cases where especially important habitat lay in private hands, the property could be purchased outright or traded for less ecologically valuable public land.

Helping Landowners

A habitat-protection approach would have to include incentives for private-property owners who protect biological resources on their land, argues Stephen Meyer, a professor of political science at the Massachusetts Institute of Technology. For instance, tax breaks could be given to landowners who enter into conservation agreements with the government or who harvest resources in a manner that allows rare creatures to thrive.

But while fresh ideas abound, few are eager to champion them on Capitol Hill. Concerned that the act will be gutted, many environmentalists are reluctant even to acknowledge publicly that the current system has flaws. And so far, the Republican Congress seems more interested in weakening the current law than devising new strategies for conserving America's biological heritage. In May, Republican Sen. Slade Gorton introduced a reform bill that would restrict the government's ability to protect crucial habitat on private property. Under the proposed bill, once a species has been listed, the secretary of interior can decide to do little to protect it except prohibit its direct killing. "Presumably, it would be perfectly legal to chop down the nesting tree of a bald eagle or bulldoze spawning habitat for salmon," says the Sierra Club's Melinda Pierce.

But rescue may still come from an unexpected quarter. At a recent hearing on the act, House Speaker Newt Gingrich warned his troops that he is "committed to our having a strong and effective environmental policy" and has a "deep concern for the biological diversity of the planet." It may fall to the man who has led Congress's antigreen forces to keep them in check. ■

Questions

1. Without human intervention, what do ecosystems do?

2. What would be the effect if a reform proposal passed in the Senate?

3. How could a new law make economic and biological sense?

Answers are at the back of the book.

21

The judicial system faces the serious dilemma of how to protect wild animals and private property rights at the same time. When property is affected, most Americans see the Endangered Species Act as an irrational restriction. The law states that one may not harm or threaten an endangered species, including the removal of a species from its habitat. However, over half of the 956 endangered species inhabit private land in the United States. Even ESA proponents admit the law has flaws. The National Academy of Sciences suggested that the creation of temporary "survival habitats" is needed and is being planned. These "survival habitats" are zones for threatened species. Regulations must be developed that simultaneously protect species while respecting private property rights.

Nature, Nurture and Property Rights

The Economist

Back in 1920, in a Supreme Court ruling about migratory wildlife, Justice Oliver Wendell Holmes pronounced that the protection of geese and ducks "is a national interest of very nearly the first magnitude." Two years later he conceded that "the general rule is that while property may be regulated to a certain extent, if regulation goes too far it will be recognized as a taking."

These paradoxical statements, from one of America's best jurists, sum up a quandary of the modern world: how to protect wild animals and private property rights at the same time. Last week the Supreme Court gave the nod to nature. By a 6-3 vote, the court ruled that the federal Fish and Wildlife Service may use the Endangered Species Act (ESA) of 1973 to protect natural life on private land. The government, for example, can forbid a landowner to cut a tree on his land because its presence is necessary for an animal or a plant of a kind close to extinction.

America has a solid record of protecting wild animals in general. With few exceptions—the passenger pigeon and the heath hen among them—endangered creatures get the government's protection, whatever it costs. DDT was banned, at great expense to farmers, because this insecticide threatened the brown pelican and the peregrine falcon. A prohibition on shooting egrets once cost some companies dearly.

The battle starts when the right to property is affected. The fifth amendment to the constitution says that no private property shall "be taken for public use without just compensation." Some intrusions on property rights—zoning laws, for example—are usually accepted without much argument. But not the ESA. Despite countless polls showing that Americans value wildlife, most of them see this law as an unreasonable restriction.

The Supreme Court's new decision focuses on the work "harm." The law says you may not harm a threatened or endangered species. The Fish and Wildlife agency interpreted this to include damaging or changing the places where such creatures live. Its opponents, paper and timber companies, contended that the law applied only to the animal itself. The court disagreed, citing the almost universal opinion of scientists that an endangered species cannot work its way off the list of doom if its habitat is at risk.

The matter now goes back to Congress. Politicians have in the past, tended to do as they pleased, regardless of what science or the courts say. In 1978, the lowly snail darter—an endangered fish the size of a stick of gum—held up completion of the $110m Tellico dam in Tennessee. The fight went to the

Supreme Court, which said that the ESA was enacted to "halt and reverse the trend toward species extinction, whatever the cost." Not to be thwarted, Congressman John Duncan of Tennessee casually dropped an amendment into an energy-and-water bill stating that the dam would be built, no matter what any law said. It was.

The ESA will not stay unchanged. Even its keenest advocates admit it has flaws. The Tellico dam provides an example. Its completion, the Cassandras predicted, would seal the fate of the snail darter. A year later, the fish was discovered living 60 miles downstream from the dam. Anyway, say hard-nosed evolutionists, perfect diversity is impossible. Evolution dictates that species should disappear. Nine-tenths of them already have. So is the law really necessary?

Absolutely, replies the National Academy of Sciences. It issued a report in May which, while admitting that changes to the law are needed, advocated the creation of temporary "survival habitats," zones demarcated by scientists while a recovery plan for the endangered species is being drawn up. This is needed because, according to a Harvard entomologist, Edward Wilson, "the rate of extinction is now about 400 times that recorded through recent geological time and is accelerating rapidly. If we continue on this path, the reduction of diversity seems destined to approach that of the great natural catastrophes at the end of the Paleozoic and Mesozoic era—in other words, the most extreme in 65m years." That ought to make even this Congress flinch.

The trouble is that most endangered species neither soar majestically on the wing nor display themselves to admiring tourists in Yellowstone National Park. Of the 10m species in the world, most are insects, plants and fungi. They live in backyards and irrigation ditches. Two-thirds of the United States is privately owned; more than half the 956 endangered species occupy private land.

Congress has already trimmed the edges of the ESA. Now a bill has been introduced by Senator Slade Gorton, a Republican from Washington state, with backing from industry, which would give the secretary of the interior the power to order a case-by-case review of threatened species. This would then permit a decision that some species are less important than others and, accordingly, may be allotted less protection.

Environmentalists back a milder bill to be submitted by Wayne Gilchrest of Maryland and Gerry Studds of Massachusetts. These Democrats hope to fine-tune the ESA, to the benefit of the small landowner. But it is industry that the law affects most. The National Endangered Species Act Reform Coalition represents 185 utilities, energy firms and water companies, mostly in the west, which want to see big changes. They suspect that Oliver Houck, a law professors at Tulane University, is right when he calls the ESA "a surrogate law for ecosystems." Such talk freezes the blood of mining firms and utilities. Ore extraction and power lines require large tracts of land. The more land you need, the higher the chance that you will run into an endangered species. Timber companies stand to lose millions.

Still, defenders of the ESA, encouraged by the Supreme Court's ruling, fight on. Steven Meyer, of the Massachusetts Institute of Technology, produced a study last year showing that states with the most endangered-species listings were also the ones with the strongest economies. He adds that a quarter of a percentage point change in the interest rate has more effect on the economy than any hardship caused by the ESA. ■

Questions

1. In protecting endangered species, when does the battle start?

2. Even though most Americans value wildlife, how do they view the zoning laws?

3. What comprises most of the ten million species in the world?

Answers are at the back of the book.

22

Marine organisms are seldom on the endangered species list. Some people think that they are resistant to human-caused extinction primarily because of their assumed large geographic ranges. Now human alterations of coastal environments and overfishing are causes of concern for marine biodiversity. The shrinking percentage of coral cover on reefs indicates a decline in marine ecosystems, but calculating marine biodiversity is difficult to do. Approximately seven percent of the world's oceans has been sampled for biodiversity and even moderately rare species can be difficult to find. If extinctions are taking place the problem is on a much larger scale than land extinctions. Also, the interconnectedness of the marine system can cause far-reaching impacts.

Is Marine Biodiversity at Risk?

Elizabeth Culotta

In the early 1970s, marine biologist Kerry Clark of the Florida Institute of Technology discovered a new species of sea slug in sea grasses in the Indian River lagoon on the Atlantic coast of Florida. Sea slugs are small, graceful creatures with brilliant coloring and many species are rare. But at first this bright green species, which Clark named *Phyllaplysia Smaragda*, the emerald sea slug, was relatively common. However, as the lagoon's shores were developed, the sea grasses shrank into the shallows and the slug's numbers dwindled. Clark hasn't seen on since 1982 and fears they are gone for good.

This tale of species found, then lost, is all too familiar to terrestrial ecologists, especially those in tropical rainforests, who have watched species vanish or hover at the edge of extinction. But it's a new story for many marine biologists. Except for large vertebrates like mammals and birds, marine organisms rarely appear on lists of extinct and endangered species. Indeed, although the fossil record is full of such extinctions, marine organisms were believed to be resistant to human-caused extinction, because many sea creatures have larvae that can drift long distances and most are thought to have large geographic ranges.

Now, a small but growing band of marine ecologists is sounding the alarm. The resilience of marine species may have been overestimated, they say, and human modifications of coastal environments, along with overfishing, may threaten marine biodiversity. These researchers are taking a new look at the question of extinctions in the sea, and some conclude that such extinctions may be common—but overlooked. "There's a perception of fewer extinctions in the ocean and I'm not sure the perception is right," says James Carlton, director of maritime studies at Williams College and Mystic Seaport (a program of marine studies in Mystic, Connecticut.) Furthermore, ecologists argue, even if oceanic species are being lost more slowly than those on land, the marine environment operates differently. Other signs of decline, such as shrinking percentage of coral cover seen on some reefs, may be more relevant indicators of deterioration in marine ecosystems.

Still, studies of marine biodiversity are in their infancy, and many biologists remain skeptical about marine extinctions. The concerns are nevertheless being taken seriously by funders and other scientific organizations, who have been sponsoring a flurry of conferences and workshops on the changing diversity of the oceans. A symposium this week at the annual meeting of the American Association for the Advancement of Science (AAAS) is one example. Four federal agencies (the National Science Founda-

Reprinted with permission from *Science*, February 18, 1994, Vol 263, pp. 918–920.

tion, the National Oceanic and Atmospheric Administration, the Office of Naval Research, and the Department of Energy) have also banded together to sponsor a National Research Council (NCR) initiative to chart a research agenda.

Counting Down

The NCR effort will likely span a wide range of research questions, but one issue sure to come up is extinction, although the numbers of recent marine extinctions are anything but dramatic. The first published account of a modern invertebrate extinction in an ocean basin, that of the Atlantic eelgrass limpet, was published only three years ago, and there are still no documented extinctions of marine fish in historic times. "Only about 14 or 15 species have been lost, mostly flightless birds, a few mammals, only one or two invertebrates. Even if there were 10 times as many extinctions, it wouldn't be a huge number," says Geerate Vermeij of the University of California, Davis. "There's just no evidence for human activity causing extinctions in the oceans in the last 300 to 400 years."

Yet the scarcity of documented marine extinctions may reflect nothing more than the difficulty of proving them, say biologists like Carlton, who cochairs the NRC committee with Cheryl Ann Butman of Woods Hole Oceanographic Institution. Only about 7% of the world's oceans has been sampled for biodiversity, and even moderately rare species are easy to miss.

In the case of the eelgrass limpet, Carlton, Vermeij, and colleagues were able to document the extinction because the literature and museum collections of the 18th and early 19th centuries indicate that the limpet was abundant on nearshore eelgrass from Labrador to Long Island—areas now reasonably well-samples by biologists. The record is much worse for most organisms. Take the emerald sea slub: Clark can't find it in its original locality or elsewhere in the Florida Keys and Atlantic Coast, but he hasn't visited potential habitats in northwest Florida. Since the slug has free-floating or planktonic larvae, it's possible that the creature is quietly grazing on seagrasses in some undeveloped lagoon. For now, *P. smaragda* won't appear on any lists of extinct species.

But many ecologists remain convinced that nearshore marine species have suffered from coastal development. Clark says that in south Florida, he's watched as human activities have dramatically changed the nearshore habitats where he's been collecting for 20 years. He believes increased sedimentation is one of the chief culprits. Sea slugs, like many marine organisms including corals, prefer clear water with low levels of silt. Coastal construction increases the amount of silt flowing into the water from the land and appears to make habitats unsuitable, says Clark. Of 33 species in one group of sea slugs in Florida, 16 are in decline, according to Clark's data (in press at the *Bulletin of Marine Science*). As a result, he says, "I have no doubt that there are more (extinctions) out there that we missed."

Following up the work on the limpet, Carlton combed the historical and recent literature looking for just such overlooked extinctions. In work published recently in *American Zoologist*, he reports two additional recent molluscan extinctions—and another "possible" extinction of a rocky shore limpet in California. That makes a grand total of four.

But the exercise persuaded Carlton of the difficulty of proving a marine extinction. He has begun to identify hundreds of molluscan species that were "missing in action"—described by 18th and 19th century systematists and never spotted again. These "species" could be extinct, have been misclassified originally, or be present but not sampled recently. Without a very time-consuming search, Carlton says, he simply can't be sure.

Of course, before researchers can estimate the amount of biodiversity that has vanished, they need to know what was there to start with—another vexing problem in the oceans. In the past few years, estimates of the number of deep-sea species alone have ranged from fewer than 500,000 to 10 million. Even on well-studied coral reefs it's tough to estimate how many species are present. "To identify every species in a reef, you'd have to blow it up, level it, to see what was living in the crevices," says reef biologist John Ogden of the Florida Institute of Oceanography. In work to be presented at the AAAS symposium this week, ecologist Marjorie Reaka-Kudla of the University of Maryland estimates that 35,000 to 60,000 described species live on coral reefs—but that this represents only 8 to 14% of reef biodiversity.

Long-range Larvae?

Even in the absence of large numbers of documented extinctions, however, some new research is beginning to question the view that marine species are resistant to extinction. This idea took hold because many marine creatures produce planktonic larvae; some may spend months in the plankton, voyaging thousands of miles. Such organisms have the potential for broad geographic distributions and large population sizes—which should be excellent insurance against a permanent vanishing act.

"There's been this overwhelming tradition of thinking of the sea as a giant bathtub in which things can slosh from India to England without any problem," says biologist Nancy Knowlton of the Smithsonian Tropical Research Institute in Panama (STRI). "But it isn't always so." Larval biologist Michael Hadfield of the University of Hawaii agrees: "The larval dispersal potential is high, but it may not be happening. Not all marine species have planktonic larvae, and some of those that do only hang around an hour or so and don't get far."

Take mantis shrimp. Many species of these crustaceans, called stomatopods, live in the crevices of coral reefs. Most species do have planktonic larvae, but work by Reaka-Kudla and colleagues has shown that many of the larvae spend only a few weeks in the water. Furthermore, they spend their days hovering close to the reef structure, venturing up into the waves only at night. So they don't travel far, say Reaka-Kudla. In general, small species are less likely than large ones to have long-lived planktonic larvae, she says. Therefore, she argues, small species—which are more numerous and also more likely to be overlooked—have more restricted geographic ranges and are more vulnerable to extinction. Sea slugs also fit that picture, says Clark. One-third of the 33 species he's studied in the Florida Keys have either no free-floating larval stage or their larvae spend only a few hours in the plankton.

For many organisms, the question of exactly how far their larvae travel has not been tested. But larvae or no larvae, researchers have evidence that at least a few marine species have relatively narrow geographic distributions—making them vulnerable to extinction due to local disturbances. For example, Michael Smith, senior research scientist at the Center for Marine Conservation in Washington, D.C., analyzed the ranges of 500 Caribbean fishes, using recently published literature. While most did have broad ranges, he found that 16% were restricted to the Caribbean Sea alone or to smaller geographical areas.

Among invertebrates, isolated cases of such narrow distributions are also trickling in, says Knowlton. For example, the Kumamoto oyster, *Crassostrea sikamea,* has planktonic larvae, but its native range was apparently restricted to the southernmost island of Japan, according to Dennis Hedgecock of the University of California. Although farmed for the restaurant trade on the U.S. West Coast—you'll pay extra for its delicate flavor—the Kumamoto now appears to be extinct in Japan, says Hedgecock.

Laws of the Sea

Despite such tales, few marine biologists claim that marine extinctions are occurring at the same rate as those on land. Most biologists agree that marine species probably do have an extra measure of resilience. But biologists such as coral reef specialist Robert Buddemeier of the University of Kansas warn that numbers of extinctions in the sea may not serve as the same kind of ecological damage indicator as they do on land. "You don't want to get trapped into a linear comparison of terrestrial and marine ecosystems," says Buddemeier. "The marine system is less extinction-prone, but if you do start getting extinctions, if means you've got a problem on a much larger scale. The rules *are* different in the sea."

For example, in January a coral reef research and monitoring panel convened by the Department of State reported that many coral reefs in close proximity to large human populations are in decline; a colloquium on coral reefs held last June in Miami came to a similar conclusion, according to the draft report of the meeting. Reef biologists suggest that increases in nutrients, intense fishing (which decreases the numbers of predatory fish), plus natural disturbances such as storms combine to put reefs under stress. Specifically, says Jeremy Jackson of STRI, these factors tip the ecological balance in favor of corals' chief competitor, algae. The end result, already seen at some locations: large, fleshy macroalgae thriving atop dead coral.

"I can't point to a single species of coral which has gone extinct in the Caribbean," says Jackson. "But that doesn't mean that corals aren't in decline. I see it everywhere: The corals look like they need a shave—they're being overgrown with algae."

The visible danger sign is not the extinction of geographically restricted species, as might happen in a rain forest, but rather a widespread change in abundance of species, explains Knowlton. "It's different from a forest. You can chop a forest down, but the pieces you don't chop are more or less OK. In the ocean, it's more interconnected. You can have the whole system slowly diving downward."

For those convinced that marine habitats are at risk, the next step is to find ways to protect them. In the Florida Keys, a new sanctuary has been set up to protect reefs. But how to manage a watery park poses a whole new set of questions, since neither currents nor organisms respect park boundaries, says Buddemeier. Although not everyone is convinced of the need for a more organized approach, many scientists are hoping that efforts like that of the NRC will outline the research needed to understand—and perhaps protect—the diversity of marine life. ■

Questions

1. Why are marine organisms thought to be resistant to human-caused extinctions?

2. What are two factors that may threaten marine biodiversity?

3. What is the visible danger sign that marine extinction is taking place?

Answers are at the back of the book.

23

The Endangered Species Act has been one of the most controversial of all the environmental laws. One reason is that it often directly conflicts with one of the most basic Constitutional rights in the U.S., the right to own private property. This occurs because about half of all endangered species live on private land. On the other hand, the Act also protects hundreds of irreplaceable species that are increasingly threatened by development. Many environmentalists have argued that the right of these disappearing species to survive must be given at least some priority over the rights of individuals to own and develop land. As often happens in conflicts, compromises must be sought. Such compromises require that clearer priorities need to made about what species and habitats to protect. Perhaps even more importantly, economic incentives for landowners to preserve land, such as tax breaks or land swaps, must be implemented.

Building a Scientifically Sound Policy for Protecting Endangered Species

Thomas Eisner, Jane Lubchenco, Edward O. Wilson, David S. Wilcove, Michael J. Bean

The primary legislative tool for protecting imperiled species in the United States is the Endangered Species Act (ESA) of 1973. The pending reauthorization of this law has sparked a fierce debate on the science, economics, and ethics of protecting vanishing species; the outcome of the debate will influence domestic and international conservation policies for years. Recent advances in our scientific understanding of biodiversity have underscored the importance of species protection for human welfare. Each species, by virtue of its genetic uniqueness, is the source of information we can learn from no other source. Species can provide us with novel molecules and new understanding of genetic capacities, which can be used to fashion new agricultural products, medicines, and other chemicals of direct benefit to humans. Indeed, prospecting for biogenetic information could well become a major scientific exploratory venture of the 21st century. Species also provide essential ecological services to humanity by regulating climate; cleansing water, soil, and air; pollinating crops; maintaining soil fertility; and performing other life-sustaining functions.[1]

Despite the importance of species to people, a significant fraction of the biota of the United States is at risk of extinction or already lost. Somewhat in excess of 100,000 native species (terrestrial and freshwater) have been described from the United States, including 22,750 vascular plants; 3110 vertebrates; and (very roughly) 75,000 insects. Within those taxa most carefully classified and studied to date, about 1.5% of the species alive at the turn of the century are now considered to be certainly or probably extinct. Extinction estimates range from 0 in reptiles and gymnosperms to 8.6% in freshwater mussels. In these groups, the overall percentage of species ranked as imperiled or rare is 22.2%, with a peak of 60.1% in freshwater mussels.[2]

Recent scientific discoveries and assessments provide valuable insights about endangered species protection.[3] We focus on three issues: (i) Does the act protect the right elements of diversity? Should the limited resources available for conservation be targeted toward the protection of higher ecological levels of diversity, such as ecosystems, rather than toward the protection of individual species? Should

Reprinted with permission from *Science,* September 1, 1995, Vol. 268, pp. 1231–1232.

protection encompass categories below the species level (that is, to subspecies and populations)? (ii) Have decisions to classify particular plants and animals as endangered been based on sound science? (iii) Can ecological and biogeographic knowledge be used to increase the efficiency of the ESA?

What Should Be Protected?

Although the stated purpose of the ESA is "to provide a means whereby the ecosystems upon which endangered species and threatened species depend may be conserved," it attempts to do so by protecting individual species, subspecies, and, in the case of vertebrates, distinct population segments. This focus on individual taxa has come under increasing criticism from those who believe it to be an inefficient and ineffective means of safeguarding biological diversity.[4] The sheer number of species present in most regions of the country and the lack of ecological information about most species are cited as the primary reasons for shifting conservation activities to higher levels of biological organization. There are four strong reasons for not abandoning the traditional focus on individual species. (i) Because ecosystems are less discrete entities, species provide a more objective means of determining the location, size, and spacing of protected areas necessary to conserve biodiversity. (ii) Population declines of individual species (for example, freshwater mussels, peregrine falcons) may indicate the presence of stress to an ecosystem before it is obvious system wide. (iii) Individual species are the source of new medicines, agricultural products, and genetic information useful to humans. (iv) Although ecological services are provided by ecosystems, individual species often play pivotal roles in the provision of these services.[1] Efforts to protect declining species are consistent with the goal of protecting ecosystems. We strongly concur with recent reports from the National Research Council (NRC) and the Ecological Society of America that emphasize the need to protect both species and habitats; neither is a complete substitute for the other.[3]

Subspecies and distinct population segments of vertebrates have been protected by the ESA since its inception and currently constitute about 20% of listed taxa.[5] Legislation to reduce protection for units below the species level has been introduced in Congress and will be debated in the forthcoming reauthorization. Advocates of this measure argue that it will reduce the number of ESA-related conflicts by reducing the number of listed taxa and will allow the federal government to focus limited resources on the protection of full species. Although sympathetic to both concerns, we believe the current policy is sound because it facilitates the protection of genetic diversity within species and encourages people to act earlier to protect declining species, rather than waiting until all subspecies or populations of a given species are imperiled. Moreover, as noted in the NRC report, there is no scientific justification for protecting populations only of vertebrates. Plants, for example, may differ chemically at the population level, reflecting genetic differences that may prove useful to humans.[6]

Criteria for Listing

Under the act, protected species are classified as either "endangered" or "threatened." The former includes "any species which is in danger of extinction throughout all or a significant portion of its range"; the latter includes "any species which is likely to become an endangered species within the foreseeable future...." These vague statutory definitions provide the Secretary of the Interior with considerable latitude in determining which taxa warrant protection. Critics of the act allege that numerous taxa have been accorded protection based on incomplete or inaccurate information.

There is, however, little evidence that the Department of the Interior has abused its authority by listing taxa that are not at risk of extinction. Since passage of the ESA, only 4 of more than 950 protected taxa have been removed from the endangered list because subsequent studies showed them to be more abundant than previously thought.[7] In fact, most species are listed when their populations are close to extinction. A recent study found that the median population sizes of taxa at time of listing were only about 1000 individuals for animals and 120 individuals for plants; at least 39 plants were listed when 10 or fewer individuals were known to survive.[5]

Where to Protect Endangered Species?

The fear that the presence of endangered species will lead to restrictions on the use of private lands has spawned much of the backlash against the ESA. It is reasonable, therefore, to ask how important private lands are to endangered species protection. Approximately 50% of listed taxa occur only on state and local public lands, tribal lands, and private lands.[8]

The current pattern of federal land ownership is imperfectly suited to protecting biodiversity. Federal lands are concentrated in the western United States, including some areas with few imperiled species. Other regions that harbor high concentrations of localized, rare species contain little or no federal land. A carefully designed program of land exchanges between the federal government, other public landholders, and private landowners could improve the federal portfolio from a biodiversity perspective while providing private landowners with relief from their endangered species obligations and compensation in kind at little or no federal cost.[9] Such a program would not negate the need to protect endangered species on private lands, but it would reduce the impact of doing so.

Improving the Process

To argue that species conservation must remain a central goal in conservation is not to say how that goal should be met. New approaches with respect to both the science and economics of protecting biodiversity could significantly improve the performance of the ESA.

Priorities for protection. Some ecosystems are more endangered than others and contain a large number of species found nowhere else. Such hot spots are critical to conservation efforts because many (but far from all) endangered species will occur within them. It is therefore most effective for the Department of the Interior to give priority to the identification and protection of such places by expediting the formal listing of imperiled species associated with them. Examples in the United States include the rain forests of Hawaii, the sand ridge scrublands of central Florida, the desert wetlands of Ash Meadows in western California and eastern Nevada, and the rivers of the Cumberland Plateau and Southern Appalachians.[10] The 50 counties in the United States with the largest number of federally listed species, which together comprise about 4% of the nation's land area, contain populations of approximately 38% of all listed species.[12]

When ecosystems are in a natural or seminatural state, the number of species, S, expected to persist is systematically related to area, A. This species-area relationship is represented as $S = cA^z$, where z is usually in the range of 0.2 to 0.4, depending on the group and place. This relationship implies that a tenfold increase in habitat area approximately doubles the number of species within it. An important consequence of the species-area relationship is that lands protected on behalf of animals with large home ranges or low population densities will provide de facto protection for numerous other species with smaller home ranges or higher densities. The Department of the Interior can maximize the efficiency of its listing duties by targeting such taxa, commonly referred to as "umbrella species".[12] Other useful criteria for determining priorities for protection include the species' ecological role, taxonomic distinctiveness, and recovery potential.[3]

Incentives for protection. The ESA relies on fines and jail sentences to punish or deter harmful conduct, but it provides no incentives to encourage or reward beneficial conduct, such as restoring habitats for endangered species. Changes in both the federal tax code and existing subsidy programs could be used to this effect.

For example, to pay federal estate taxes, inheritors of large land holdings are occasionally forced to sell, subdivide, or develop the property, resulting in loss of wildlife habitat. In cases where the property contains endangered species, the heirs could be given the opportunity to defer part of the estate taxes by entering into an endangered species management agreement with the Department of the Interior. In other cases, endangered species will persist on a site only if the habitat is actively managed on their behalf. The expenses associated with habitat management (for example, prescribed fire) are not currently tax deductible; if they were, more landowners would likely participate in efforts to recover endangered species.

The federal government funds a number of incentives programs aimed at encouraging farmers, ranchers, and small woodlot owners to protect wetlands, forests, soils, and water quality.[13] To date, no effort has targeted these programs to areas where endangered species are likely to benefit. This could be done by simply modifying the criteria for eligible lands or by paying a premium for lands harboring endangered species. Such measures cannot wholly supplant the regulatory requirements of the ESA, but they can limit the need for, and increase the flexibility of, such requirements.

The ESA is scientifically sound. A continued focus on species protection is necessary and appropriate and complements the protection of ecosystems. The effectiveness of the act can be improved by emphasizing the protection of hot spots and umbrella species, by protecting disappearing species and ecosystems earlier, and by supplementing the law's regulatory requirements with economic incentives.

References and Notes

1. C. Perrings, C. Folke, K-G. Maler, C. S. Holling, B-O. Jansson, Eds., *Biodiversity Loss: Ecological and Economic Issues* (Cambridge Univ. Press, Cambridge,1994); H. A. Mooney, J. Lubchenco, R. Dirzo, O. Sala, Eds., *Biodiversity and Ecosystem Functioning*, Sections 5 and 6, in R. Watson *et al*, Eds., *Global Biodiversity Assessment* (Cambridge Univ. Press, Cambridge, in press).
2. Data on numbers of species at risk of extinction or recently extinct were taken from the National Heritage Central Databases of The Nature Conservancy, June 1992. The count for the total number of known species is only an estimate, due to the imperfect nature of summary databases for U.S. biota. Our estimate is a conservative one, especially in view of the lack of clear estimates for groups such as fungi and microorganisms. See also E. T. LaRoe, G. S. Farris, C. E. Puckett, P. D. Doran, M. J. Mac, Eds., *Our Living Resources: A Report to the Nation on the Distribution, Abundance and Health of U. S. Plants, Animals and Ecosystem* (U. S. Department of the Interior, National Biological Service, Washington, DC, 1995).
3. National Research Council, *Science and the Endangered Species Act* (National Academy Press, Washington, DC, 1995); C. R. Carroll *et al.*, *Ecol. Appl.*, in press.
4. J. F. Franklin, *Ecol. Appl.* **3**, 202 (1993); G. Easterbrook, *A Moment on Earth* (Viking, New York, 1995).
5. D. S. Wilcove, M. McMillan, K. C. Winston, *Conserv. Biol.* **7**, 87 (1993).
6. T. Eisner, *BioScience* **42**, 578 (1992).
7. The four taxa include three plants (*Tumamoca macdougalii*, *Astragalus perianus*, and *Hedemoa apiculatum*) and one amphibian (Florida population *of Hyla andersonii*).
8. B. A. Stein, T. Breden, R. Warner, in *Our Living Resources* [see 2].
9. An example is the Collier land exchange approved by Congress in 1988, in which the Department of the Interior will trade 111 acres in Phoenix, Arizona, for 122,000 acres of private land in Florida. The Florida land will become part of Big Cypress National Preserve and two national wildlife refuges and will protect habitat for the endangered Florida panther. Land exchanges are complex; for recommendations, see National Research Council, *Setting Priorities for Land Conservation* (National Academy Press, Washington, DC, 1993).
10. D. W. Lowe, J. R. Mathews, C. J. Moseley, *The Official World Wildlife Fund Guide to Endangered Species of North America* (Beacham, Washington, DC,1990).
11. Data on county occurrences of threatened and endangered species obtained from U.S. EPA's Office of Pesticide Programs' Endangered Species-By-County List, 31 March 1995. Our analysis is based on known occurrences of federally listed species in each county.
12. The northern spotted owl (*Strix occidentalis caurina*) is an example of an umbrella species. One study concluded that 280 species of plants (vascular and nonvascular) and vertebrate animals closely associated with old-growth forests in the Pacific Northwest would be protected in a reserve system established for the owl. See J. Thomas *et al.*,"Viability assessments and management considerations for species associ-

ated with late-successional and old-growth forests of the Pacific Northwest" (USDA Forest Service, Washington, DC, 1993).

13. Examples of such programs include the Conservation Reserve Program, Wetlands Reserve Program, Water Quality Incentives Program, and Forest Legacy Program,

14. We thank M. McMillan for data analyses and A. R. Blaustein, T. Janetos, S. A. Levin, T. E. Lovejoy, G. Orians, D. Policansky, G. B. Rabb, J. Reichman, and C. D. Trowbridge for helpful comments. ■

Questions

1. According to the species-area relationship, a tenfold increase in area will increase the number of species within it by how much?

2. Name some useful criteria, besides "umbrella species," for determining priorities for species protection.

3. Give two reasons why it may be useful to continue listing subspecies and populations instead of listing only full species.

Answers are at the back of the book.

24

Beavers, the largest rodent in North America, are making an astonishing comeback. Until recently, they were trapped for fur and eradicated as "pests" into extinction in many areas of the United States. But their protection by law, their rapid reproduction, and the extinction of wolves and other major predators, has led to a dramatic increase in beaver abundance. They are not only numerically more common, but beavers are geographically expanding into areas where they have not been found for centuries. By some estimates, more than 12 million beavers now live in the United States. But this rapid increase has led to a classic and increasingly common dilemma in conservation biology: What to do about native species that become too common because their predators have been exterminated by humans? This is especially problematic with beavers because of their drastic impacts on ecosystems. By building dams, beavers alter the courses of streams, create wetlands, and cause flooding. These activities not only cause many millions of dollars in damage each year, they can threaten other species.

Back to Stay

Jon R. Luoma

Once nearly wiped out by trapping, beavers are booming across the country—and land managers must cope with the only creatures that can alter an ecosystem nearly as efficiently as humans themselves.

Dick Foster's little corner of retirement heaven lies down a narrow lane near a sweeping oxbow of the Oswegatchie River, so far north in the state of New York that the nearest metropolis is the Canadian capital, Ottawa. On a sunny day last summer, I pulled up to the place he calls the End-of-the-Road Farm just as Foster, a retired gym teacher and coach, was coming out the door of his 19th-century stone farmhouse.

The place hasn't been a farm for decades. Like much of the rest of the former agricultural land here in New York's North Country, Foster's 300 acres reverted to forest decades ago. And that's just how he and his wife, Sandra, like it. The "farm" is now a place to hike, to bird, to canoe and garden and cross-country ski, and to log out a few of the valuable northern hardwoods, like sugar maple and hard cherry.

Of course, the view from the house used to be prettier, with a row of fat old black oaks hanging their boughs over the Oswegatchie. There is also the matter of a little bog, once located on state-owned land a few yards beyond the property, where Foster used to take grade-school kids to look at insect-eating flora like sundews and pitcher plants. The oaks, it seems, were girdled by busy teeth and died of thirst; the bog was inundated into oblivion. In fact, if the Fosters allowed nature to take its course, much of their stream-laced forest and meadow—a third of their land—would soon be under water.

And that's part of the story of why Dick Foster, nature lover, finds himself at war, not with the usual suspects—polluters, drainers, ditchers, and developers—but with booming populations of beavers, the only living creatures with enough engineering skill to alter ecosystems nearly as efficiently as humans themselves.

Foster is not the only one in conflict with these, the largest of North America's rodents. Farmers, woodlot managers, and homeowners are being driven to distraction by the result of an environmental success story. Once fantastically abundant, beavers had by the end of the 19th century, been pushed to the brink of extinction by trapping. Now they are back. And with their major natural predator, the wolf, largely absent, it looks as if they are back to stay.

In fact, they are breeding prodigiously in some regions, accomplishing with busy efficiency precisely what they evolved to do: cutting down trees, incessantly gnawing on bark to keep their chisellike teeth razor-sharp, stopping running water from running, and flooding dry land into ponds in which to feed and shelter and breed.

As a species, the creatures are simply attempting to follow a biological imperative that leads them to manipulate large swaths of the American landscape back to what it once was: a great patchwork of beaver-made ponds and wetlands, rich with related species. The rub is that, in the interim between their near extinction in the late 1800s and their present recovery, we humans have taken over, bringing with us our roads, basements, backyards, wells, septic tanks, woodlots, and golf courses—artifices that respond poorly to, say, a sudden inundation by thousands of gallons of water.

Massachusetts Department of Fisheries and Wildlife biologist Tom Decker says that in the past five years alone his state, with perhaps 29,000 of the animals, has seen a doubling of beaver-related complaints. The good news, he says, is that "beavers are now found throughout almost all of their traditional range in Massachusetts." The bad news: "They're trying to share that landscape with 6.2 million people."

Mike DonCarlos, a wildlife-management specialist with the Minnesota Department of Natural Resources, says, "It's reached the point where there are too many beavers to avoid conflict with people." In Minnesota there are, in fact, too many of them to count. Instead, the state samples a series of standard reference areas, which, DonCarlos says, show "dramatic increases since the late 1980s," with a population peak in 1992 three times as high as that of the early 1970s. Extrapolating from those surveys, he believes that there could be as many as 1 million of the animals in the state. In fact, he says, the population has boomed so much in the past half-decade that "we've pretty much lost control of it."

In northern New York's St. Lawrence River valley, where Foster's home lies, the number of beaver families—which average seven animals each—has doubled in 10 years, to the point where the animals now occupy about 40 percent of all potential habitat. State wildlife biologist Joe Lamandola says that with the population boom has come a boom in complaints: about 500 each year in the valley. "I try to tell people that beavers have value," he says. "But when a guy bought a young woodlot twenty years ago to provide for his retirement and he finds that it's now up to his nose in water and all the trees are dying, it's hard to tell him to appreciate them."

No one knows precisely how many beavers now thrive across the United States, but estimates run to 12 million and beyond. Nor does anyone know precisely why the numbers have soared so dramatically in so many places. One theory is that the phenomenon follows a principle of biology, that populations grow or recover gradually until breeding-age adults reach critical mass, and then the numbers begin to soar exponentially. At least as important is the fact that trapping, which once limited population growth, has declined across North America as fur has fallen out of favor and pelt prices have plummeted.

Foster says the booming beaver population on his land has compelled him to trap the animals. "I was thrilled the first time I saw beavers on this property," he says. His retirement haven, like many present-day dairy farms and villages here-abouts, probably spent much of its presettlement history under the waters of beaver ponds. He took me back along rutted roads and woodland trails to a pair of stick-and-mud dams for a look at how beavers here are attempting to establish their domain once again.

For Foster, the conflicts are evident. "I don't think trapping's wrong or immoral," he says. In fact, he markets the pelts from the animals he traps, so that at least some use is made of them. "But I don't happen to like killing these animals. I don't like it one bit. I got so tired of it that I got my family together and asked whether we should just give the place over completely to the beaver. I got a resounding no."

According to the New York Department of Environmental Conservation (DEC), in 1994 beavers caused about $6.2 million in damage to crops, commercial trees, and human structures in the state. Minnesota's DonCarlos says his state doesn't compile such figures, but he suggests that the numbers there also run into the millions—money spent to repair roads or to cover the costs of preventing even more damage.

Comprehending the efficiency and energy with which beavers can build and dam, particularly in the autumn as they prepare for winter, takes direct observation. During a year when I lived in the St. Lawrence River valley, not many miles from Foster's house, I marveled at how, in a matter of days, a solitary beaver had mowed down and hauled into a culvert virtually all the alder shrubs in a seasonally damp wetland behind our house. At the other end of the culver—an extremely long one that ran perhaps 75 yards under an adjoining yard and diagonally across a road—another neighbor was more distressed than amazed. In a matter of a few more days, the creature had logged out about a third of a windbreak of aspen trees. Still, the animal was doomed: There would be virtually no autumn or winter water in which it could submerge the food cache it would need to survive the frozen months. Water would come, but only with the spring snowmelt—too late to save the beaver, but just in time to inundate the neighbor's yard and cellar.

Of course, there's another side to this animal's story. "Beavers," says Ray "Bucky" Owen, Maine's commissioner of inland fisheries and wildlife, "are the number one wetland managers in the state." Surveys show that about half of that state's inland ospreys nest and feed in beaver flowages. Similar studies in Massachusetts show that about two-thirds of its great blue heron nesting sites are in the flowages. Beaver ponds also provide habitat for scores of other species, from muskrats to ducks. During their clear, unsedimented early years, the ponds support brook trout. They help buffer flooding and erosion on stream courses and recharge groundwater. Beavers are, in fact, so effective at this sort of ecological good work that experiments by ranchers and scientists in some arid western states have proven beaver damming can restore eroded, damaged stream courses and adjoining parched and livestock-beaten lands, with benefits ranging from newly abundant songbird habitat to a burst of vegetation for cattle and wildlife.

When the first white settlers arrived on this continent, North America was up to its ecological keister in the creatures the Ojibwa called *amik* and taxonomists have come to call *Castor canadensis*. From Maine to Oregon, from the Arctic tundra to the Sonoran Desert, there were once 60 to 400 million beavers across the North American landscape—meaning that there were beavers on nearly every suitable pond, lake, stream, and rivulet. Their effect on the landscape was often profound: Early explorers in Oregon reported having to make their way only along ridgetops, so flooded and swamped were the valley floors—now farmland and towns—by beaver activity. Biologists Robert Naiman, Carol Johnston, and James Kelley concluded in a 1988 article in *BioScience* that lands so blessed with beavers become a "shifting mosaic" of habitats, with flooding, more flooding, and the expansion of forest openings as trees are beaver-cut or pond-drowned. Wetlands transition back to young forest and then older forest as one pond site is abandoned for another. The biologists reported that between 1834 and 1988, the United States lost perhaps 100,000 square miles of wetlands, much of that former beaver pond.

Plants and animals generally live in habitats to which they have adapted. Beavers, on the other hand, adapt habitats to suit their needs. In fact, says Thomas Eveland, a biologist from Pennsylvania's Luzerne College who has served as an adviser to several states struggling to manage human-wildlife conflicts, "Next to man, the beaver probably does the most of any animal to manipulate its own environment."

That environmental manipulation occurs mainly through the construction of spectacularly engineered dams of sticks, logs, and mud, resolutely knit together on a foundation of logs braced against rocks or trees, or forced into bottom sediments at angles that impel the current to wedge them even deeper. Ponds, and sometimes virtual lakes, swell into existence behind the dams, leaving waters deep enough for a submerged entrance to a beaver lodge, safe from predators, and for a winter food cache of sticks; beavers eat the cambium, the nutritious, sapfilled inner bark.

The animals use their famous flat tails not only to flail upon the water, warning of danger with a great wet kaboom, but also to prop themselves upright while eating or working. Their hind legs, huge and powerful, are webbed for propulsion. Their front paws, by contrast, are surprisingly delicate—almost tiny hands, with which they can pick at food or pack mud into crevices.

Perhaps the beaver's most remarkable feature, though, is its chisellike teeth. To keep them sharp, the animals gnaw on trees even when they're not hungry. The gnawing peels off a layer of fast-growing dentin on the back of the teeth and sharpens the bright-orange enamel on the front. (It also means that a pondful of beavers, by debarking living trees, can kill acres of trees beyond those they drown or cut down.)

Although revered by Native Americans, the beaver was valued by early white settlers not for its ecological good works but rather, as Eveland puts it, only as "living gold"—for its valuable pelt, the "mining" of which reached extraordinary levels. Between just 1630 and 1640, for instance, some 80,000 beavers were trapped annually from the Hudson River and what is today western New York.

Tom Decker, the Massachusetts biologist, worries that increasing conflicts will cause an overreaction. "People's tolerance can rapidly change from a sense that it's a value to have these animals around to a sense that they're just pests, like cockroaches," he says. "If their main experience with these animals is property damage, that their basement is flooded, or that they can't use their septic system, they're going to start seeing them as vermin. So our ability to keep beavers and their wetlands around depends on our ability to control beavers."

By way of example, Decker tells the story of the township of Chelmsford, now a suburb of Boston but once a pioneer fur-trading community. In 1988 Chelmsford residents, at the urging of animal rights activists, voted to ban trapping on the wetlands and streams that lace the area. Four years later, after two public wells had been flooded and homeowner complaints about beaver damage had soared from an average of about 1 per year to 30, voters rescinded the ban. "Four years before," Decker says, "trappers had been portrayed as evil people. But trapping was the only thing that was controlling the beaver population. After they lifted the trapping ban, people were literally inviting the trappers into their homes, saying, 'Please come through our yard and set your traps back there by the stream.'"

Indeed, in 1994 Massachusetts liberalized its trapping rules, doubling the take of beaver pelts. In North Carolina, where beavers were a rare sight as recently as 1950, the legislature two years ago set up a trapping season to help suppress a population now estimated to be causing more than $1 million in damage each year.

But trapping as a management tool raises plenty of hackles. In New York, a 1995 attempt to change trapping laws, intended to suppress beaver populations by as much as one-third in some areas, ran into a wall or opposition not only from animal rights groups but also from mainstream conservationists, including both the Sierra Club and the state's Audubon Council. "Beavers are the only source of new wetlands in the state," says council president Andrew Mason. "If the population is cut by a third, that's going to have an effect on the number of wetlands." On the other hand, Mason says, his group does not oppose changes in the law that would make it easier for a local official or a homeowner faced with a beaver nuisance to obtain a permit to have the animal trapped or shot.

But even some traditionally trained wildlife managers question whether trapping can again be the management tool it once was. Europe, for example, is the greatest (albeit declining) market for pelts; however, the European Union may soon act to ban the importation of pelts from countries that use methods it defines as inhumane. As the market collapses, dragging prices with it, the number of knowledgeable trappers is falling. Thus, purely practical questions remain as to how to increase trapping.

In the face of what he calls his state's "out-of-control" beaver population, Minnesota's DonCarlos has begun to promote the use of Clemson Beaver Pond Levelers, drainage pipes developed at Clemson University. Enclosed in wire mesh and filled with holes, the gadgets are designed to eliminate any sound of running water or trace of suction, so the ever-vigilant beavers won't bury them in sticks and mud. The devices allow wildlife managers to target areas where the animals present problems and either to drain a pond completely or to maintain one deep enough for beavers but small enough to prevent conflicts with people.

New York's DEC has criticized similar gadgets as too costly and prone to failure. But DonCarlos says that experiments with more than 100 of the Clemson levelers in his state show that they almost always work. Although similar devices have been used successfully in Maine and Massachusetts, they hardly solve all the problems. They don't work in shallow, low-flow streams. And they wouldn't have prevented the loss of Dick Foster's oaks, girdled by "bank beavers," which live in lodges along the sides of deeper rivers like the Oswegatchie.

As conflicts between beavers and humans accelerate, the dilemma goes to the heart of our relationship with nature. There may be some merit in the practice of sticking sewer pipes into beaver dams in order to out-engineer these natural engineers. But there is hardly aesthetic merit, or even great environmental merit. Unlike beaver dams, polyvinylchloride pipes do not degrade. Similarly, plastic-based fake fur or synthetic clothing stuffed with sheets of fossil fuel-based polyester insulators might keep us warm and feeling virtuous. But it's possible we'd do well to learn from the native people who once lived in harmony with this creature. They saw no contradiction between reverence for the beaver and a warm fur hat on a winter day. ■

Questions:

1. Why do beavers build dams?

2. When was the beaver population in the U.S. at its lowest? What was the original North American population size?

3. What do beavers use their tails for?

Answers are at the back of the book.

25

Farmers, environmentalists and anti-hunger activists are working together to change the way food is raised and distributed. Food security helps to bridge gaps and sets plans for environmental and environmental justice groups. These groups conceive much more than just providing food for people. They visualize a food system that nourishes the environment and local economy. By supplying locally grown food, communities help themselves economically. They also conserve energy and lessen pollution by not having to transport food from across the country. As an added benefit, organic and sustainable agriculture reduces the amount of chemicals on food and chemical seepage into groundwater. Some programs concentrate on the garden itself as a basis for employment and education. Subjects other than environmental issues and sustainable agriculture are taught. The program tutors instruct the youths in basic science, math and economic development. Most importantly, these groups are learning that through group action and organization they have power.

Common Ground

Barbara Ruben

Farmers, environmentalists and anti-hunger activists join forces to redefine the way food is grown and distributed.

Fourteen miles beyond the White House and ten miles from the dilapidated rowhouses of Washington's poorest neighborhoods, Pennsylvania Avenue gives way to exurbia's newest pastel housing developments, interspersed with rolling fields. A blue heron spreads its wings against the sky above a seven-acre organic farm plot, where fledgling sprouts of broccoli, carrots, collard greens and beets push through the soil in the late-April sun.

Alesia Dickerson strides down the neat rows, straightening the cloth batting covering the cabbage and broccoli to help protect the crops from an invasion of maggots. Last summer, she worked as a volunteer on the farm, lived in a Washington homeless shelter and could barely distinguish between kale and Swiss chard. Today, she makes $5 an hour at the farm, lives with her mother and aerates compost like an expert.

"I didn't know anything about farming before," she says. "I do now. I like to watch things grow."

The farm is tended by employees and volunteers of From the Ground Up, a project of the Capital Area Community Food Bank that links sustainable agriculture with low-cost produce for inner city residents. The program also employs two other people living in shelters.

From the Ground Up is one of a growing number of programs that join the environment, access to nutritious food and the inner-city poor under the idea of "community food security." Whether through farmers' markets in low-income areas, community gardens at public housing or community supported agriculture—in which consumers directly pay farmers for a share of the crop—anti-hunger groups, farmers and environmentalists are sowing the seeds for changing the way food is raised and distributed.

It's a coalition of groups that has rarely joined forces in the past. "It's as if the environment is one box here, welfare and hunger in a box there," says Robert Gottlieb, coordinator of environmental analysis and policy in the Department of Urban Planning at UCLA, whose students did one of the first studies of community food security in the wake of the Los Angeles riots. "Food security is the kind of issue that builds bridges and sets agendas for environmental and environmental justice groups, which have touched on pesticides or farmworkers, but not in conjunction with access to the actual food produced."

Barbara Ruben, Common Ground, *Environmental Action* magazine. Summer 1995, *pp. 26–28.*
Reprinted with permission.

Kate Fitzgerald, executive director of Austin's Sustainable Food Center, says that she has worked smoothly with local and national environmental groups, but that, "Historically there's been a tension between sustainable agriculture and hunger groups. Agribusiness put forth myths saying that sustainable grown food would be more expensive," she says. "It prevented a dialog between farm groups and anti-hunger groups."

But that may be changing. The key, says Mark Winne, executive director of the Hartford Food System, is to change traditional views of agriculture as merely a money-making business. "I get the sense sometimes that if agriculture wasn't producing food, it wouldn't make a difference to the [Connecticut] Department of Agriculture. If farmers in the state stuck to high-end stuff like oysters or mushrooms—multi-million dollar enterprises—people would still say you have farming going on.

As grocery stores flee from what owners consider unsafe and unprofitable inner-city locations, residents are often left with less nutritious and more expensive food. In such cities as New York, Los Angeles and Hartford, low-income residents pay 10 to 40 percent more for groceries than those with high-incomes and access to large supermarkets, according to a study by the Community Food Access Resource Center in New York. A national study by the Second Harvest Food Bank in 1994 found that one in every 10 Americans has used an emergency food pantry.

But more than just providing food for people, community food security advocates envision a food system that also nourishes the environment and local economies. For example, the average food item travels 1,400 miles before it reaches a consumer, says Andy Fisher, coordinator of the Community Food Security Coalition, which formed last year and includes a number of sustainable agriculture, hunger and environmental groups. At that rate, it takes about 10 energy calories to deliver one food calorie to the dinner plate. By providing locally grown food, communities not only get an economic boost, but save the energy—and resulting pollution—from trucking food in from across the country.

In addition, the organic and sustainable agriculture practices advocated by community food security proponents reduce the amount of chemicals on food and washed into groundwater. Urban gardening and farmers' markets can also brighten the often bleak landscape of the inner city.

"Some people use our farmer's markets as a way to escape the Bronx or Brooklyn," says Tony Mannetta, assistant director of New York City's Greenmarkets, which organizes markets in 20 locations, many of them in lower-income areas. "I've seen elderly people just taking a walk through them, smelling a mound of basil here, admiring the bunches of flowers there."

Although only a portion of the produce in the markets is organic, Mannetta says once farmers get into the program they start using fewer chemical pesticides and fertilizers. Cosmetic perfection isn't required. Customers will tolerate an apple that's misshapen, or even one that has a worm hole," he says. "We've found the chemical salesmen even start complaining about the change."

The farmers' market run by the Ecology Center in Berkeley, California, at three locations, including one low-income neighborhood, boasts 70 percent of its produce as certified organic, the highest rate in California and perhaps the country, according to its co-manager Kirk Lumpkin. The Ecology Center also forbids any produce grown in soil using the biocide methyl bromide, which has been shown to destroy the ozone layer.

Produce not sold at the Ecology Center and some other markets is donated to local food banks and shelters. Seattle's community gardening program donates more than two tons of produce a year through a program called Lettuce Link.

Although farmers' markets around the country are traditionally found in upscale neighborhoods, groups like the Ecology Center and Greenmarkets say they are committed to bringing them to lower-income sections of the city. In Austin, for instance, the only farmers' market was located an hour and 25 minutes by bus from a low-income, mainly Latino community. Last year, the non-profit Sustainable Food Center organized a farmers' market for the east side, in which more than 40 percent of the population lives below the poverty level.

In addition to the farmers' market, the Sustainable

Food Center runs a community garden at the site and a food school. "Some of the people will initially look at an eggplant and say, 'I won't eat that. I don't know how to cook it,'" says Nessa Richman, who started the farmers' market. The center's education efforts extend to fliers posted at the gardens to inform residents that, for instance, the cilantro has purposely been left to grow unwieldy to attract lady bugs that will eat the aphids preying on other vegetables.

The farmers' market accepts both food stamps and special Women's Infants and Children's (WIC) farmers' market coupons that are available, in addition to the regular benefits. Interest in using food stamps was so high that in the two weeks after the center put out a pamphlet on the subject, the office was deluged with 7,000 calls for more information.

But it can be difficult persuading farmers to accept the food stamps and coupons—or to even set up a stand in a lower-income neighborhood.

At a farmers' market organized by the Ecology Center, farmers have had their cash boxes stolen and once a man confronted two women with a knife. The Ecology Center then provided a cellular phone for emergencies, but no one so far has had to use it. "The market creates its own positive atmosphere," says Lumpkin.

Greenmarkets in New York offers a $100 credit to farmers toward renting stall space if they will accept food stamps. But in many states, including New York, food stamp transactions have entered the high-tech age. Benefits are accessed through an electronic card, much like an automatic teller machine card. But since most farmers' markets have no electricity or even access to a phone to confirm food stamp eligibility, the few farmers who would accept food stamps many times aren't able to adapt to the new system.

"We're looking at cellular phone links for EBT (electronic benefits transfer), but all I keep thinking is there would be a $5 call for a $2 purchase," says Manetta.

Some programs focus on the garden itself as a source of employment and education. The Seattle Youth Garden Works, started this year, will employ five to 10 homeless youth in organic gardens to raise produce for a farmers' market. In addition to teaching about sustainable agriculture and environmental issues, the program tutors the youth in basic science and math and provides a forum to learn about economic development.

"It seems sometimes that recycling and conservation and environmental righteousness are for the rich, and that the poor are people who litter and eat junk food and don't buy recycled because it costs more. And I think that's wrong," says Margaret Hauptman, who began the program. Hauptman says she joined her love of gardening and working with children in planning the program. "I experience a lot of joy and healing working in the soil. It's a kind of transcendent experience, and I wanted to share that."

The Homeless Garden Project in Santa Cruz, California employs about 20 homeless people part-time in its 5-acre organic garden. Food is sold at a farmers' market and directly from a community supported agriculture (CSA) program. Since the project began in 1989, more than 100 homeless have worked at the garden.

The Hartford Food System has run low-income community gardens as well as a CSA farm since 1978. The farm provides produce for both middle-income residents and low-income groups, which distributes the food to about 900 of their clients. Food recipients help out at the farm.

"Hartford is radically divided between rich white suburbs and a poor inner city made up largely of Latinos and African Americans. Our biggest education effort is on the local food supply," executive director Mark Winne says. "We have teen mothers who have never seen a carrot growing before. They can't get over that they can eat it right there and make lunch out of what they've harvested."

Winne's group is now studying the feasibility of the Hartford School System purchasing more locally grown food. About 25,000 children are enrolled in the schools, 80 percent of whom qualify for the subsidized school lunch program. Almost no food used now is locally grown, and none of it is organic.

Back at From the Ground Up's farm in suburban Washington, director Leigh Hauter talks about weeding with the Civilian Conservation Corps and

Americorps. The acres the program tills are part of a larger sustainable agriculture center owned by the environmental group the Chesapeake Bay Foundation. Farmed primarily for tobacco and corn for 200 years, the land hasn't yet recovered from centuries of monoculture, but Hauter is still expecting a fair crop.

Part of the 45,000 pounds of produce he estimates will be raised will be sold for half price to community groups and churches, which in turn sell the food to their members. When Hauter talked to parishioners at a church in Washington's impoverished Anacostia, several members told him they had been share croppers in North Carolina decades before and recalled spreading ashes on the soil to serve as fertilizer, a practice Hauter emulates using lime today.

"I think we have a lot to learn from each other," he says. "We teach people that by group action they have power. That lesson is what we're trying to get across everywhere—organize to make your life better. Maybe that can translate beyond buying vegetables." ■

Questions

1. What does community food advocate?

2. What does the Seattle Youth Garden Works focus on and what are other areas of learning that they include in their program?

3. What has the concept of food security done for bringing diverse groups together?

Answers are at the back of the book.

---

26 *Food shortages due to population increases have been predicted for centuries. In general, however, they have not occurred. Between 1950 and 1988, the world's population doubled, yet the food supply kept pace with food demand. Genetically altered strains of high-yield cereals called the "green revolution" helped raise food output. Other factors affecting food supply and demand were better farming methods, more irrigation, and the by greater usage of chemical fertilizers. Every year the population of developing countries expands approximately 90 million. UN estimates that by the year 2020 the world population will have increased to over eight billion. Even though crop growth in many areas is slowing, the author argues that wide use of the most basic of technological techniques could still expand the food supply.*

Will the World Starve?

The Economist

For centuries wise men have been predicting global famine. Thomas Malthus argued two centuries ago that the growth of population must outstrip that of food supplies. In 1968 Gunnar Myrdal, a Swedish Nobel laureate, declared that India would have trouble feeding more than 500m people. One 1960s book "Famine 1975!" turned Malthusian predictions into a best-seller.

Today's leading Malthusian is Lester Brown, president of the Worldwatch Institute, a Washington-based environmental group. He argues that China's growing demand for grain imports could trigger food price shocks, in turn causing starvation for hundreds of millions around the world.

Next week, the world's top agricultural experts and various politicians will gather in Washington to debate the global food prospect until 2020. Doomsters will be asked why anyone should believe today's Malthusian predictions, when earlier ones have been proved so wildly wrong.

The world's population doubled between 1950 and 1988, yet food supply kept pace with demand. Pessimists had failed to anticipate the "green revolution" of agricultural productivity, as scientists devised strains of high-yielding cereals. In Asia, wheat yields rose five-fold between 1961 and 1991. Output was boosted further by better farming methods, more irrigation and more chemical fertilisers. Apart from some blips in the 1970s, food prices continued their long-term decline.

Certainly, the challenges are daunting. Every year the population of the developing countries expands by almost 90m—as it were, another Mexico. UN estimates suggest that by 2020 world population will exceed 8 billion, up 45% from today. Food demand will rise faster still: as people are lifted out of poverty, they eat more.

There are some worrying signs, too, on the supply side. Yields of rice and wheat in Asia are still rising, but much more slowly than in the 1960s and 1970s. Growth in the use of fertilisers has slowed worldwide. In many African countries government spending on agricultural research fell in the 1980s, after rising for decades. Rich countries too are cutting back. The green revolution, argues Ian Carruthers, of London University, has brought a false sense of security.

It has brought problems too. In Asia many irrigated areas have become saline or waterlogged. Pests have developed resistance to chemicals. In many countries, fertile areas still uncultivated are often precious habitats for wildlife as well. In some areas, such as northern China, irrigation has led to water shortages. "We're pressing against the limits of lots of natural systems," says Mr. Brown.

Yet for every headline-grabbing prediction of

doom, there are sober reports predicting the opposite. A forthcoming study from the UN's Food and Agriculture Organisation, for example, argues that Mr. Brown has miscalculated China's productive capacity. In Washington, the International Food Policy Research Institute advocates more investment in agricultural research, but sees no immediate constraints on food supply. "Our estimates show that the world is perfectly capable of feeding 12 billion people 100 years from now," says Per Pinstrup-Andersen, the institute's director-general.

The growth in crop yields in many areas may be slowing. But wider use even of basic modern techniques could still boost food supply: in Africa, for example, fertiliser use is only a quarter of Indian levels. And technology can still promise dramatic advances. Scientists have recently devised crops suitable for the acid soils of Latin America's huge *cerrado*—previously infertile plains covering over 200m hectares (nearly 800,000 square miles). Biotechnology may yet revolutionise farming worldwide.

The environmental problems brought by modern farming are real enough (though the likeliest alternative, traditional methods spread to every pocket of cultivable land, might well have been worse). But it is unlikely that the world will let them seriously restrain food production.

Arguably, the true issue is not the risk of global food shortage in the next century, but the real food shortages that specific areas and classes suffer today. The UN reckons that more than 700m people in poor countries are chronically undernourished. In 1993 10m–12m children aged under five died from malnutrition and related illnesses. They are victims not so much of food scarcity as of poverty. And the solution to that problem—witness the dramatic falls in undernourishment over recent years in East and South-East Asia—are not just those of farming technology but the macro and micro policies, social and economic, that lead to overall economic growth. ■

Questions

1. What does Lester Brown predict?
2. Population expands in developing countries by how much each year?
3. In poor countries, what is the UN estimate of chronically undernourished people and how many children under the age of five died from malnutrition and related diseases in 1993?

Answers are at the back of the book.

27

Air, soil, water, and rice and beans have for thousands of years been accessible to most everyone. Recently, advances in biotechnology have made issues of ownership and control of commercial and wild germplasm very important. Biotechnologists can create plants to tolerate herbicides, counter insects, or thrive in drought or heat. The companies they work for are claiming certain types of cotton, soybeans, and vegetables as their own. Some are succeeding in patenting concepts which could hinder corporate research. However, efforts are being made to reward farmers and indigenous communities for their contributions to biological diversity. Biotechnology has made genetic resources even more valuable to future generations.

Who Owns Rice and Beans?

Fred Powledge

The Patent Office's seeming shift on broad biotechnology patents still leaves unanswered questions about plant germplasm.

For thousands of years, the fundamental elements of agriculture have been dirt cheap. Everybody knew seeds were important, but hardly anybody claimed to own them. The plants that a researcher produced were usually available to other scientists for experimentation and research. Rice and beans, wheat and soybeans were part of the global commons, along with air, soil, and water.

The promises of biotechnology have made ownership and control of commercial varieties and of wild germplasm—even that with no proven agricultural and medicinal potential—increasingly important. Legal battles, often involving multinational companies, are being fought over proprietary claims to certain forms of cotton, soybeans, other vegetables, and their progeny. At the same time, farmers, government officials, and nongovernmental organizations in many cash-poor, germplasm-rich countries are demanding a place at the international bargaining table. They say they have a legitimate claim to the germplasm that serves as the raw materials of biotechnology. Representatives of these countries want control of, or at least compensation for, commodities that traditionally have been there for the taking.

Classical breeders have always been able to produce new plants by patiently crossing related varieties in experimental fields. But the genetic manipulation possible with biotechnology allows researchers to freely traverse species lines, to insert just one desired quality into a new plant rather than an uncontrolled number of traits, and to create plants that will do just what their designers want them to do—tolerate herbicides, resist insects, or prosper in drought or heat.

Until biotechnology came along, seeds and plant tissue were covered by a set of laws and international conventions that were considerably less strict than the utility patents that apply to mechanical inventions. Protected germplasm could be used freely by researchers other than the original breeder to stimulate further innovation, and there was at least rhetorical recognition of the contribution of farmers to the preservation and improvement of plant varieties.

But as government-sponsored basic research on food crops declines and industrial research grows, control of plant breeding is passing rather quickly into the hands of multinational chemical companies. These companies insist that genetically manipulated

Fred Powledge, Who Owns Rice and Beans?, *BioScience* 45: 440–444.

germplasm be fully covered under the utility patent system, just like a new kind of light bulb or automobile starter. As intellectual property, rice and beans are big business for corporate lawyers and international trade negotiators who never need go near a wheat field or research station. This enormous change, which can profoundly affect the staple foods enjoyed by consumers around the world, is taking place with little notice by the general public.

One recent event captured the attention of the plant breeding community and advocates of Third World rights, if not of the general public. It is the Agracetus case.

In 1991 and 1992, Agracetus, a Middleton, Wisconsin-based subsidiary of the multinational specialty chemicals company W. R. Grace, was granted two patents on all cotton created in the United States through any technique of genetic engineering. The company had asked the US Patent and Trademark Office (PTO) for the patents in 1986. In the words of an Agracetus background sheet, the more comprehensive of the claims "covers all cotton seeds and plants which contain a recombinant gene construction (i.e., are genetically engineered)." Utility patents do not automatically give other scientists the right to use patented material for research purposes or as a basis for newer varieties. In this case, Agracetus has said it would make free research licenses available "to all academic or governmental researchers upon request."

News of the amazingly broad patents, which run for 17 years, shook fellow members of the biotechnology community and outraged advocates of farmers from the Southern Hemisphere. Five requests were made to PTO to reexamine and negate the claims. The first step on the established route for challenging patent grants is a review by the patent examiner.

Two entities requested a reexamination of Agracetus's broadest claim. One is an anonymous requester represented by a law firm. The other is the plant patent arm of the US Department of Agriculture (USDA), which claimed that the method described by Agracetus for transforming cotton was not new, having been previously described in several papers. (To be granted, a request for a utility patent must he novel, or not anticipated by what patent experts call "prior art"; it must have utility; and it must be nonobvious to someone with "ordinary skill in the art.")

Some of Agracetus's critics raised other objections that had little to do with the mechanics of patenting. It is "morally unacceptable," said one of them, to control the manipulation of life on such a magnificent scale. Furthermore, the patent would give a single multinational corporation unprecedented control over one of the world's staple crops. But the opponents knew it is not the patent office's job to consider such arguments, and they framed their objections in technical terms.

When the examiner announced, last December, that she had reviewed the claims and was now rejecting them, there was considerable speculation in the industry that PTO had not realized how broad the patent was and now was attempting to backpedal. *The New York Times* noted that it was the third time since Bruce A. Lehman became Commissioner of Patents and Trademarks that PTO "had retroactively rejected the claims made on behalf of a highly visible, controversial patent after it had been issued." Both previous rejections are now under appeal.

While opponents of broad species patents celebrated and *The Wall Street Journal* proclaimed "a potentially crippling blow" to the broad patents, Agracetus's parent company, W. R. Grace, dismissed the decision as a mere "office action," which it referred to as "a routine preliminary step in a lengthy reexamination process." The company also reminded competitors that its patents "continue to be valid" pending the outcome of the process.

Russell Smestad, Agracetus's vice president for finance and commercial development, said in an interview in April that "rejection," in the language of the patent office, does not mean the same thing as "rejection" in lay language. "What it means, basically," he said, "is that they cannot permit or allow or issue the claims without more information. They use the word 'rejection,' which sounds drastic, but that is a normal part of the patent process."

The PTO's "Office Action in Reexamination" in the Agracetus patents uses the phrase, "Claims 1–16 [or 1–7, or 13–15] are rejected under 35 U.S.C. 103 as being unpatentable. . ." or ". . . as being anticipated" by what the office calls "prior art references."

And: "Patentee believes that his is the 'first biochemically verifiable genetically engineered' cotton plant. This is not persuasive because being first is an indication of diligence, not of obviousness."

Smestad said the company had filed a response to the PTO action. "Certainly at this point in time we think we would have a valid case to proceed further," he said.

Patent experts say it is not at all unusual for reexaminations to end in rejection, because it may not be until the claim is issued that challengers can marshal their opposition. More than 90% of such rejections remain rejected, according to a PTO spokesperson. What sometimes happens is a rejection will result in a revision of the application so the patent will be accepted; applicants naturally seek the broadest possible coverage and are not surprised when they must tone down their claims.

Far fewer rejected patents—approximately 1%—enter the formal appeal process, which starts with a PTO review and can extend to the US Circuit Court of Appeals. Of that 1%, says PTO, patents are reissued in approximately half the cases.

In 1994, Agracetus obtained a similar cotton patent from the European Patent Office. A challenge was filed to the patent during Europe's nine-month opposition period, and the company is awaiting a decision by the patent examiner there. Agracetus has also received a patent on genetically modified soybean in Europe. The multinational company is conducting research on transgenic rice, maize, peanuts, and beans. Smestad said his company was "not currently pursuing commercial opportunities ourselves" in those crops.

Industry Wants Patents

Although biotechnology companies are in favor of patents for plants, some elements of industry were clearly disturbed at the granting of broad species patents to Agracetus/Grace. Simon G. Best. is the chief executive officer and managing director of Zeneca Plant Science of Wilmington, Delaware, a part of the British-owned multinational Zeneca group of pharmaceuticals, agrochemicals, and specialty chemicals. "Broad patents for broad and pioneering inventions can be legitimate," he said in an interview. "But that doesn't mean necessarily that patents like those granted to Agracetus have met or did meet those standards" required by the US patenting system. The Monsanto Company, which has invested heavily in agricultural biotechnology, has filed opposition papers to Grace's European patent on transgenic soybean, arguing that the patent will stifle research and, in the words of a spokesperson, is "simply too broad."

Others, however, say, if they had filed first, they would feel entitled to the patent and its rewards. William Tucker is manager for technology transfer at DNA Plant Technology of Oakland, California, which includes in its repertoire techniques for making transgenic tomatoes that ripen properly even when picked early. As for broad patents on food crops such as Agracetus's, he said recently, "The view is from where you sit. If we'd got something as broad as that on a patent, then we'd be very happy. But coming from the other side, it's clearly hard to deal with a patent as broad as that."

Tucker says PTO's initial granting of the cotton claim may have been a fluke. "I don't think those things will be issued any more," he said. "It's one of those quirks of the Patent Office [from] back in the days when they really didn't know much about biotechnology." As patent examiners become more familiar with the prior art that is around, he added, it is going to get harder for applicants to secure broad patents.

Like others in the industry, Tucker sees no moral or ethical problem with the protection of genetically manipulated plants under the utility patent system, as opposed to the less restrictive plant breeders' rights that have prevailed for the last few decades. "A plant variety has been created because of the inventiveness of a set of people," he said. "And it has certain properties that have certain commercial utility. I think you need to be able to protect that invention.... Just because it's biological and self-reproducing doesn't to me make it any different from a piece of machinery that you manufacture from nuts and bolts and screws."

Spokesmen for industry associations agree with this position. "Patenting plants is no different than patenting any other kind of product," said David R. Lambert, executive director of the American Seed Trade Association, which represents approximately

600 active members. Richard Godown, senior vice president of the Biotechnology Industry Organization, another Washington group, which has 585 members, sees the cotton case as an example of a smoothly functioning system of protection: "Agracetus was granted these two broad-range patents, and the industry recognized it, and so the industry is dealing with Agracetus."

Patents on Concepts?

The university research community, which has become closely affiliated with the biotechnology industry itself, has been relatively quiet in discussions about the broadness of patent claims. One exception is Peter J. Day, the director of the Rutgers, New Jersey, Center for Agricultural Molecular Biology and chair of the National Research Council's committee on managing global agricultural genetic resources. Before going to Rutgers, Day was director of the Plant Breeding Institute in Cambridge, England.

"I think it's perfectly legitimate to patent a particular method of producing transgenic plants," said Day in a recent interview. "What I object to in the Agracetus patent is that it patents any genetically engineered cotton produced by any means whatever. So that means that if someone develops a novel method for genetically engineering plants, including cotton, and they apply it to cotton, say, next year, they are unable to market that genetically engineered cotton without an agreement with Agracetus, because Agracetus have patented the concept of genetically engineered cotton. That, I think, is absurd." Such a patent, he added, would certainly stifle corporate research.

John Barton, a professor at the Stanford University law school and recognized expert on global intellectual property rights, is also concerned about patents of the sort claimed by Agracetus. Barton said in an interview: "I think that kind of patent is a mistake. At the same time, I think that the fact that there are some errors in the way the patent system has been implemented doesn't mean it isn't going to be useful. I do think it's going to be important to build some international arrangements such that patents don't become too powerful as tools of monopolization for the existing major companies."

More Seats at the Table

Industry tends to think of patents as a matter between the company holding the patent and its competition. Others argue that the discussion about ownership of germplasm actually should involve many more people, including those farmers whose stewardship has kept the plants alive for centuries. Under the utility patent system, farmers and indigenous communities have no formal role unless they can successfully argue that they perfected germplasm that is novel, has utility, and is nonobvious—a most unlikely possibility.

The biotechnology industry considers genes from undomesticated plants, animals, and microorganisms the strategic raw materials for the creation of new agricultural, pharmaceutical, and industrial products. The great majority of these genetic materials exist now in the tropics and other parts of the nonindustrial world. "But these genes are seldom 'raw materials' in the traditional sense because they have been selected, nurtured, and improved by untold numbers of farmers and indigenous peoples over thousands of years," said Hope Shand at the 1994 National Agricultural Biotechnology Council conference "Agricultural Biotechnology and the Public Good." Shand is the research director of the Rural Advancement Foundation International (RAFI) in Pittsboro, North Carolina. RAFI has formally challenged the issuance of broad species patents, including Agracetus's 1994 transgenic soybean patent in Europe.

Organizations in the Southern Hemisphere frequently accuse northern industries of collecting seeds and other germplasm from the traditional centers of genetic diversity (which are concentrated in the South), changing them a bit through plant breeding, and then selling them back to their original curators at an unconscionable markup. Norah Olembo, of the Kenyan Ministry of Science and Technology, put it this way: "What went freely now comes back with a price tag." This objection is widely made whether the collected germplasm comes from farmers' fields, weedy wild relatives of the agricultural crops, foods and medicinal products from forests, or collecting forays into tiny village markets.

Suman Sahai, head of the Gene Campaign, an organization based in New Delhi and dedicated to protecting the genetic resources of the Third World and the rights of its farmers, says: "God didn't give us 'rice' or 'wheat' or 'potato.' There were wild plants in the forests, which forests happened to be located in the tropical countries in the South today. But it was a very concerted effort to convert, refine, breed, select, cross, and create economically valuable crops—food crops, cash crops, everything. Who did all of that work?"

Demands have increased in recent years for greater recognition for the native stewards of germplasm—not only for the varieties they have developed but also for the job they have been doing of conserving it in the face of rapid global overdevelopment and population growth. These demands contributed to the adoption, as part of the 1992 United Nations Conference on Environment and Development, of a Convention on Biological Diversity. The convention is a binding treaty affirming that "the conservation of biological diversity is a common concern of humankind," but it lacks real teeth to support the efforts of farmers and indigenous communities. There are ongoing efforts by nongovernmental organizations and developing countries to counter the treaty's weaknesses (which were largely imposed by the United States and some other industrialized nations) and to make it a force in ensuring intellectual property rights for the less-affluent world.

The point that is frequently made by these advocates is that wild or indigenously cultivated germplasm is just as valuable as finished seed that is used in commercial agriculture. Breeding programs must continue because farmers need ever-new varieties to stand up against changing insect or fungal populations and as farmland spreads to more marginal areas. And those programs, whether they involve classical plant breeding or sophisticated biotechnology, draw constantly upon the unfinished seeds of the Andean slopes, the African savannas, and the tropical peasant's backyard garden. The stewards of this germplasm deserve recompense, say the advocates.

For many in the northern biotechnology community, the arguments for southern compensation are not persuasive. They see a vast difference between finished and wild germplasm. Simon Best, of Zeneca, says: "By and large, wild material historically was perhaps a small portion of the finished genome of a hybrid family or a varietal family that took 20 or 30 years to develop: hundreds or thousands of man-years...My view, and I think most people's view, is in reality that the value of the wild material is much less than most people think—and that sort of discrepancy is artificially inflaming the debate right now. "

Henry L. Shands, who oversees the USDA's genetic resources program, tends to agree, and he adds that the "ownership" of wild and indigenous germplasm is difficult to pin down. "In the case of agricultural plants," he said in an interview, "there is no country that is not dependent on at least one other country for the genetic resources it grows...

"If you say, 'Well, I own this and you own that,' then I could say, 'Well, you own that but I own this.' We're interdependent. We need to exchange material for our own benefit. It's an 'I win, you win' situation then."

Furthermore, said Shands, "There's nothing that says that [southern farmers] have to buy this material. But farmers are going to buy what yields well."

Meanwhile, to insulate its holdings from potential ownership claims on germplasm, last fall the Consultative Group on International Agricultural Research (CGIAR) placed its enormous genebank collections under the control of the Food and Agriculture Organization (FAO) of the United Nations. CGIAR, which took the action to protect what it often calls "the common heritage of humankind," is a collection of 17 research centers that are situated in the South but receive most of their funding from northern governments and organizations such as the World Bank. Currently, any researcher can request and get seeds from its genebanks.

Now, partly with John Barton's help, the CGIAR centers are exploring the possible use of material transfer agreements to control the use of their germplasm by other interests. Such agreements, which are currently used by many commercial groups, set the rules for transferring genetic material. The US National Cancer Institute (NCI), which annually obtains samples of approximately 6000 plants and marine organisms, has drafted such an agreement that, according to the Institute, "recognizes the need to compensate source country organizations and peoples in the event of commercialization or a drug derived from the organism collected." NCI has signed such agreements with approximately a dozen nations, including Madagascar, Indonesia, Belize, and the Philippines, and with a foundation representing indigenous people who live predominantly in Ecuador. Even in those areas where an agreement has not yet been signed, NCI's four plant and one marine collectors do their work under the terms of the agreement, said an official in NCI's Natural Products Division, and leave copies of the agreement behind.

A Reward System

Despite industry views on ownership, there is an effort under way to devise a system to reward farmers and indigenous communities for their ongoing contributions to biological diversity. The trick, all agree, lies in ensuring that the rewards actually get to the people who deserve them and do not get filtered out by bureaucracies. There is an even more basic problem of deciding who really made the contributions and deserves the credit. Even agricultural plants in the rich centers of biological diversity have ancestors from foreign lands.

The World Resources Institute (WRI), a Washington policy and research organization, has been studying the problem, particularly in connection with germplasm from tropical countries that is desired by northern pharmaceutical firms. Walter V. Reid, a WRI vice president, says he thinks the "most realistic mechanism" would be to collect royalties. Otherwise, says Reid, "you're going to rely on the good will of governments to return these benefits locally. And that's a big 'if.'" The royalties from germplasm would go to a fund, which would be used to encourage biological conservation by distributing it through networks of nongovernmental organizations that work with local communities.

The framework for such a fund already exists through the Fund for Plant Genetic Resources, which was established in 1987 by the FAO's Commission on Plant Genetic Resources. But contributions to it are voluntary, and they have been minimal.

Another way for the biologically rich nations to take advantage of what they have, thinks Sahai, is to take control of biotechnology itself "The raw material for biotechnology is in the South," she said during an interview in New Delhi. "The technological expertise is there." (Indeed, research institutions in the southern India city of Bangalore have been hailed as hotbeds of technological brainpower. Both homegrown firms and branches of US corporations have established what many call the Silicon Valley of India.)

"And, most of all, for us there is a very important factor: Unlike all the other technologies so far, this is not a capital-intensive technology. It's a skill-intensive technology. The fact that we have the skilled manpower at very, very competitive costing makes India a potential player in the field of biotechnology. And therefore it becomes even more important for us—and it's always important as an environmental goal— to preserve the resources. . . .

"Because of the advent of biotechnology, the fact is that the most lucrative resource on Earth right now is genetic resources. There are big bucks out there," says Sahai. "Any one who gets control of genetic resources is going to dominate the world economy." ■

Questions

1. What does genetic manipulation with biotechnology allow researchers to do?

2. How does industry tend to think of patents?

3. What is the single most important factor that has made genetic resources the most lucrative resource on Earth right now?

Answers are at the back of the book.

Section Three

Problems of Environmental Degradation

28

The decline of many long-lived species due to the exposure of man-made chemicals is a very serious and alarming problem. Researchers have been looking at environmental pollutants only as a carcinogenic threat. Now they are looking at these pollutants as a cause of damage to the reproduction, immunity, behavior, and growth systems of a body. In the mid-1970s exposure to the chemical DDT seemed to be correlated with an abnormally low number of males in a California gull population. In the early 1990s there were similar findings in alligators at Florida's Lake Apopka. The populations of certain animal species are in decline, but how does this affect the human population? In 1991, an analysis was done of many smaller studies of global human sperm counts over the past 50 years. The sperm count had declined by half between 1940 and 1990. Studies show endometriosis, testicular cancer, and possibly other types of cancers are increasing due to environmental pollutants.

The Alarming Language of Pollution

Daniel Glick

On California's Channel Islands in the mid-1970s, an ecologist found an abnormally high ratio of female gulls to male gulls. In Florida in the early 1990s, a team of endocrinologists discovered abnormally small penises in alligators near a former Superfund site. And in Great Britain, biochemists have noticed in the last few years that something in wastewater effluent appears to be creating hermaphroditic fish.

Sound like bizarre episodes of *Wild Kingdom*? Actually, these observations are all clues to a far-flung scientific sleuthing saga. Over the last few years, experts from a dozen disciplines have been piecing together field and laboratory evidence that environmental pollutants may be doing far more damage to wildlife and humans than previously suspected, in ways no one had imagined possible. For starters, by sending various false signals to endocrine (or hormonal) systems in the body, pollutants could be harming vertebrate reproduction worldwide. All of this evidence could comprise one of the most alarming messages wildlife has ever sent our way. "If we don't believe that animals in the wild are sentinels for us humans, we're burying our heads in the sand," says Linda Birnbaum, director of the environmental toxics division of the Environmental Protection Agency (EPA).

Endocrine-disrupting chemicals are associated with problems ranging from developmental deficiencies in children, to smaller penises in pubescent boys, to infertility. "Every day, I get more concerned," says John McLachlan, chief of the laboratory of reproductive and developmental toxicology at the National Institute of Environmental Health and Science (NIEHS).

Implicated are huge numbers of products—including some pesticides, industrial solvents, adhesives and plastics. A very few, such as PCBs and the pesticide DDT, have been banned or are more heavily regulated in this country than in the past—though they persist in the environment. But thousands have never been regulated. Much of the stuff is deposited worldwide by the atmosphere and has been found in both the Arctic and Antarctica.

Until the last few years, the biggest question for regulators has been: Does a given chemical cause cancer, and if so, at what exposure level? (And very few chemicals have even been tested for carcinogenicity.) Now some researchers are also asking: Does a chemical harm reproduction, immunity, behavior or growth?

Also, regulators have long assumed each chemical to be innocent until proven guilty. But researchers are growing increasingly concerned at evidence that related chemicals may be able to harm the body in similar ways. For example, DDT and dioxins (often commonly referred to in the singular) are members of a group of similar chemicals called organochlorines. They are not to be confused with the chlorine we safely use to disinfect swimming-pool water and bleach our clothes. While DDT is a deliberate product, dioxins are unwanted byproducts of industrial high-temperature use of chlorine.

A 1994 National Wildlife Federation report, *Fertility on the Brink: The Legacy of the Chemical Age*, concluded that there is enough evidence to warrant phaseouts, at the very least, of certain chemicals released into the environment. The list includes dioxins, some pesticides and hexachlorobenzene. Federation counsel Elise Hoerath argues that the problem has become "a significant public health threat."

Others warn that hormonal activity is so complicated and poorly understood that costly action to ban certain chemicals is uncalled for until we know more. "As a citizen, I would like to see some of these chemicals banned," says Carlos Sonneschein, professor of cellular biology at Tufts University School of Medicine. "As a scientist, I would like to have more data."

Still, the data have been steadily adding up, thanks largely to the work of zoologist Theo Colborn, a senior scientist at the World Wildlife Fund and director of its wildlife and contaminants program. In late 1987, Colborn began sifting through studies of declining wildlife populations in the Great Lakes region. On the left side of a piece of paper, she listed species with steep population drops: bald eagle, Forster's tern, double crested cormorant, mink and river otter, among others. On the right, she listed their health problems, including organ damage, eggshell thinning, hormonal changes and low birth survival rates.

Each of the animals depended on a fish diet. Fish in the notoriously polluted Great Lakes were known to contain high concentrations of various synthetic chemicals, especially in fatty tissue, and Colborn wondered if the pollutants were causing the disorders. Were toxics tinkering with the immunity, behavior, growth or behavior of fish eaters? Colborn began searching the scientific literature. "I was really concerned," she recalls. "It was very obvious that these chemicals were developmental toxicants." Yet for the most part, testing had only looked for cancer. "We've been blinded," she says. "We never tested for developmental effects."

Even so, some studies did find those effects. Researchers had found in the mid-1970s that exposure to DDT seemed to be correlated with an abnormally low number of males in a California gull population. In the late 1970s, toxicologist Michael Fry of the University of California at Davis was able to cause "feminization" of male gull embryos (they developed abnormal testes containing ovarian tissue) in his lab by injecting uncontaminated eggs with DDT.

Many years later, in the early 1990s, University of Florida comparative endocrinologist Louis Guillette started finding similar problems in alligators at Florida's Lake Apopka. The area was a former Superfund site that had been contaminated in 1980 with the chemical dicofol, an organochlorine that also contained some DDT. The lake also contained a mix of agricultural chemicals from farm runoff.

Working with colleague Timothy Gross and other researchers, Guillette found that alligator eggs were barely hatching, teenage males had abnormally small penises and the level of the male hormone testosterone was far below normal. Later, Guillette conferred with a researcher who had produced remarkably similar results in lab rats by exposing them to a compound similar to DDE, a breakdown product of DDT. "Oh my God," Guillette said after seeing the data. "I think we have a major problem here."

As Colborn compiled evidence from wildlife biologists, toxicologists and the medical literature, she realized that other scientists were asking some of the same questions. So, in 1991, she helped bring a group of them together to compare notes for the first time. After another meeting last year in Washington, D.C., 23 wildlife biologists agreed that "populations of many long-lived species are declining .... Some of these declines are related to exposure to man-made chemicals and their effects on the development of embryos."

Their reasoning is based on the knowledge that sex differentiation is determined by tiny amounts of male and female hormones interacting in the developing fetus. Contrary to what we've all been taught in introductory biology classes, animals do not exhibit male or female traits simply because they possess or lack a Y chromosome. If a hormone impostor shows up during fetal development, sexual function can go akimbo. "Very, very low levels of contaminants can have an effect on developing embryos," says the University of Florida's Guillette. "A dose that wouldn't bother an adult can be catastrophic to an embryo."

Soon after Fry's discovery that DDT injections could "feminize" gull eggs, biologist David Crews of the University of Texas discovered in 1984 that he could control the gender of slider turtles with minute quantities of the female hormone estradiol. For many turtles, the temperature of the eggs' environment determines gender. Heat produces a female; cold yields a male. But in the lab, Crews could coax embryos incubating at a male-producing temperature to become female with just a drop of estradiol on the eggs.

Estradiol is an estrogen, and Crews' study fits a scary pattern. A number of synthetic substances are so-called "environmental estrogens," acting like the hormone Crews used to bend the turtles gender. In recent work, he and colleagues have found they can create sexually mixed-up turtles with "cocktail" mixtures of certain PCB compounds. Some of the turtles have testes and oviducts. Others have ovaries but no oviduct. Most alarming, these effects occurred at extremely low doses. Somehow, the combination of several PCBs is far more disrupting than one PCB compound alone.

Of course, not all estrogens are bad; when they occur naturally, they play critical roles in the body. Deliberate therapeutic doses even help women through and beyond menopause, in part by protecting bone density and cardiac health. Environmental estrogens, however, are a different story. NIEHS researcher McLachlan, who calls estrogen the "Earth Mother of hormones," has shown that certain chemicals can bind to or block estrogen receptors, which may in turn cause developmental deviations.

Think of the estrogen receptor as a lock on a cell, and natural estrogen as a perfect key. Scientists believe that literally hundreds of compounds have a chemical structure that also fits the lock—and which could produce similar responses. But then, these chemicals may "fit" into estrogen receptors without producing the cascade of cellular events that follow exposure to actual estrogen—and no harm may be done. Still, even if that's so, when the impostor key is in the lock, the real key may not be able to enter.

Since the number of chemicals that fit into the estrogen lock, or receptor, are so numerous, no one can clarify all the effects of these multiple exposures. "If there are so many estrogens out there, how can anybody figure out which one is doing what?" asks Thomas Goldsworthy of the Chemical Industries Institute of Toxicology. "Some of the mechanisms aren't clear yet."

Some of the effects, however, are becoming clearer. Toxicologists Earl Gray and Bill Kelce of the EPA reported last year that the common fungicide vinclozolin, used on many fruits and vegetables, can block receptors for the male hormone androgen and cause sexual damage in male rats. At certain doses, rats exposed to vinclozolin do not develop normal male traits even though they do produce testosterone. At high exposures, male rats develop severely abnormal genitalia. Gray thinks fruit treated with the fungicide does not contain enough residue to harm humans, but he is looking into the question. And he is sure of one thing: "There are clearly other environmental anti-androgens we haven't discovered yet," he says.

The findings of field work like Guillette's and laboratory analysis like Gray's have been bolstered by studies of inadvertent human exposures to endocrine-disrupting compounds. In 1979, women in Taiwan who ate rice oil contaminated with polychlorinated biphenyls (PCBs) and polychlorinated dibenzofurans (PCDFs) offered an ideal if tragic laboratory to track long-term effects in humans. Researchers have followed 118 children of the women and an identically sized control group. Members of the exposed group have suffered developmental delays, growth retardation and slightly lower IQs. Many of the boys, who are now reaching puberty, have abnormally small penises.

Between the 1940s and 1970s, diesthylstilbestrol, or DES, was given to an estimated two million to six million women during pregnancy to help prevent miscarriage. In children of DES mothers, the drug caused a range of developmental and health problems, some of which only surfaced in the process of creating the next generation. Among males, researchers have noted abnormalities in scrotums, an unusually high prevalence of undescended testicles and decreased sperm counts. Among DES daughters, clinical problems include organ dysfunction, reduction in fertility, immune-system disorders and other difficulties.

The DES example leads to an alarming hypothesis: If some endocrine-disrupting pollutants act like DES, which had effects long after birth, perhaps we won't see the consequences until exposed offspring themselves begin trying to have kids. And that raises the question: What actual harm to humans have scientists found from exposure to the sea of chemicals released into the environment over the past 50 years?

Enter Niels Skakkebaek, a Danish researcher in Copenhagen. In 1991, he published a meta-analysis of many smaller studies of global human sperm counts over the past half century and found that the counts declined by half between 1940 and 1990. Other more recent European studies sought to disprove Skakkebaek's results, but ended up corroborating them. If sperm counts have indeed dropped, one clue to the reason may come from lab tests in which estrogen-mimicking compounds have affected the Sertoli cell, which is related to sperm production.

Research has also implicated environmental toxics in the rise of endometriosis, testicular cancer and possibly other cancers as well in recent decades. In one study that went on for 15 years, 79 percent of a rhesus monkey colony exposed to dioxin developed endometriosis (the development of endometrial tissue in females in places it is not normally present). Dioxin is not thought to imitate estrogen, but is clearly an endocrine disrupter in at least some animals. In the monkeys, the endometriosis increased in severity in proportion to the amount of dioxin exposure.

What should the rest of society do while the researchers compare notes? "The tough call isn't for the scientists now," says Devra Lee Davis, a top scientific advisor at the U.S. Department of Health and Human Services. "It's for the regulators." There are signs that the federal government is beginning to pay heed. In the EPA draft dioxin reassessment report, now under review, dioxin is characterized as a potent toxic "producing a wide range of effects at very low levels when compared to other environmental contaminants."

The International Joint Commission, a bilateral organization that advises on environmental issues along the U.S.-Canada border, has repeatedly called for virtual elimination of toxic substances in the Great Lakes region. And a little-noticed amendment to the Clean Water Act proposed by the Clinton administration (the reauthorization died in the last Congress) would have required regulators to look at "impairments to reproductive, endocrine and immune systems as a result of water pollution." Even skeptic Goldsworthy of the Chemical Industries Institute of Toxicology says, "We are changing our environment. There's no question about that."

The World Wildlife Fund's Colborn says she welcomes scientific skepticism and even has days when she hopes she is imagining the whole thing. "We admit there are weaknesses, because we are never going to be able to show simple cause-and-effect relationships," she says of the complicated theory. Still, she adds, "The research has reached a point where you can't ignore it any more, and new evidence is coming in every week." For visitors to her Washington, D.C., office, Colborn lets a pesticide manufacturer have the last word: On the wall hangs a 1950s label from a one-pound package of a substance called DuraDust, 50 percent of which was pure DDT. The label promises, "Its killing power endures." ■

Questions

1. How can pollutants harm vertebrate reproduction?
2. What component determines gender in turtles?
3. What potent toxic does the EPA characterize as producing wide range effects at very low levels?

Answers are at the back of the book.

29

Beachgoers, commercial fishing fleets, military crafts, cruise ships, and pleasure boaters discard tons of garbage into the ocean. In 1993, the national beach clean-up collected more than seven million pieces of trash. Oil and invisible pollutants, such as disease-causing pathogens, contaminate beach waters and estuaries. Nutrient chemicals pumped into coastal waters can set off ecological chain reactions that destroy a huge number of fish and invertebrates. The coastal population of the U.S. in 1990 was 80 million. Estimates show that it will be 127 million by the year 2010. If the current trend of beach pollution continues, beaches will become more polluted with the increasing population.

What Is Polluting Our Beaches?

Joe Brancatelli

Everything from cigarette butts to raw sewage. But it's what you don't see that really hurts.

When you stand on the gently sloping hillside overlooking Fleming Beach on the rugged western shore of Maui it's hard to imagine there could be anything wrong with any beach anywhere in the United States.

The powdery, white sand is clean and pristine. The blue-green Hawaiian waters really do sparkle in the afternoon sun. Off in the distance, perched on craggy rocks, fishermen cast *upena kiola*, their hand-thrown nets, into the sea.

The tragedy of this perfect scene is that Fleming Beach is an exception. Bluntly put, many of our beaches are under attack. Pollutants as exotic as giardia—a single-celled protozoan—and as prosaic as the cigarette butt foul our shores, damage fragile ecosystems, and profoundly threaten the simple pastime of spending a day at the beach.

The most visible form of beach pollution is "marine debris," what the layman prosaically calls "garbage." In 1993, the national beach clean-up sponsored by the Washington, D.C.-based Center for Marine Conservation, collected more than seven million items of trash. In Texas alone, volunteers collected more than a ton of debris for every mile cleaned. In Connecticut, they collected 1,840 cigarette butts per mile of beach (in December 1994, the Hanuama Bay beach in Honolulu, Hawaii, became the first "no-smoking" beach). CMC volunteers also gathered more than 40,000 rubber balloons, 25,000 plastic six-pack holders, 300,000 glass and plastic beverage bottles, and 200,000 metal beverage cans.

Beachgoers are not the only culprits. Commercial fishing fleets, military vessels, cruise ships, and pleasure boaters dump tons of galley waste, plastic fishing lines and nets, and other garbage into the ocean. A survey conducted on the beaches of Amchitka Island in Alaska determined that commercial fishermen left behind more than 950 pounds—almost half a ton—of trawl webbing for each mile of beach.

Marine debris is not only harmful to wildlife (a sperm whale found dying on the New Jersey shore in 1985 had a mylar balloon lodged in its stomach, and three feet of purple ribbon in its intestines); it also is not easy to eradicate and occasionally has caused beach closures. Yet its most visible effect on beaches is aesthetic.

Another visible polluter is oil, principally because of the sensational aspect to spillages. Yet oil rarely affects beach-goers. Not so other, more insidious types of pollution. Invisible pollutants known as "pathogens" contaminate beach waters and estuaries, reaching us via raw sewage, sludge, and wastewater effluent, or from storm drains discharging into coastal waters and rivers.

As a nation, we spent S76 billion between 1972 and 1992 to build or expand sewage treatment plants, but it clearly has not been enough. For example, more than 2.5 billion gallons of untreated waste is flushed into Narragansett Bay, Rhode Island, every year, and eight coastal states don't even monitor pathogenic activity regularly.

These pathogens can bring disease, either from our swimming in infected waters or our consuming contaminated seafood or shellfish. Among the principal types of pathogenic organisms are viruses such as hepatitis A, the bacteria responsible for cholera and gastroenteritis and giardia which can cause chronic diarrhea and even death. During the last decade, cholera has been detected in shellfish beds in Mobile, Alabama, and directly traced to an outbreak in South America. Shellfish beds were closed and fortunately no one contracted the disease.

The effect of pollutants also can be measured by beach closures. During 1992 and 1993, according to the National Resources Defense Council, 23 states issued almost 5,000 beach closures or advisories, and many other polluted beaches went undetected because states did not have monitoring systems.

Still another form of dangerous pollution is a category known as "nutrients"—chemicals such as nitrogen and phosphorous, and even small amounts of excess nutrients pumped into coastal waters can set of£ a horrifying ecological chain reaction. The nutrients overfertilize seabeds and cause massive increases in the blooms of algae and phytoplankton. The blooms die, and decomposing bacteria depletes oxygen from the water, leading to mass kills of fish and invertebrates.

Where do excess nutrients come from? From land, usually in the form of sewage, fertilizers, and sediment. Large doses of excess nitrogen also reach our beaches from the atmosphere, and once again people are to blame: Automotive and smokestack emissions are a main source of airborne nitrous oxides.

If nutrient-based pollution sounds exotic and unfamiliar to you, perhaps you know it by more colorful names. The "brown tide" that wiped out bay-scallop beds and eelgrass bays in Long Island Sound several years ago was caused by nutrient pollution. And part of the blame for the periodic "red tides" that plague seashores may lie with nutrients (other factors, such as weather conditions, contribute as well).

There is also one unavoidable conclusion: Things are going to get worse before they get better. The coastal population of the U.S. was 80 million in 1990, but will grow to more than 127 million by the year 2010. Given our dismal record in tending to the aesthetic and ecological imperatives of beaches and coastal waterways, it seems inevitable that more people will mean more pollution. ■

Questions

1. Where do excess nutrients come from?

2. What are invisible pollutants and by what means do they contaminate coastal waters and rivers?

3. What types of diseases can pathogens bring?

Answers are at the back of the book.

30

Many urban areas have utilized most of the available water resources. New sources of drinking water need to be found. Now communities nationwide are studying the safety, economics and feasibility of inserting treated sewer water into the ground to supply decreasing aquifers. This also includes those tapped for drinking water. This is known as artificial groundwater recharge, a process that involves injecting treated city wastewater directly into aquifers or leaching it into the ground. Two methods for returning the treated wastewater to the aquifer are: (1) the use of absorption layers of soil and percolation to filter viruses and contaminants from the water; and (2) direct injection into the aquifer after being treated thoroughly. It is likely that recycled wastewater will be increasingly relied upon in coming years.

Drinking Recycled Wastewater

Ginger Pinholster

Can groundwater recharge safely address the drinking-water needs of rapidly growing urban areas?

As the 21st century approaches, communities around the world, faced with population growth and increased urbanization, scramble to find new sources of drinking water. "Most cities have already fully exploited the readily available water resources," EPA warns.[1] The California Department of Water Resources braces for annual water shortages ranging from 3.7 million to 5.7 million acre-feet (1.2 million to 1.8 million gallons, or 4.6 billion to 7 billion cubic meters) by 2020.[2] To meet growing population needs, U.S. planners traditionally have built new dams, levees, and canals. But public willingness to pay for new water facilities "has declined dramatically from the dam-building heyday of the 1950s and 1960s," according to a recent report by the National Research Council.[3]

Consequently, communities throughout the United States are studying the safety, economics, and feasibility of directing treated sewer water into the ground to replenish dwindling aquifers—even those tapped for drinking water. The practice, known as artificial groundwater recharge, typically involves injecting treated city wastewater directly into aquifers or spreading it onto the ground to infiltrate the surface. It is unclear exactly how many regions practice groundwater recharge, but James Crook, director of water reuse at Black & Veatch (Cambridge, MA), estimates that "hundreds" of U.S. cities are recycling wastewater for nonpotable purposes from crop irrigation in arid western states to landscaping at Florida's Walt Disney World. A half-dozen cities, including El Paso, TX, and Los Angeles are recharging potable aquifers; a dozen more communities are considering similar projects, says Crook.

Though reclaimed wastewater has been used to augment drinking-water supplies in Los Angeles County since 1962, mounting public concern about the safety of recycled water is sparking renewed debate among the scientific community. Debate has focused on whether treated wastewater can be clean enough to drink and is free from viruses and hazardous substances. But economics also is a concern: Is it cheaper to treat wastewater to replenish aquifers or "import" water from other sources?

A National Research Council (NRC) report titled "Ground Water Recharge: Using Waters of Impaired Quality," released in September 1994, offers a promising, though qualified, endorsement of groundwater recharge practices. Treated city wastewater can be used to boost potable aquifers under

Reprinted with permission from *Environmental Science & Technology*, Vol. 29, No. 4, April 1995, pp. 174A–179A.

certain conditions, when no "better quality" water exists, according to the report. The committee was careful to add, however, that recharge technologies are "especially well-suited to nonpotable uses such as landscape irrigation."

For water managers facing public opposition to recharge projects using treated wastewater, the report provides new ammunition. Coincidentally, it was released the same week that the Miller Brewing Company filed a lawsuit to block a $25 million recharge effort in Upper San Gabriel Valley, CA. If successful, the lawsuit "would really give water reuse a black eye," says Crook, who served on the NRC groundwater recharge committee.

Miller's lawsuit charges that use of treated wastewater "will irreversibly pollute the basin," possibly damaging its product. Cass Luke, a spokeswoman for the Upper San Gabriel Valley Water District, says Miller's objections are based solely on public relations concerns, rather than scientific evidence of a health hazard.

The brewery supports the principle of groundwater recharge, but the company has a problem with this specific case, says Miller spokesman Victor Franco. The spreading ground for the treated wastewater, adjacent to the brewery, is a very porous soil, according to Franco, and the filtering effect as water percolates through the soil would not occur because the water would "cascade through."

The lawsuit underscores recurring questions about the safety of treated wastewater. Given proper treatment, reclaimed wastewater can be as safe or safer than traditional drinking-water sources, according to Herman Bouwer, chief engineer for the U.S. Department of Agriculture's Water Conservation Laboratory in Phoenix, AZ. After all, he says, wastewater slated for use in aquifers typically is subjected to primary, secondary, and sometimes advanced cleanup procedures—from settling of solids to biological oxidation of organics to salt extraction via reverse osmosis and adsorption of synthetic organics by granular activated carbon.

"About half the people in our country use groundwater [for drinking]," says Bouwer, an NRC committee member, "but not all groundwaters are pure and pristine." Surface water supplies from rivers, streams, and lakes frequently are polluted by sewage effluent upstream. Runoff from cattle farms can contaminate surface waters with potentially deadly parasites such as *Cryptosporidium*, which can resist disinfection.

But other researchers insist that uncertainties related to the chemical composition of reclaimed wastewater could jeopardize public health. "Human sewage is a mish-mash of complex organic materials, only some of which have been identified," says Henry Ongerth, former sanitary engineer for California's Department of Public Health. "Of the substances that have been identified, only a fraction have been studied for their toxicity.... Why add to the risks we already face?"

Is It Safe to Drink?

Nonpotable aquifers frequently are recharged without fuss or fanfare. "For car washes, street sweeping, and golf-course irrigation, no one is opposed to this type of recycling," says Forest Tennant, a physician in West Covina, CA, who is head of the grassroots group Citizens for Clean Water. Yet Tennant and others staunchly oppose refilling potable aquifers with treated wastewater. Tennant hopes to thwart the recharge project in Upper San Gabriel Valley.

When it comes to groundwater recharge, the controversy most often hinges on whether treated city wastewater is safe to drink. Questions focus mainly on two constituents of treated wastewater: viruses and disinfection byproducts.

According to a 1993 analysis of virus-monitoring data collected over a 10-year period from six California wastewater treatment plants, secondary-level treatment effectively removes 99.8% of detectable viruses.[4] Large quantities of treated wastewater, averaging 275 gallons per sample, were collected monthly. Only one of 590 samples tested positive for enteric viruses, reports study author William A. Yanko, a laboratory supervisor for the Los Angeles County Sanitation District (LACSD). Yanko says his data show that California's existing treatment requirements ensure "essentially virus risk-free effluents" for recharge projects.

But another 1993 study of viral risks was less conclusive.[5] In that report, author David K. Powelson of the University of Arizona at Tucson notes that although soil can strip remaining viruses from treated wastewater as the water infiltrates an aquifer, studies have shown that "virus removal is dependent on virus type and environmental conditions." Removal or inactivation of two types of viruses (MS2 and PRD1) from treated effluent directed into test basins composed primarily of sand and gravel varied depending on sample depths. At a depth of 4.3 meters, for example, virus removal ranged from 37% to 99.7%.

Earle Hartling, water recycling coordinator for the LACSD, says viruses require a host and therefore are inactivated quickly in underground aquifers. "All a virus does is invade a cell, take over the DNA-replicating machinery of the cell, and cause disease. It does not survive long periods [in an aquifer] because it has no opportunity to replicate itself," says Hartling, who has been known to delight news photographers by chugging vials of treated wastewater. As proof that viruses pose no threat in recharged aquifers, Hartling and others frequently cite a landmark 1984 health effects study prepared by Margaret H. Nellor and colleagues.[6] Over a 5-year period, the authors say, "No viruses were detected in groundwater or chlorinated reclaimed water samples" collected from three sites in California.

Bob Hultquist, senior sanitary engineer for the California Department of Health Services, counters that "viruses don't live very long, but it's all relative because they can live in excess of a year." Therefore proper pretreatment of recycled wastewater is critical, he adds.

Disinfection Hazards

To eliminate viruses, wastewater usually is disinfected with chlorine and less frequently with alternative disinfectants such as ozone, monochloramine, or ultraviolet radiation. Unfortunately, a host of chlorine disinfection byproducts (DBPs) are thought to be hazardous to human health, Philip C. Singer, professor of environmental science and engineering at the University of North Carolina–Chapel Hill, notes in a 1994 study.[7] In 1976, for example, the National Cancer Institute deemed chloroform, a trihalomethane (THM), to be carcinogenic. Hundreds of other DBPs have been found since in drinking water, including dichloroacetic acid, which "is believed to be a more potent carcinogen than any of the THMs, based on animal studies," says Singer.

"There's very clear evidence that these compounds are carcinogenic, but the doses must be very high," says Richard J. Bull of Washington State University in Pullman, author of a major study on DBP health effects.[8] "Now, whether they're dangerous at the concentrations you would find in drinking water is a question."

Nellor's data indicate "no increased rates of infectious diseases, congenital malformations . . . or all cancers combined," and thus don't link DBPs to cancer, the National Research Council report notes. However, cancers may remain latent for more than 15 years—longer than the period of Nellor's study, it adds. Chlorine DBPs in reclaimed water could be minimized by expanding the use of alternative disinfection processes, the report notes. But it cautions that little is known about the byproducts of ozone, monochloramine, and ultraviolet disinfection.

Potential problems associated with alternative disinfectants were discussed in a recent one-year study of DBPs at Jefferson Parish, LA.[9] A treatment strategy of preozonation and postchlora-mination produced the lowest levels of 18 halogenated DBPs, based on total organic carbon and total organic halide. Author Benjamin W. Lykins, Jr., points out that ozonation results in increased assimilable carbon, which must be controlled to prevent microbial regrowth. Ozonation byproducts such as aldehydes, ketones, and acids also are a concern, says Lykins, chief of the EPA's Systems and Field Evaluation Branch in Cincinnati, OH.

In general, the National Research Council report says the body of existing health effects studies "do not suggest a health concern" associated with treated wastewater. For example, in a 20-year study in Denver, 344 rats showed no toxicologic, carcinogenic, or reproductive effects after drinking samples of treated water, even when concentration levels were amplified 500 times.[10] But study author William C. Lauer notes that "as little as 10% of the measurable total organic carbon [in treated wastewater] has been estimated to be amenable to identification."

Entering the Aquifer

Two recharge methods, surface infiltration and direct-well injection, are available to water managers when importing water is not an option. These methods can also be used to prevent seawater from contaminating a coastal fresh water aquifer.

The Montebello Forebay Groundwater Recharge Project in south-central Los Angeles County, CA, established in 1962, is an off-channel surface infiltration-type system. Wastewater at Montebello Forebay is subjected to advanced treatment, including chemical coagulation, dual-media filtration, and chlorine disinfection, Hartling explains. Then it is combined with other sources, such as storm water runoff and surface waters from northern California and Colorado. Spreading basins are flooded and dried alternately to prevent clogging and mosquito breeding and to promote aerobic soil conditions. Reclaimed water currently represents 20–30% of the inflow to the Montebello Forebay aquifer.

The other groundwater recharge method, direct-well injection, has been practiced at Water Factory 21 in Orange County, CA, since 1976. Construction of the facility began in 1972 to prevent seawater intrusion into four overtapped groundwater aquifers, says Mike Wehner, health and regulatory director for the Orange County Water District. Today, highly treated wastewater is mixed with deep well water and injected into threatened aquifers through a series of 23 multiple-casing wells located about 3.5 miles from the shore.

The direct-well approach enables Orange County to penetrate aquifers through a thick layer of clay. The same method lets the city of El Paso recharge a 350-foot-deep aquifer. At both locations, wastewater is extensively treated prior to injection. Pretreatment of activated sludge secondary effluent at Water Factory 21 begins with lime clarification, which removes suspended solids, heavy metals, dissolved minerals, and many viruses by elevating the water's pH level. Recarbonation then normalizes the pH, and mixed-media filtration removes more suspended solids. In a final step, some water is fed through an activated carbon adsorption system where organic molecules latch onto complex carbon pores before the remaining flow is rechlorinated. Another stream is forced under high pressure through a reverse-osmosis membrane, which removes salt and organics.

Soil infiltration recharge systems may offer the advantage of scrubbing additional contaminants from treated wastewater. According to Hartling, the soil infiltration system at Montebello Forebay reduces total organic carbon by as much as 90% (though this could be either by removal or dilution) and removes 50% of all nitrogen. Soil infiltration also can remove parasites that tend to be resistant to disinfection. Given the right soil conditions, "parasites like *Cryptosporidium* and *Giardia* can be filtered out mechanically by nature," says Herman Bouwer of the USDA's Water Conservation Lab.

Recharge Price Tags

Recharge may be an increasingly attractive option as competition for water grows and public funding for new water facilities shrinks, says Henry Vaux, Jr., professor of resource economics at the University of California–Riverside. Costs associated with artificial recharge "are going to be quite variable," he says, but reclaimed water can be less expensive than imported water.

When the Orange County Water District completed an internal economic feasibility study in 1993, Vaux says, the price tag for nearby surface water was $600 per acre-foot.[3] Transporting this surface water to spreading grounds would have cost the district an additional $82.40 per acre-foot, for a total of $682.40 per acre-foot. Conversely, reclaimed water is readily available for recharge purposes and the cost of advanced treatment ranges from $251 to $387 per acre-foot.

Upper San Gabriel Valley has proposed building a new nine-mile pipeline to funnel treated wastewater from the San Jose Creek Water Reclamation Plant to nearby spreading grounds. Water District Manager Bob Berlien says the project would be cost-effective over the long term because the region currently imports water for $235 per acre-foot, whereas reclaimed water would cost just $200 per acre-foot.

Groundwater recharge costs also are affected dramatically by the level of pre- and post-treatment required to ensure compliance with drinking-water standards. At Lake Buena Vista, FL, where the

Reedy Creek Improvement District is studying the feasibility of discharging treated wastewater into a recreational lake, researchers predicted the costs of four treatment options.[11] Estimated capital costs ranged from $11.3 million for a treatment system based on biological "deep bed" denitrification (rather than reverse osmosis) to more than $23 million for a system that includes a full suite of treatment technologies. Because direct-well injection recharge provides no soil-aquifer treatment, reclaimed water tends to require more advanced, and more costly, pretreatment, Vaux notes.

Regulatory Issues

Water drawn from recharged U.S. groundwater aquifers must comply with maximum part-per-million type standards set by the Safe Drinking Water Act. Aside from that, however, no federal regulations directly govern recharge practices such as the level of monitoring or pre- and post-treatment needed to ensure water quality.[1]

Many states have regulations governing recharge. According to a 1993 survey, 36 states have regulations or guidelines related to water recycling, but only 10 directly address the use of reclaimed water for purposes other than irrigation.[12] In states without regulations, EPA's *Guidelines for Water Reuse* manual serves as a blueprint for water reuse projects, says James Crook, a principal author of the publication.

"There's no formal agency position saying that we're strongly encouraging or discouraging [recharge]" said Robert Bastian, an environmental scientist with EPA's Office of Wastewater Management, but EPA typically has been "very positive toward water reuse practices," he adds. Michael Cook, director of EPA's Office of Wastewater Management, recently offered a qualified endorsement of the proposed recharge project in Upper San Gabriel Valley. In a letter accompanying an Environmental Impact Report commissioned by the water district, Cook says that "comprehensive research and demonstration projects as well as monitoring of existing operating systems have documented the environmental safety and lack of public health hazards associated with well-run water reclamation and reuse practices in the United States, including projects involving the recharge of potable surface water reservoirs and groundwater aquifers." He softens his support, however, by adding, "concerns are often raised when new reuse projects [such as the San Gabriel project] are proposed." Public concerns may be based on a "lack of knowledge," a fear of plummeting property values, or "the negative consumer image of drinking reclaimed water," Cook says.

California and Florida are revising their recharge regulations. Title 22 of the California Administrative Code defines a minimum treatment level for reclaimed water slated for nonpotable reuse (a cycle of coagulation, flocculation, sedimentation, filtration, and disinfection). A brief 1978 amendment to Title 22 (Article 5.1) notes that requirements for potable reuse will be established "on an individual case basis." A committee directed by the California Department of Health Services is refining a more specific set of regulations that include minimum performance standards, based on water quality analysis, for potable recharge systems. The proposed regulations also would require that reclaimed water be held underground for at least six months prior to reuse—"long enough to get a high percentage of [virus] die-off," says Bob Hultquist. He expects the draft regulations to be approved in early 1995. Florida is revising Chapter 62-610 of its Administrative Code, which deals mainly with nonpotable water reuse systems. The state Environmental Regulation Commission will review revised regulations in May, reports Dave York, reuse coordinator with the Florida Department of Environmental Protection. The draft regulations would set maximum contaminant levels for groundwater recharge.

While dealing with more stringent maximum contaminant levels in the future, recharge managers also will have to comply with new EPA limits on DBPs, according to Philip Singer.[7] EPA's revised rules probably will include requirements for collecting data on raw water quality, tougher treatment safeguards against *Giardia* and *Cryptosporidium*, and maximum contaminant levels for total THMs and other DBPs, Singer says.

Russell Christman, professor of environmental sciences at the University of North Carolina–Chapel Hill, fears that existing federal regulations may not ensure public safety when potable aquifers are recharged. "When you have a [water] source that can't be entirely characterized—if only 10 or 20% of its organic content can be identified—of what comfort is the fact that the water meets existing drinking water standards?" asks Christman.

A Critical Test Case

Concerns about public safety are driving the San Gabriel Valley lawsuit, according to Miller Brewing Company. To prove its case, attorneys for the company have collected expert declarations from researchers including Daniel A. Okun, Kenan professor of environmental engineering, emeritus, at the University of North Carolina–Chapel Hill. No long-term health hazards have been linked to the 32-year-old Montebello Forebay recharge project, Okun notes. But he also claims that "San Gabriel doesn't have quite the same aquifer situation [because] the soil there is more permeable and less likely to remove trace organics." Planners have proposed tertiary treatment based on conventional granular filtration. More advanced technologies, such as reverse osmosis, would be needed to remove organics, Okun says.

Recharge into San Gabriel aquifers might disturb a Superfund site containing a plume of industrial contamination, groundwater activist Tennant charges. But the contamination is located well upstream from the proposed recharge sites, countered the water district's Cass Luke.

In an August 9, 1994, letter to water district manager Bob Berlien, a representative of the California Regional Water Quality Control Board wrote that questions about water quality were "satisfactorily addressed" by a draft 1993 Environmental Impact Report.[13] "We believe the District's project will have positive regional and statewide benefits," wrote Robert P. Ghirelli, an executive officer for the Control Board.

U.S. water managers are following the Miller lawsuit closely. "We are worried that it might have negative impacts because public perception of water reuse is very, very important," says Bahman Sheikh, water resources and reuse policy specialist for the West Basin Municipal District. Already there have been two delays in getting to court, and the first hearings were delayed until February, Victor Franco said.

In Los Angeles County, Earle Harding began receiving calls from concerned citizens shortly after the Miller lawsuit was filed. "The reverberations and repercussions are starting already," he says. "This is scary, given the fact that reclaimed water will be critical to compensate for impending water shortages."

References

1. "Guidelines for Water Reuse"; U.S. Environmental Protection Agency: Washington, DC, 1992; EPA/625/R-92/ 004.
2. "California Water Balance: California Water Plan Update"; California Department of Water Resources, 1994; Bulletin 160-93.
3. "Ground Water Recharge: Using Waters of Impaired Quality"; National Research Council: Washington, DC, 1994.
4. Yanko, W. A. *Water Environ. Res.* **1993,** 65, 221–26.
5. Powelson, D. K.; Gerba, C. P; Yahya, M. T. *Water Res.* **1993,** *27*(4), 583–90.
6. "Summary: Health Effects Study Final Report"; Nellor, M. H.; Baird, R. B.; Smyth, J. R. County Sanitation Districts of Los Angeles County: Los Angeles, CA, 1984.
7. Singer, P. C. *J. Environ. Eng.* **1994,** *120*(4), 727–44.
8. "Health Effects of Disinfectants and Disinfection By-products"; Bull, R. J.; Kopfler, F. C. American Water Works Association Research Foundation: Denver, CO, 1991.
9. Lykins, B. W.; Koffskey, W. E.; Patterson, K. S. *J. Environ. Eng.* 1994,*120*(4), 745–58.
10. Lauer, W. C.; Wolfe, G. W.; Condie, L. W. *Toxicol. Chem. Mixtures* **1994,** 63–81.
11. "Advanced Wastewater Reclamation Program Final Report"; CH2M Hill: Gainesville, FL, 1993; prepared for the Reedy Creek Improvement District, Lake Buena Vista, FL.
12. Payne, J. F. et al. *Proceedings of the Water Envi ronment Federation 66th Annual Conference.* **1993,** *9*, 137.
13. "Draft Environmental Impact Report"; CH2M Hill: Santa Ana, CA, 1993; prepared for the

Upper San Gabriel Valley Municipal Water District, El Monte, CA.
14. Lauer, W. C. et al.; "Denver's Direct Potable Water Demonstration Project: Final Report"; Denver Water Department: Denver, CO, 1993.
■

Questions

1. Name two recharge methods used when importing water is not an option.

2. What have U.S. planners traditionally done in order to meet growing population needs for water?

3. With what substances is wastewater usually treated to eliminate viruses?

Answers are at the back of the book.

31

Future growth opportunities in the pollution control industry are predicted for companies and individuals seeking employment. Air pollution control systems markets are expected to expand around the world as developing nations industrialize. In the United States, the Clean Air Act (CAA) of 1990 will continue to create new technologies. A considerable near–term market is anticipated in retrofitting city solid waste combuster plants. The pollution control industry in Canada and Mexico may increase by 50 percent or more in the near future. These opportunities illustrate once agian how the marketplace, jobs, and the environment do not have to be in conflict.

Particulate Control: *The Next Air Pollution Control Growth Segment*

Robert W. McIlvaine

Where should environmental firms direct their research, money, and work force? Where will the next growth spurt in the industry occur? How should firms prepare for the next decade? In what direction should environmental professionals take their careers as they search for the field's next market boom? Robert W. McIlvaine of The McIlvaine Company makes his predictions for companies and job seekers as he forecasts future growth segments. He draws a road map for how and where companies can benefit from the next environmental hot topic.

Summary

Expect markets for air pollution control systems to balloon, thanks to a resurgence of heavy industry in the United States. The Clean Air Act (CAA) of 1990 will also have a major impact in forcing the creation of new technologies, such as cleanable HEPA filters and other novel devices. A substantial near-term market in retrofitting municipal solid waste combustor plants is anticipated, and the market for thermal gas treatment will rise substantially. In addition, although the Canadian-Mexican market is relatively small compared to the United States, the pollution control industry in each country will expand by 50% or more within five years.

The recent documentation of health risks associated with small particle inhalation forms the basis for tighter particulate control regulations. This is likely to result in rapid growth in sales of particulate control equipment.

Hello Mr. Greenfield

There is a new face in the neighborhood. In fact, Mr. Greenfield has not put in many appearances over the last fifteen years so his name may be unfamiliar to some readers. But back in the '70s when basic industry was expanding in the United States and lots of new plants were being built on what were formerly green fields, the term "greenfield project" was quite familiar. Now industrial America Is making a big comeback and this will have an incredibly large effect on the air pollution industry in the United States and therefore the world. More than $600 million of air pollution control projects have been started in just six weeks as part of tens of billions of dollars in new plants committed by manufacturers of steel, castings, semiconductors, chemicals, pharmaceuticals, automobiles, engines, pulp mills, refineries, and municipal wastewater treatment plants.

The U.S. philosophy of focusing on the bottom line for next month, rather than looking at the long-term has proven a weakness in the country's industrial outlook generally and a major detriment to the air pollution industry. This had resulted in continuous downsizing. The steel industry output has shrunk by 33%. There were twice as many foundries some years ago as there are today. Five years ago, the United States was running far behind Japan in the race to dominate the semiconductor industry.

Today the picture has been dramatically transformed. The United States leads the world in supplying semiconductors and has become a low-cost producer of steel. Demand is soaring for both products. As proof, the automobile companies just negotiated purchases of steel at prices 8% higher than their last contracts. Industrial America has been running at an unprecedented capacity rate and there is no way to increase output except by building new plants. A secondary effect can be expected from these expansions. The power industry and the municipal wastewater treatment plants have not anticipated this kind of expansion for industrial America. The Gross National Product (GNP) is headed for a 3% increase this year. Because electricity demand follows GNP, expect a need for 21,000 MW of power plants just to accommodate the burgeoning market. And this brings us to the next question.

What About the Power Industry?

The increase in demand for electricity is just one of the factors that will impact air pollution control expenditures. Utilities in the United States are receiving confusing signals on which direction to take in complying with both future and present regulations not only for air pollution control, but for transmission of power. The next two years could be a watershed period of overwhelming change for the North American electric power industry. The changes might include open access transmission and premature closure of nuclear power plants.

Competition through open access transmission could result early on in overcapacity. Areas that are short of electricity will receive power from those that have sufficient reserves. This could have a negative impact on air pollution control expenditures over the next year or two. The long-term effect, however, could be a huge increase in air pollution control expenditures as part of a massive construction program for mine mouth, coal-fired power plants. The CRSS/Philips mine mouth power plant in Mississippi represents an indicator of the future. The newly announced power plant will supply electricity to several states in the Southeast and can be expected to compete or cooperate with local utilities. Mine mouth power plants in Wyoming and Montana could supply the West Coast with coal-fired electricity. Nuclear power plants face a competitive disadvantage in this new environment. The California Public Utilities Commission expects full retail wheeling of electricity by 2002.

The March and August 1994 issues of the *Journal of the Air & Waste Management Association* provided extensive coverage of the controversy relative to particulate emissions from utility boilers. Recent studies strengthen the contention that utility emissions are both greater in quantity and more harmful than previously thought.

The emphasis on fine particulate control is opening the door for suppliers of wet precipitators. Research-Cottrell, Joy, and Lurgi have enjoyed success in the mining industry. Beltran, Sonic, and Belco have a number of units on industrial applications such as incinerators. One of the most significant advances is the application to coal-fired boilers. Southern Environmental is completing installation of a counterflow, two-field wet precipitator with four-point suspension on each field for one of the boilers at the Sherburne Station of Northern States Power. Presently, only the FGD scrubber on this plant exists. In the new arrangement, the FGD scrubber will be followed by the wet precipitator making it unique in the United States, where typically a dry precipitator is followed by a wet FGD system.

Southern Environmental, EPRI, Southern Services, and Southern Research have embarked on a project to analyze particulate removal at a 1 MW pilot combustor installed at Southern Research. In addition to determining particulate and air toxics removal performance, investigators will examine the use of additives for mercury control and the ability to operate in an unsaturated state.

Several new developments are directed at fine particulate. Wahlco Environmental Systems has a patented, new in-duct flue gas conditioning system.

The system simplifies the conditioning process by generating sulfur trioxide directly from the sulfur dioxide existing in the flue gas stream. Externally supplied feedstock (either sulfur or sulfur dioxide) is no longer needed, nor are auxiliary process heaters, which were formerly required. The new in-duct system precisely controls the generation of SO_3 by controlling the exposure of catalyst blocks positioned in the duct to the gas flow. Deflectors placed over the catalyst blocks adjust the contact between the gas flow and the catalyst surface, controlling the catalyzation of SO_2 in the gas flow to SO_3.

Other Markets Are Strong as Well

Delays in implementing Title I and Title III of the Clean Air Act have also resulted in increased potential for air pollution control equipment. McIlvaine is in the midst of an extensive analysis of the post-combustion nitrogen oxide (NO_x) market for power plants, municipal incinerators, and other applications. Close to $10 billion will be spent on NO_x control over the next decade. Nonattainment areas in the East will be particularly active.

The continuous emissions monitoring market has continued at a high level because of NO_x nonattainment continuous emissions monitors (CEMS). There will also be a positive impact on the market from the enhanced monitoring rule. The rule, slated for April 1995, will require enhanced monitoring at some 18,000 sources, where an individual emission point has the potential to emit 30 tons per year of a regulated pollutant. This drops to 15 tons per year in serious nonattainment areas and 7.5 tons per year in severe nonattainment areas, and only 3 tons per year in extreme areas. The operators must collect sufficient data to demonstrate continuous compliance. Continuous emissions monitors are the most accurate and also the most expensive. A plant faces fines and other sanctions if it cannot demonstrate compliance.

The mass particulate emissions monitoring offers the fastest potential growth. This is a market waiting for the technology. The beta gauge tape sampler available from F.A.G. and Environment is one option. The micro balance is another. The biggest hurdle is getting a representative sample out of the duct work and providing a transport system which allows consistent results. Graseby Andersen has developments in all three areas including a micro balance mass determination system, a sophisticated software program to calculate any losses in the transport system, and a new shrouded probe which is shown in the attached photograph. This unit operates at a fixed flow rate and collects a representative particulate sample even with changes in free-stream velocity. The shrouded probe was originally developed to provide continuous representative sampling of radionuclide particles at U. S. Department of Energy facilities. More recently, the shrouded probe has demonstrated the capability to provide continuous single-point aerosol samples from industrial sources.

Another particulate measurement technique has been overlooked in the United States, but has been used extensively in Europe. The typical transmissometer application in Germany, for example, is installed and individually calibrated as a particulate monitor. Operators collect stack measurement data simultaneously with a manual method for dust density and with a transmissometer setup to measure optical density. When a regression line is established to define this relationship between optical density and dust density, the instrument output can be linearly translated into dust density. In Germany most stacks are regulated in dust density established from such transmissometer measurements, or more recently from optical side scattering instruments. Monitor Labs LS541 transmissometer, which is normally used to monitor opacity in the United States, can be set up with the slope and offset of the above regression line entered into the control unit, in which case an output is available to provide dust density directly in mg/nM^3. As long as some degree of consistency exists in the particle size distribution and composition, which is typical after a baghouse or electrostatic precipitator, the optical measurement proves a reasonably accurate measure of dust density.

Expect substantial business for air pollution control companies in dealing with Department of Defense cleanup. A $450 million incinerator system is under construction at the Tooele Army Depot and is designed to destroy mustard and nerve gas. Incinerators are planned in seven other states, including

Alabama, Arkansas, Colorado, Indiana, Kentucky, Maryland, and Oregon.

Cleanable HEPA filters have been a boon for industries handling hazardous fine dusts. The first cleanable absolute filters introduced by several companies in Europe three years ago had similar designs comprising a box-shaped cartridge of glass media cleaned by an indexing tube delivering reverse pulse air. Since that time, MAC Environmental has redesigned the cleanable HEPA filter into the Miasmactic™ which incorporates a cylindrical cartridge made of reinforced HEPA media. Cleaning is accomplished off-line with medium pressure pulsing. This cleanable HEPA has found a waiting market in manufacturing processes for computer circuit boards, television tubes, inorganic paint pigments, and pharmaceuticals. In the making of television tubes, lead powder is sprayed on the inside of the screen to prevent radiation emissions. A closed Miasmactic filter system captures overspray powder and pulses it from the cartridges directly into a recycle system for return to the spraying process. Problems with hazardous dust and cartridge disposal are avoided.

Big Incinerator Retrofit Program

A massive investment in air pollution control equipment will be required at existing municipal waste combustor sites around the country. The U.S. Environmental Protection Agency (EPA) has proposed rules and must, under a statutory deadline, finalize standards by September 1995. The proposal would require these combustors to reduce acid gas emissions, including sulfur oxides and hydrogen chloride. It would also require substantial reductions in particulate matter and heavy metals including mercury. A very low limit would be set on dioxins and furans, and NO_x would also be reduced.

The proposal calls for most plants to add additional CEM equipment and increase intermittent stack testing, particularly for the various heavy metal emissions. Emission requirements are more stringent for one hundred nineteen large municipal waste combustor plants with two hundred thirty-five individual combustor units than they are for the sixty smaller plants with one hundred thirty-seven units.

$1.7 Billion Thermal Treatment Market

Volatile organic compounds (VOCs) and odors are effectively removed by either thermal oxidation or adsorption. Regulations to reduce ozone and air toxics are rapidly expanding the market for pollution control systems. In 1991, the annual purchases of thermal gas treatment and adsorption systems in the United States totaled just over $700 million. By the year 2000, the annual purchases are projected to increase $1.7 billion, resulting in an incremental increase of over $1 billion per year.

There are several types of thermal treatment equipment. One type is recuperative thermal treatment, which uses heat exchangers to capture heat generated by the incineration of the organic pollutants and the extra fuel necessary to ensure combustion. Another category, catalytic thermal treatment, involves the use of a catalyst to accomplish the oxidation at lower temperatures thereby using less fuel.

A third technology is called regenerative thermal oxidation (RTO). It treats the gas in multiple combustion chambers filled with ceramic packing. Because these units are capable of operating at high temperature and utilize low quantities of outside fuel, they have been the most popular segment of the market.

Whereas only a handful of companies offered RTOs in 1987, presently more than thirty companies in the United States manufacture these devices. Of the four largest companies, one was started as an environmental company offering conventional thermal oxidizers, and another was the original developer of the RTO. The other two largest suppliers are companies furnishing complete surface finishing systems. Large Fortune 500 companies are entering this field. One is primarily a catalyst supplier offering the RTO as an alternative technology. Another is a major chemical company with a pollution control division and a number of plants which will utilize the RTO technology. Oriented strandboard VOC reduction has the largest present application for these systems. Another large purchasing segment has been the automobile industry, which is using the RTO technology for automobile paint facilities.

Semiconductor plants have become major RTO purchasers. The chemical industry and wastewater treatment plants are major longer-term markets. Market growth will not be limited to industrial applications. Bakeries and other small emitters in nonattainment areas will be installing this equipment.

50 Percent Growth for NAFTA Air Pollution Control Market

Canada, Mexico, and the United States will experience air pollution control industry growth of 50% or more over the next five years. Orders for air pollution control systems in Canada are expected to rise from $285 million in 1994 to $460 million in 1998. Mexican air pollution control system sales will rise from $60 million in 1994 to $90 million in 1998.

The 1994 total North American Free Trade Agreement (NAFTA) air pollution control market including hardware, testing, and services was just under $4 billion. The United States accounts for over 91% of the present NAFTA market with Canada and Mexico accounting for 7% and 2%, respectively. In contrast, the Mexican market compares to that of an average U.S. state. The Canadian market is equivalent to larger U.S. states, such as Ohio and California.

The types of equipment being sold in the three markets differ. In the United States, measurement and reduction of nitrogen oxides represents a major potential. In Mexico, the largest potential can be found in dust control in basic industries. In Canada, pulp mills and mining are much more important than in the United States. In all three countries, enforcement responsibilities are shared between federal and state/provincial governments.

Conclusions

Future markets will be greatly affected by new plant expansions with opportunities in Mexico, Canada, and more importantly in the developing Asian nations. New advances in technology will both change and increase the market. ■

Questions

1. By what methods are volatile organic compounds removed?

2. What two changes could the North American electric power industry see in the next two years?

3. What is the expected growth in the pollution control industry in Canada and Mexico?

Answers are at the back of the book.

32

The Scripps Institution of Oceanography in California plans a new way to assess global warming. They will measure temperature below the surface of the Pacific Ocean by using low-frequency sound waves. Such data is crucial because the ocean stores most of the heat that powers the climate. However, marine mammal experts and environmentalists are concerned for the underwater inhabitants. Sound waves may harm whales and other creatures. A test is planned to determine if sound sources harm marine mammals.

The Sound of Global Warming

Stuart F. Brown

Cold War spy technology could become the world's largest ocean thermometer to measure global warming. But will the residents object?

A program that uses sound waves to measure ocean temperatures could improve global climate models but may be scrapped if whales and sea lions don't like it.

The Scripps Institution of Oceanography in La Jolla, Calif., hopes to inaugurate a $35 million program to measure temperatures beneath the surface of the Pacific Ocean. Knowing whether ocean temperatures are rising or falling over time would greatly increase the accuracy of computer models designed to predict climate change.

But the Scripps program—called the acoustic thermometry of ocean climate (ATOC)—uses low-frequency sound waves to determine ocean temperature. This has some marine mammal experts and environmentalists concerned because sea creatures such as blue whales, elephant seals, and sea lions either have hearing ranges that can detect ATOC sound waves or occasionally dive to depths where the sound-generating devices would operate. "This sound will travel through more than a quarter of the whole Pacific," says whale biologist Linda Weilgart of Dalhousie University in Nova Scotia, who triggered the controversy in a flurry of Internet exchanges last year. "You've got to be sure you're not affecting the long-term welfare of marine mammals, such as fertility rates, growth, and mortality" before going ahead with the program, she says.

The controversy over the Scripps ATOC program is especially ironic, since it pits animal-protection advocates such as the Sierra Club and the Natural Resources Defense Council against scientists who believe the program could help us learn more about one of the biggest mysteries of the environment—global warming. The ocean plays a bigger part in global climate than most people realize. "Most of the heat that powers the climate is stored in the seas," says oceanographer Walter Munk, the project's principle investigator. "You won't get the atmospheric climate prediction straight unless you get the ocean climate prediction straight."

Measurements of ocean-surface temperatures can be made by satellites, but their radar or infrared instruments are only capable of sensing the top few millimeters of water. Deep waters, which contain most of the ocean's heat, can also be sampled by oceanographic research ships dispensing sensing instruments. This process is expensive, however, for the amount of data gathered. ATOC's proponents argue that acoustic thermography's ability to sense the average temperature of an immense volume of water makes it the most practical method.

Using sound waves to measure water temperature has its origins in a 1944 experiment by Maurice

Ewing of Columbia University, who discovered a natural "channel" in the sea that could carry sound waves across vast distances. This sound channel, at a depth of about 2,800 feet, is a seam of medium-temperature water that's isolated by a layer of warmer surface water above, and a layer of cooler, deep water below. The thermal and pressure barriers on either side of the middle-depth layer act as a wave guide, keeping certain sound frequencies bouncing between its two "walls." Ewing's original experiment proved this when an underwater explosion was detected 900 miles away.

The U.S. Navy was quick to realize the implications of Ewing's work. It developed sound fixing and ranging (SOFAR), a method that uses hydrophones, or submerged microphones, to pick up noises underwater. For example, the noise of vibrating machinery aboard Soviet submarines was detected from one thousand miles away by SOFAR monitoring of the oceanic sound channel. "SOFAR was a mainstay of U.S. security during the Cold War," says Munk. "This method was our main source of information about one of Russia's most threatening activities."

Curiously, a Soviet scientist had independently discovered the sound channel in a 1946 experiment, but for some reason the Soviet navy didn't capitalize on the findings. Soviet strategy changed abruptly in the 1970s, when spies revealed the success of SOFAR. Much quieter Soviet submarines were soon in service.

By the late 1970s, Munk was pioneering the techniques that would lead to acoustic thermography. He performed acoustic tomography experiments, using multiple sound emitters and receivers to produce three-dimensional temperature maps of areas of the sea. The method exploited a natural phenomenon: Sound travels faster in warm water than it does in cold water. Therefore, the average temperature of the water can be determined by clocking the travel time of low-frequency sound from its source to a receiver. For example, water warmer by a mere five one-thousandths of a degree decreases sound travel time across a 6,000-mile path by 150 milliseconds.

In 1991, Munk led a test program which showed that an underwater sound source exploiting the sound channel could be broadcast over great distances. A signal transmitted underwater from Heard Island in the southern Indian Ocean near Antarctica was detected by sensors 11,000 miles away. This experiment opened the door for ATOC, which uses a pair of sound emitters and an array of sensors to measure temperatures across vast areas of the Pacific.

ATOC sound emitters produce low grumbles of 75 hertz that are transmitted in coded 27-second sequences and are repeated 43 times for a total of about 20 minutes. The coding identifies the signal, which weakens and becomes buried in the ocean's random noise before it arrives at a receiver. Repetition provides replacements for individual signals that may be canceled out by background noise en route. Each chunk is tagged with its exact departure time. Computers then correlate clearly received chunks of different sequences into a precise measurement of sound travel time.

"Actually detecting climate change in the oceans will take at least a decade, and 20 years would be even better," says ATOC program manager Andrew Forbes. "If you only measure for five years, you may find yourself tracking a trend that is just riding on the back of an El Niño," Forbes says. "So you have to get beyond those known periodicities in the ocean and atmosphere and measure at least a couple of normal cycles."

The mission plan calls for one sound source to be anchored in the waters at Pioneer Seamount off central California, and another to be placed eight miles off the north shore of Kauai, Hawaii. An assortment of acoustic receiving devices located 3,000 to 6,000 miles distant will collect sound waves and convert them into average temperature measurements along 18 paths through the Pacific. The Navy has agreed to allow researchers access to part of its once-secret network of submarine-detecting hydrophones.

Once the ATOC network is established in the Pacific, the equipment will gather temperature measurements for 24 months. Forbes hopes the early insights gained into the thermal characteristics of the Pacific will lead to the development of a long-term acoustic thermometry network monitoring large ocean areas, particularly in the Atlantic, where major flows of cold, polar bottom water enter the global ocean.

But the entire program hinges on the outcome of a six-month test aimed at determining if ATOC sound sources harm marine mammals. If federal and state permits are issues—perhaps as early as this summer—researchers will gradually bring the omnidirectional sound emitters up to full power for 20-minute transmissions for four-day periods, alternating with week-long periods of silence. Time-depth recorders will be attached to marine mammals in the neighborhood, and their swimming speeds and patterns, along with heart and respiratory rates, will be monitored.

Sound intensity from the ATOC microphones, which at 195 decibels roughly equals the noise of a large container ship at close range, diminishes rapidly as it radiates. At 2,700 feet from the source, the sound level decreases to 136 decibels, the equivalent of breaking waves, according to program officials. "We don't expect to see distress in these animals," says Daniel Costa, a marine biologist at University of California, Santa Cruz, who is heading the $2.9 million study on the effects of ATOC sound waves on sea mammals. "What we expect to see, if anything, is that the animals would express annoyance and avoid the site."

Program managers at Scripps have given the marine mammal group control of the sound emitters and the authority to modify or halt the entire program if they detect harm to sea life. Ocean temperature measurements will begin only if the system is found to be safe. Still, marine biologist Christopher Clark, head of the ATOC marine mammals study, is embittered by the controversy. He feels that opposition to the program is grounded in emotion rather than scientific fact. "This is environmental activism gone completely astray," Clark says. "They should be focusing on the real acoustic pollution in the sea, which is the barrage of noise from supertankers and other shipping traffic."

If ATOC clears regulatory hurdles, Walter Munk and his acoustic oceanographers will transform a Cold War submarine snooping technique into the biggest thermometer in the sea. Scripps scientists plan to share data with researchers at NASA's Jet Propulsion Laboratory, who are now receiving ocean surface-height measurements from the orbiting *Topex-Poseidon* spacecraft. "The satellite altimetry gives you the temperature of the upper oceans, because when they are warmer the water level is higher," Walter Munk explains. "With ATOC, we get the temperature structure of the interior ocean. We will use these two views to help produce a good climate-prediction model, which we think does not now exist." ■

Questions

1. Why is there a controversy between animal-protection advocates and scientists?

2. How can the average temperature of water be determined?

3. Where is most of the heat that powers the climate stored?

Answers are at the back of the book.

33

Debate over global warming continues. Climate scientists are searching for a greenhouse fingerprint in order to determine whether natural forces or human actions are responsible for global warming. Since the late 19th century, the annual average global temperature has risen about 0.5 degrees Celsius. Recent research finds only 1 chance in 20 that natural forces were responsible for the temperature rise in the last century. However, uncertainties remain. But scientists warn that delaying action until we have a conclusive greenhouse fingerprint will increase potential dangers of future warming caused by greenhouse gas pollution released into the atmosphere now.

Dusting the Climate for Fingerprints

Richard Monastersy

Has greenhouse warming arrived? Will we ever know?

After lurking in the back pages amid the ads for the past few years, the topic of global warming has once again clambered onto the front pages of newspapers around the world.

In recent months, an iceberg nearly as large as Rhode Island broke off an Antarctic ice shelf, apparently because of rising temperatures there. A statistician declared that the seasons have slipped out of sync with the calendar, perhaps because of greenhouse gas pollution. And just in time for a climate summit in Berlin 2 months ago, a German research team reported finding an abnormal pattern of change in climate records that does not correspond to any known natural causes.

Although the annual average global temperature has risen by about 1.5ºC since the late 19th century, investigators have had difficulty determining whether natural forces or human actions deserve the blame. But in late February, Klaus Hasselmann, director of the prestigious Max-Planck Institute for Meteorology in Hamburg, Germany, stepped forward to point a finger.

The Max-Planck researchers find it highly improbable—only 1 chance in 20—that natural forces caused the temperature to rise during the last century.

Environmental groups attending the Berlin climate summit rallied around the recent findings. Proof of greenhouse warming has arrived, many trumpeted.

Not quite, say Hasselmann and other researchers intimately involved in the hunt for human influences on climate. Recent developments do indeed bolster the theory that greenhouse gases have reset Earth's climate dial. But scientists acknowledge that uncertainties continue to plague studies aimed at detecting the human fingerprint in climate change—a distinctive change attributable only to human activities. Proof remains elusive.

What's more, researchers warn that the public should not hold its breath waiting for the unambiguous detection of human-caused greenhouse warming. Studies aimed at such a detection, while illuminating, will never give a definitive answer. "From the scientific point of view, I think it's just a sport, frankly. I really think it's wrong to pin all one's hopes on this proof," says Hasselmann.

Sport or not, the search for a greenhouse fingerprint has become the rage among climate scientists. As the name suggests, such efforts resemble the methods employed by police trying to crack a crime—especially the new technique of DNA fingerprinting, which has figured prominently in the current trial of O.J. Simpson.

In the forensic process of DNA fingerprinting, investigators compare a suspect's complex DNA pattern with samples found at the crime scene. In the case of global warming, the purported crime scene is the environment, and green house gases are the prime suspects. Clues such as out-of-sync seasons (*Science News*: 4/8/95, p. 214) or giant icebergs (*Science News*: 4/29/95, p. 271) do not, by themselves, implicate any one culprit. Researchers must sift through climate records for specific patterns that could only arise from the effects of greenhouse pollution.

What does the fingerprint of greenhouse warming look like? Hasselmann's group and others use computer climate models to calculate the kind of abnormal signal that human tinkering with the climate might create. The German team generated an anthropogenic, or human-caused, climate change by slowly boosting the amount of greenhouse gases in the model's atmosphere, starting from 1935 values. As the simulation progressed through the decades, a greenhouse fingerprint emerged: a specific pattern of global changes with temperatures increasing most dramatically in the interior of continents.

The team then compared this pattern with available temperature records for the period 1854 through 1993. Using statistical analyses, they tested how closely their computer-generated fingerprints matched the observations.

To solidify their case, Hasselmann's team had to address the same concerns that confront experts who use DNA fingerprinting in a criminal trial. Just as prosecutors must convince juries that the forensic technique will not implicate the wrong person, climate researchers must argue that their fingerprinting methods do not yield false matches.

Because natural factors can alter climate, Hasselmann and his colleagues needed to estimate the extent to which conditions can vary on their own. Ideally, this information would come out of the available climate data. But useful regional records of surface temperatures reach back only 140 years, not nearly long enough to give a full picture of the kind of changes that nature alone can produce. Moreover, the records are probably compromised because they contain changes wrought by greenhouse gases and other pollutants.

So the researchers again turned to computer climate models. To estimate Earth's own variability, they let the climate model run for 1,000 years, with greenhouse gas concentrations locked in at modern values. They also removed the estimated contribution of greenhouse gases from the climate observations to obtain another gauge of natural variability.

Pulling the entire case together, Hasselmann and his co-workers estimate a 95 percent chance that the observed climate changes exceed the range of natural variability. In a paper to appear in the *Journal of Climate*, they declare the recent warming abnormal. But they do not take the next step, which would be to attribute the warming to greenhouse gases.

"We simply think the probability is very high that we can see a human impact. But we're not saying that we have detected it. That's a too deterministic statement," Hasselmann says.

Several hurdles stand in the way of making clear-cut declarations about climate change. Hasselmann notes that the fingerprinting study relies on estimates of natural variability taken from a computer model. If these calculations are incorrect, then the German group may have underestimated the chance that Earth itself caused the observed pattern. Moreover, the model estimates of natural variability don't include volcanic eruptions and solar fluctuations, two features that can alter climate and skew the detection tests.

The study currently in press also suffers because it left out the influence of sulfur pollution, which exerts a cooling effect on industrial regions. The German researchers have included this factor in more recent computer runs. That addition tempers the simulated warming over the United States and other regions, bringing the model fingerprint more in line with actual observations and increasing the statistical significance of the detection work, Hasselmann says.

As the techniques used in these detection efforts improve, researchers have edged closer toward identifying a signal of human-induced warming in the climate record. Aside form the German study, teams at the Lawrence Livermore (Calif.) National Laboratory and at the Hadley Center for Climate Prediction and Research in Bracknell, England, are also searching for a human signal.

"It would be premature to claim unambiguous detection and attribution of some human influence on climate. But preliminary evidence suggests that if you believe the model signal and the model noise, something unusual is going on," says Benjamin D. Santer of Lawrence Livermore, who participated in the German study. Santer also served as lead author for a chapter on detection that will appear in an upcoming report of the Intergovernmental Panel on Climate Change (IPCC).

Thomas R. Karl of the National Climatic Data Center in Asheville, N.C., also sees current experiments pointing toward successful detection.

"The bottom line on a number of these studies is that if one considers both the greenhouse [factor] and the sulfur aerosol [factor], then one comes much closer to seeing an anthropogenic effect in the climate record," he says. "And a number of these studies are very close to the 95 percent significance level. But if you really wanted to be unequivocal, you'd probably like to see that there is less than 1 chance in 100 that you could be misled."

That level of certainty—99 percent statistically significant—might take a while, however. In an article published in the Jan. 21, 1994 *Science*, Stephen H. Schneider of Stanford University wrote: "For the detection of anthropogenic climate signals, we must recognize that a goal of 99 percent statistically significant signal detection over the next decade or two is unrealistic."

For scientists, the recognition of greenhouse warming will never come from a specific detection study or any single observation. Rather, it grows from a steady accumulation of evidence. "Detection and attribution will be an evolutionary, not a revolutionary process," says Livermore's Santer.

What's more, some scientists now voice the conclusion that they misled the public by giving false hope for unequivocal detection of changes caused by greenhouse gases. Researchers always strive for more certainty by ruling out alternative explanations. But they can never reach the 100 percent level, even in establishing a link between smoking and cancer.

"I am beating very hard on the community to drop forever from our vocabulary that pernicious term 'unambiguous detection,' " Schneider told *Science News*.

In the case of greenhouse warming, scientists warn that built-in delays in the climate system compound the potential dangers of waiting for certainty. Greenhouse gas pollution emitted into the atmosphere this year will take decades to warm the atmosphere and oceans. So society has already committed to a certain amount of warming.

If government leaders want to head off the potential for greater changes, they will have to act before ever seeing the major effects of greenhouse gas pollution. At the climate summit in April, 115 countries agreed to negotiate a treaty on emissions reductions by 1997, but they did not specify the extent of such cuts (*Science News*: 4/29/95, p. 271).

Santer, Schneider, and others draw a distinction between the needs of scientists and policy makers when it comes to seeking answers. "We're talking about being very certain about things when we talk about 99 percent or 95 percent certainty," says Santer.

But politicians often take actions in the face of uncertainty, balancing risks on one side versus risks on another. "A politician usually takes a decision at a far lower level of confidence about things," Santer says. "They don't wait until they are 99 percent confident. Decisions are made against a background of substantially greater risk." ■

Questions

1. Describe a greenhouse fingerprint.

2. What percent does observed climate change exceed the range of natural variability?

3. How would scientists recognize greenhouse warming?

Answers are at the back of the book.

34

Since the discovery of the Antarctic ozone hole, the atmospheric science community has increased its understanding of stratospheric ozone. This has been done by a series of field observations, laboratory experiments, and computer modeling. During this time, countries from around the world have joined the Montreal Protocol on Substances That Deplete the Ozone Layer to phase out chlorofluorocarbons and halons. Over time, evidence has accumulated that the Arctic winter stratosphere has the same chlorine species as the Antarctic ozone hole. Scientists believe that an unusually long Arctic winter could initiate severe ozone destruction. When Arctic stratospheric temperatures hit new lows in 1995, reports indicated that the ozone had diminished dramatically. For the ozone layer to recover, nations must abide by the Montreal protocol.

Complexities of Ozone Loss Continue to Challenge Scientists

Pamela S. Zurer

Severe Arctic depletion verified, but intricacies of polar stratospheric clouds, midlatitude loss still puzzle researchers.

This past winter brought record ozone loss to the Arctic polar regions, scientists at an international conference confirmed last month. While reluctant to call the severe ozone destruction a "hole" like the one that develops each year over Antarctica, European researchers presented a convincing case that the depletion in the far north resulted from the same halogen-catalyzed chemistry that triggers the Antarctic phenomenon.

But few other issues addressed at the weeklong International Conference on Ozone in the Lower Stratosphere, held in Halkidiki, Greece, could be resolved with as much certainty. Atmospheric scientists are still struggling to grasp the exact nature of the polar stratospheric clouds that are so crucial to the fate of ozone in the polar regions. The dynamics of the stratosphere remain imperfectly understood. And questions persist about the mechanism of the gradual ozone-thinning trend over the midlatitude regions of North America, Asia, and Europe, where most of the world's people live.

Without a doubt, a broad outline of the complex behavior of stratospheric ozone is in focus. And—as the research presented at the conference attests—scientists are working diligently to fill in the details. Until the remaining uncertainties are clarified, however, the ability to quantitatively predict the condition of the ozone layer will remain elusive, especially as chlorine levels are expected to peak over the next few years and then slowly decline over several decades.

The goal of the recent meeting was to bring together U.S. and European scientists to share their results. "It is vital [that] information is exchanged as widely as possible within the world scientific community," said University of Cambridge chemist John L. Pyle, one of the organizers. Designed to provoke discussion, the conference featured a handful of invited talks and more than 200 poster presentations. Its sponsors included the European Union (EU), the World Meteorological Organization (WMO), the National Aeronautics & Space Administration (NASA), and the National Oceanic & Atmospheric Administration (NOAA).

For some of the 300 scientists from 40 nations who attended, the meeting marked a return trip to the

Reprinted with permission from *Chemical & Engineering News*, June 12, 1995, *73*(24), pp.20–23.

lush Halkidiki Peninsula in northern Greece. Eleven years ago they had gathered to discuss ozone depletion at the very same beach resort.

Back then, depletion of stratospheric ozone by chlorine and bromine from man-made chemicals was only a hypothesis. But the discovery of the Antarctic ozone hole the following year brought theory to life, kindling a surge of research and political activity. In the decade since, the atmospheric science community has greatly improved its understanding of stratospheric ozone through a combination of field observations, laboratory experiments, and computer modeling. Meanwhile, the nations of the world have joined in the Montreal Protocol on Substances That Deplete the Ozone Layer to phase out production of chlorofluorocarbons (CFCs) and halons, the major sources of ozone-depleting halogen compounds in the stratosphere.

On the minds of many researchers as they arrived at the conference was the question of Arctic ozone depletion during the winter just past. For some years, evidence has accumulated that the Arctic winter stratosphere is loaded with the same destructive chlorine species believed to cause the Antarctic ozone hole. Only the normally milder northern winters have prevented massive ozone loss.

Atmospheric scientists have been predicting that a prolonged cold Arctic winter—more like those usually experienced in Antarctica—could unleash severe ozone destruction. Arctic stratospheric temperatures hit new lows in the early months of 1995, and preliminary reports indicated that ozone had decreased dramatically (C&EN, April 10, page 8).

Indeed, WMO's network of ground-based ozone-monitoring instruments observed record low ozone over a huge part of the Northern Hemisphere, averaging 20% less than the long-term mean, Rumen D. Bojkov told the conference. Bojkov, special adviser to the WMO secretary general, said the period of extreme low ozone began in January and extended through March.

"In some areas over Siberia, the deficiency was as much as 40%," he said. "It's only because of normally high ozone in that region that we do not have what you would call an ozone hole."

Measurements from ground-based instruments alone are not enough to prove that Arctic ozone has been destroyed chemically, however. Changes in ozone could also result from the constant motion of the atmosphere transporting air with different ozone concentrations from one place to another.

"It's difficult to tell if the variations are chemical or dynamical," mused NASA atmospheric scientist James F. Gleason. "Could an ozone high over Alaska have been balancing the low over Siberia?" Unfortunately, there have been no daily high-resolution satellite maps of global ozone to help answer that question since NASA's Total Ozone Mapping Spectrometer (TOMS) aboard Russia's Meteor-3 satellite failed in December.

Even if TOMS data were available, it would be extraordinarily difficult to untangle chemical destruction of ozone from dynamical fluctuations. The exception is the evolution of the Antarctic ozone hole each September. The stratosphere over Antarctica in winter is isolated by a circle of strong winds called the polar vortex, so ozone concentrations there are normally at a stable minimum before the ozone hole begins to form. The dramatic ozone depletion that occurs as the sun rises in early spring is unmistakable.

The Arctic polar vortex, in contrast, is usually much weaker, and ozone concentrations in the north polar regions are constantly changing. They normally increase in late winter and early spring, bolstered by waves of ozone-rich air from the tropics. So even if ozone amounts hold steady or increase somewhat in the Arctic, ozone still may have been destroyed: The concentrations may be significantly less than they would have been had there been no chemical depletion. Researchers have resorted to ingenious methods of calculating Arctic ozone loss indirectly—for example, by studying the changing ratio of the amount of ozone to the amount of the relatively inert "tracer" gas nitrous oxide in a given parcel of air.

This past winter, however, the Second European Stratospheric Arctic and Midlatitude Experiment (SESAME) generated a wealth of data that is helping to overcome the complications posed by ozone's tremendous natural variability. The 1994–95 EU-sponsored research campaign employed aircraft, balloons, and ground-based instruments to study the stratosphere. European scientists eagerly presented their latest findings at the Halkidiki conference—all of which point to widespread chemical destruction of Arctic ozone.

One elegant experiment used coordinated balloon launches to measure ozone loss directly during the Arctic winter. The trick is to measure the amount of ozone in a particular parcel of the air, track the air mass as it travels around the Arctic polar vortex, and then measure its ozone content again some days later. The approach was described by physicist Markus Rex, a doctoral student at Alfred Wegener Institute for Polar & Marine Research (AWI) in Potsdam, Germany.

In a previous experiment, Rex and his coworkers used wind and temperature data from the European Center for Medium-Range Weather Forecasts to identify which of some 1,200 ozonesondes—small balloons carrying electrochemical ozone sensors—launched during the winter of 1991--92 intercepted the same air mass at two different times. From the matches they identified, they estimate over 30% of the ozone at 20 km within the Arctic polar vortex was destroyed during January and February of 1992 [*Nature*, **375**, 131 (1995); C&EN, May 15, page 28].

Rather than again rely on chance matches as they had in 1991–92, the researchers coordinated releases of more than 1,000 ozonesondes from 35 stations during this past winter's SESAME campaign. After launching a balloon, the scientists used meteorological data to forecast its path. "As the air parcel approached another station, we asked that station to launch a second sonde," Rex said.

The scientists observed that ozone was decreasing throughout January, February, and March 1995. And they found the decline in a given air parcel was proportional to the time it had spent in sunlight, consistent with photochemical ozone depletion catalyzed by halogens.

Rex and his coworkers calculate that ozone was being lost at a rate of about 2% per day at the end of January and even faster by mid-March, when the sun was flooding a wider area. Those depletion rates are as fast as those within the Antarctic ozone hole.

"The chemical ozone loss coincides with, and slightly lags, the occurrence of temperatures low enough for polar stratospheric clouds," Rex said. Such clouds provide surfaces for reactions that convert chlorine compounds from relatively inert forms to reactive species that can chew up ozone in sunlight.

The results of other SESAME experiments add to the conclusion that the Arctic suffered severe ozone loss last winter. For example:

- Temperatures in the Arctic stratosphere in winter 1994–95 reached the lowest observed during the past 30 years, reported Barbara Naujokat of the Free University of Berlin's Meteorological Institute. The north polar vortex was unusually strong and stable, breaking up only at the end of April.
- Profiles of the vertical distribution of ozone within the polar vortex revealed as much as half of the ozone missing at certain altitudes compared with earlier years, reported AWI's Peter von der Gathen.
- An instrument carried by balloon into the stratosphere above Kiruna, Sweden, recorded high concentrations of chlorine monoxide (ClO) in February, said Darin W. Toohey, assistant professor of earth systems science at the University of California, Irvine. Chlorine monoxide, the "smoking gun" of ozone depletion, forms when chlorine atoms attack ozone. Simultaneous ozone measurements showed substantial amounts missing.

These and many other findings coalesce into an "incredibly consistent" picture of Arctic ozone depletion, said Cambridge chemist Neil Harris. But was there actually an Arctic ozone hole this year?

"I wouldn't call it a hole," said Lucien Froidevaux of California Institute of Technology's Jet Propulsion Laboratory (JPL).

"At most we've got half a hole," said Cambridge's Pyle.

"We don't have to be so hesitant," said NOAA research chemist Susan Solomon. "The data show ozone was not simply not delivered but actually removed. This is exciting confirmation of substantial Arctic ozone depletion. It's never going to look exactly like an Antarctic ozone hole, but so what?"

Whatever one chooses to call what happened in the Arctic earlier this year—one wag suggested "Arctic ozone dent"—key questions remain unanswered.

Will the severe depletion return in subsequent winters, intensify, or increase in area? How are the dramatic ozone losses in the polar regions affecting stratospheric ozone over the rest of the globe?

"Yes, I believe there have been statistically significant changes in Arctic winter ozone that we can observe," said NOAA chemist David Fahey. "But where do we go from here? Can we predict future changes?"

One issue hampering atmospheric scientists' ability to make qualitative predictions of future ozone changes is the difficulty in understanding the exact nature of polar stratospheric clouds (PSCs). For several years after their critical importance to the Antarctic ozone hole was discovered, researchers thought they had a good grasp of the situation. But reality has turned out not to be so neat, said Thomas Peter of Max Planck Institute for Chemistry, Mainz, Germany.

It is the presence of PSCs that makes ozone in the polar regions so much more vulnerable than it is in more temperate regions. The total amount of chlorine and bromine compounds is roughly uniform throughout the stratosphere. The halogens are carried there by CFCs and haloes, which break down when exposed to intense ultraviolet light in the upper stratosphere.

In most seasons and regions, the halogen atoms are tied up in so-called reservoir molecules that do not react with ozone—hydrogen chloride and chlorine nitrate ($ClONO_2$), for example. However, PSCs—which condense in the frigid cold of the stratospheric polar vortices—provide heterogeneous surfaces for reactions that convert the reservoir species to more reactive ones.

The most important reaction is between the two chlorine reservoirs:

$$ClONO_2 + HCl \rightarrow Cl_2 + HNO_3$$

The molecular chlorine produced flies off into the gas phase, where it is photolyzed easily by even weak sunlight to give chlorine radicals—active chlorine—that can catalyze ozone destruction.

Equally important is the fate of the nitric acid (HNO_3) produced by the heterogeneous chemistry. It remains within the PSCs, effectively sequestering the nitrogen family of compounds that would otherwise react with active chlorine to reform chlorine nitrate. That process, called denitrification, allows the photochemical chain reactions that destroy ozone to run efficiently for a long time without termination.

Laboratory and field experiments have confirmed the importance of heterogeneous reactions in the winter polar stratosphere. What is at question now is the actual composition of the PSCs, how they nucleate and grow, their surface area, and their chemical reactivity. All those factors affect the interconversion of halogen compounds between their active and inactive forms, and thus the rate and amount of ozone depletion.

Peter described the "happy period" in the late 1980s when scientists were confident they understood just what PSCs are. Type I PSCs were thought to be crystals of nitric acid trihydrate (NAT) that condensed on small sulfate aerosol particles once temperatures cooled below about 195 K. Frozen water ice (type II) appears once stratospheric temperatures plunge lower than about 187 K, which generally happens only in Antarctica.

Now it appears type I PSCs are not so simple. "Say 'bye bye' to the notion PSCs must be solid," Peter said. "They can be liquid. Both types of particles are up there."

Margaret A. Tolbert, associate professor of chemistry and biochemistry at the University of Colorado, Boulder, explained that "Everybody assumed type I PSCs were NAT, which condenses about 195 K. But observations show nothing actually condenses until about 193 K. That doesn't prove the particles aren't NAT. But they may instead be supercooled ternary solutions" of water, sulfuric acid, and nitric acid.

Such supercooled solutions could develop from small sulfate aerosol particles. (The stratosphere contains a permanent veil of sulfate aerosol droplets, which form when sulfur dioxide from volcanic eruptions and carbonyl sulfide emitted by living creatures are oxidized to sulfuric acid.) As the stratosphere cools in winter, the sulfate aerosols could take up water and nitric acid, growing larger but remaining liquid.

Whether type I PSCs are solid or liquid can make a significant difference, said NOAA chemist

A. J. Ravishankara. He noted that chemistry could occur not just on the surface of supercooled solutions, but also in the interior of the droplet, which he likened to a little beaker. That implies scientists would have to consider not just the surface area of the particles but their volume in calculating the rates of chemical processes involving PSCs.

Furthermore, he said, the supercooled solutions may persist over a larger temperature range than solid NAT particles. That would allow transformation of halogens to their active forms to take place over a wider temperature range than previously recognized.

The implications of chemical processing taking place on and in supercooled ternary solutions extend beyond the polar regions, where PSCs appear, to more temperate regions of the globe. Although changes in ozone in the midlatitude stratosphere are nowhere near as dramatic as in the polar regions, they are real and substantial. The latest United Nations Environment Program study "Scientific Assessment of Ozone Depletion," concludes that ozone in the midnorthern regions, for example, has been decreasing at a rate of about 4% per decade since 1979.

Atmospheric scientists have been struggling to quantitatively explain that decrease, which is predominantly in the lower stratosphere. Chlorine and bromine radicals are clearly implicated, but in the absence of PSCs they destroy ozone most voraciously at much higher altitudes where there simply isn't all that much ozone to begin with. Modelers have been plugging every known ozone destruction cycle into their calculations but still have not been able to account for all of the ozone loss observed below about 20 km.

Roderic L. Jones, of Cambridge's chemistry department, noted that chemistry involving supercooled ternary solutions could have significant effects on ozone trends at midlatitudes. "Look at areas in the Northern Hemisphere that are exposed to temperatures just above the NAT [condensation] point, about 197 or 199 K," he said. "It's a big area, extending as far south as 50° N."

Supercooled solutions may turn out to play an important role in explaining what's going on at midlatitudes. But JPL modeler Ross J. Salawitch said he thinks the problem may arise from the way scientists approximate the dynamics of the stratosphere in their models.

Two-dimensional models treat ozone as if it diffused out uniformly from the tropical stratosphere where it is produced. "The real world is more like the 'tropical pipe'" paradigm, Salawitch said, in which air rises high into the stratosphere in the tropics and moves downward again at higher latitudes, like a sort of fountain.

"The issue of whether the global diffusion or the tropical pipe model better represents reality is not just an academic discussion," Salawitch said. "The question of a midlatitude deficit may disappear when the models handle dynamics better."

Answers to some of the issues that continue to bedevil atmospheric researchers may become clear as more data accumulate. Participants in the SESAME campaign have barely had time to think about what they observed. And NASA has just begun its new three-year Stratospheric Transport of Atmospheric Tracers mission.

But the stratospheric ozone layer may reveal even further complexities in the coming years. As Christos S. Zerefos, local organizer of the conference from the laboratory of atmospheric physics at Aristotle University of Thessaloníki, Greece, pointed out in his closing remarks, the ozone layer will only begin to recover if the nations of the world continue to comply with the Montreal protocol. If not, atmospheric scientists may have even greater puzzles to contend with. ■

Questions

1. What are the major sources of ozone-depleting halogen compounds in the stratosphere?

2. What has prevented massive ozone loss?

3. What is used to study the stratosphere and what were the results?

Answers are at the back of the book.

35

Depletion of the atmospheric ozone layer is generally considered as one of the major environmental threats facing humanity and indeed all life on earth. The ozone layer shields the earth from high levels of ultraviolet and other dangerous solar radiation so that the loss of even a few percent of the ozone layer would lead to increased mutation rates, skin cancer, blindness, and many other health hazards. Harmful effects on plant and animal life, including plankton in the ocean would also occur, with potentially widespread impacts on many ecosystems. Because of these threats, the Montreal Protocol of 1987 was signed by many nations to phase out production and use of chlorine-containing chemicals such as CFCs which deplete the ozone layer. Recent evidence suggests that the Protocol has been successful in slowing the rate of CFC in production. However, developing nations have been less effective in reducing production, for economic reasons.

Ozone-Destroying Chlorine Tops Out

Richard A. Kerr

The title of the session at the American Geophysical Union's fall meeting last month was "The Montreal Protocol to Protect the Ozone Layer: Has It Worked?" The answer was "yes, so far at least." The prime evidence for the success of the 1987 international agreement to restrict the use of ozone-destroying chemicals—most of which contain chlorine—is that atmospheric chlorine has peaked and is on the way down.

The chlorine decline, reported by atmospheric chemist Stephen Montzka of the National Oceanic and Atmospheric Administration in Boulder, Colorado, bodes well for the stratospheric ozone layer. One computer model of ozone destruction presented at the meeting predicts that the chlorine controls imposed by the recently strengthened protocol should allow the ozone layer to begin recovering before the end of this decade. But some new observations suggest that the recovery may be delayed. Losses to another ozone-killer, bromine, may be temporarily offsetting the gains from the chlorine reduction.

The good news on chlorine comes after decades when the annual release of one million tons or more of chlorofluorocarbons (CFCs), used for refrigeration, blowing foams, and cleaning electronics, drove chlorine concentrations up several percent each year. If no controls had been adopted until 2010, calculates Michael Prather of the University of California, Irvine, stratospheric chlorine would have topped out at levels more than three times higher than today's, the Antarctic ozone hole would appear well into the 22nd century, and springtime ozone losses over middle northern latitudes would soar above today's 8%, possibly to well above 30%.

Instead, by 1990 the Montreal Protocol restrictions began to slow the rise in chlorine noticeably (*Science*, 1 January 1993, p.28). By the end of 1994, as production of CFCs plummeted, chlorine peaked, reported Montzka, at least in the lower atmosphere where it can be readily monitored. And now the amount of chlorine in CFCs and other halocarbons seems to be headed down, Montzka said. Using a computer model, Charles in Greenbelt, Maryland, found that if the chlorine controls hold, ozone should begin recovering by 2000 and return to 1979 levels—the year the Antarctic ozone hole became obvious—by around 2050.

Recent atmospheric data reported by Ronald Prinn of the Massachusetts Institute of Technology suggested, however, that potential gaps in the existing protocol could slow the recovery. Although chlorine—the major ozone destroyer—seems to have peaked, says Prinn, the total ozone-destroying power of humanmade chemicals in the atmosphere has "perhaps reached a plateau, but certainly is not decreasing." This measure also includes bromine, an ozone destroyer that is scarcer than chlorine but at least 40 times more potent on an atom-for-atom basis. Prinn finds that the increasing concentrations of halons—bromine-containing compounds released from fire-protection systems—have for the moment sustained the ozone-destroying power of the atmosphere.

Production of halons was halted in 1994 under the protocol, but halons in existing firefighting systems are a continuing source of emissions. And because replacements for some halons have not yet been identified, users may be tempted to go on consuming existing industrial stocks. Methyl bromide, a soil and stored-grain fumigant, could also contribute to a future bromine increase. Meeting in Vienna last month, the parties to the protocol decided to phases out use of methyl bromide by 2010, but as yet unspecified exemptions may be granted.

Even chlorine, although under control, is far from eliminated as a threat. Rich, technologically sophisticated countries have found it relatively easy to move away from CFC use, notes Prather, but developing countries, which have laxer schedules to meet under the protocol, have yet to follow suit. Even in wealthy countries, the hydrogenated CFCs used in place of traditional CFCs are still weak ozone destroyers. Clearly, the Montreal Protocol has some more milestones to reach. ■

Questions

1. According to the model by Charles Jackman, when should the ozone layer return to its 1979 levels, if current controls hold?

2. Is bromine more destructive than chlorine? How much? Where does bromine come from?

3. What are halogenated CFCs? Do they destroy ozone?

Answers are at the back of the book.

36

Congressional debates on Superfund usually focus on liability and cleanup standards. Recently, the debates are concerned with whether the excess money in the Superfund trust fund should be used to clean up hazardous waste sites or used to lower the federal deficit. Three proposals have been discussed to reform Superfund. They include remedy selection, liability for cleanups, and funding. More flexibility in the choice of cleanup remedies could save 35% per cleanup. One of the industry groups, Chemical Manufacturing Association (CMA) estimates that at the current pace of cleanups it will cost $2.4 billion a year. There is approximately $3 billion in the Superfund trust now, with an expected increase to $4 billion by fiscal 1996. Many members in Congress would like to see those funds used for deficit reduction. CMA contends taxes collected in support of Superfund should be used for cleanups only.

Superfund Renewal Expected to Be Rocky

Lois Ember

The tortuous path to Superfund renewal has shifted direction yet again. Last year's congressional debates focused on cleanup and liability standards. This year, the spotlight is on money—whether an excess in the Superfund trust fund should be used to clean up hazardous waste sites or to lower the federal deficit.

This switch in focal point has pulled in a more diverse array of congressional players—many more interested in spending cuts, tax relief, and deficit reduction than in hazardous waste cleanup. This diversity will make the programs reauthorization more problematic this year. The program's taxing authority expires on Dec. 31.

Also making renewal more difficult is the absence of broad-based support for specific reform measures. Last year's environmental-industry coalition "has not been reconstituted, and I don't see a coalition in any formal structure in the near term," says Dell Perelman, senior assistant general counsel for the Chemical Manufacturers Association (CMA).

Further, no Superfund bills have yet been introduced, so there is nothing that stakeholders can rally around. But in an effort to spark and shape debate, CMA and other industry groups have come out with proposals to reform Superfund comprehensively.

For these industry groups, the three principal issues are remedy (cleanup) selection, liability for cleanups, and funding. But for CMA, the pivotal issue is remedy selection. Says CMA's vice president for regulatory affairs, Morton L. Mullins, "Remedy selection is where it's at, but we always get distracted by liability."

More flexibility in the selection of cleanup remedies—and a reduction in the program's administrative overhead—could save an average of at least 35% per cleanup. Savings can be achieved if the law's preference for permanent cleanup solutions and treatment over containment and its requirement that cleanups comply with other environmental laws are eliminated, CMA contends.

The trade association additionally calls for site-specific, risk-based remedies that take into account current and realistic future land use. Risk-based solutions are also supported by the Business Roundtable, a New York City based lobbying group made up of chief executive officers from major U.S. corporations, in its position papers on Superfund reform.

Mullins says CMA advocates "a more robust role for the public," including "early and often public participation . . . so we can avoid getting way down the road with too much invested in an outcome before the public is included." The Business

Roundtable also calls for increased community involvement in Superfund cleanups.

One of the most contentious provisions of the original law is its strict joint and several liability scheme. This system forces all potentially responsible parties to pay cleanup costs, regardless of fault. Critics argue that the scheme is inherently unfair, and there is a major push in Congress this year to repeal it.

The Business Roundtable recommends repeal of retroactive liability prior to Dec. 11, 1980, when Superfund was first enacted. But that industry group places several caveats on this course of action. Remedy selection should be substantially reformed, the existing tax burden on the economy should not be increased, and responsible parties that have paid for cleanups in the past should be reimbursed for their payouts.

CMA essentially agrees with the roundtable's position, but says it could also support a fair-share allocation liability scheme or a hybrid system, given the right conditions. However, CMA will "not support any proposal that slows the pace of cleanups," Perelman insists.

If retroactive liability is repealed, CMA estimates that it will cost about $2.4 billion a year to maintain the current pace of cleanups. The Business Roundtable pegs a maintenance level at about $2.2 billion. Both groups are calling for no increase in taxes. So the difference between the amount Congress appropriates for the program—now about $1.4 billion and what will be needed if retroactive liability is eliminated should be made up from funds salted awayin the Superfund trust fund, they argue.

There's roughly $3 billion in the trust fund now, a figure expected to reach $4 billion by fiscal 1996. Many members of Congress eye these funds as a source for deficit reduction. CMA argues that such use of trust fund monies is inappropriate. Taxes collected from industry to support cleanups should only be used for cleanups, so the trade association suggests that Congress place the trust fund off budget.

Superfund Reform '95, a group that includes mainly insurance companies, state and local governments, and community groups, says retroactive liability before 1987 can be repealed without slowing the pace of cleanups or increasing Superfund taxes. This can be achieved through appropriation of all funds in the trust fund.

CMA argues that trust fund revenues should come from a mix of taxes that includes insurance companies. Insurance companies do not now contribute to the trust fund.

The House has been holding a series of hearings on Superfund issues, but the timing of the law's reauthorization remains unclear. The House Commerce Committee has primary jurisdiction over the law and is expected to take the lead in introducing a bill. The Transportation Committee has sequential referral, but can introduce its own bill. At press time, Sen. Robert C. Smith (R-N.H.), chairman of the Senate Environment & Public Works Superfund Subcommittee, was slated to release his Superfund reform principles.

Smith's concept paper could form the basis for a Senate bill. The senator would like to see markup of a bill in July, but September appears more realistic, his staff says.

At a recent briefing CMA stressed the urgency of Superfund reform, stating it would not support simple reauthorization of the program. The apparent congressional disarray on Superfund renewal may not resolve itself before the program expires. CMA's Jon C. Holtzman, vice president for communications, says Congress' "knee-jerk fallback position may be to let the program die for a while." ■

Questions

1. What is the pivotal issue for the Chemical Manufacturing Association?

2. How could clean-up remedies save an average of 35 percent per clean up?

3. What is the controversy surrounding the Superfund renewal?

Answers are at the back of the book.

37

According to a 1994 report by an Environmental Protection Agency workgroup, animal factories are the largest contributors to polluted rivers and streams. This includes storm sewers and all industrial sources combined. Most of the two billion tons of manure that are produced by farm animals per year comes from factory farms. Super-wastes that come from these factories have devastating impacts on the local environment. The enormous waste problems that these factory farms have created have not been solved, but some states are campaigning to keep animal farms.

Fowling the Waters

Jim Mason

Animal factories are the biggest contributors to polluted rivers and streams—bigger than storm sewers, bigger than all industrial sources combined.

Roger Brown joined Citizens Against Corporate Hog Factories (CACHF) because he believes his rural Missouri community is turning—literally—to manure.

Brown ought to know. The former manager of a poultry complex for Arkansas-based Hudson Foods parted ways with the company in a dispute over how to safely get rid of the manure-soaked litter from the company's turkey "growout" buildings, or "factories," as he and the locals call them. Brown and other Missourians are fighting a half-dozen agribusiness corporations that want to bring in hog factories. "If they come in, we'll have one big stink around here," he says.

"Factory" farming is a billion-dollar industry in America. Over 90 percent of the chicken and eggs we eat get produced in crowded, windowless buildings, each holding as many as 100,000 animals. Beef cattle are raised not in bucolic fields, but in huge, dusty feedlots. Farmers slaughter more than one million veal calves—raised in tightly confining crates—every year. Seven-hundred-pound hogs stand crowded together in spaces barely larger than their own bodies.

People in North Carolina, Iowa, northern Missouri, Illinois, Tennessee, and almost a dozen other states want to keep factory farms out. These states are becoming battlegrounds as the corporate takeover of poultry and hog farming is rapidly changing the face of animal agriculture. Where once farm animals were dispersed among the nation's grainfields and pasturelands, allowing their wastes to go back to the soil slowly and evenly, today the mobility of corporate money and technology allows Tyson, Hudson, Conagra, Cargill, Continental Grain, Premium Standard Farms, Murphy Farms, and other agribusiness firms to out-compete traditional farmers.

Brown and other critics feel that big agribusiness wants to cash in on the public's seemingly insatiable demand for cheap meat. The corporations go to regions that have poor environmental laws, an anti-regulatory culture, low wages and a docile, anti-union labor pool. There they set up huge factories in which most caretaking chores are mechanized, and any manual labor is so routinized that workers are easily trained and replaced.

The trouble is, such schemes have created huge waste problems, which corporations have been neither quick nor eager to solve. A 60,000-bird egg factory full of "layer" hens jammed into rows of cages stacked four-high, for example, produces 82 tons of manure every week. In the same week, 50,000 hogs in a megafactory can produce over 3,000 tons of manure. In all, America's farm animals produce about two billion tons of manure each year—about 10 times that of the human population—and most of it now comes from factory farms. The environmental hazards of animal manure were brought dramatically home last June 24 when 25 million gallons of hog waste spilled into the New River in Haw Branch, North Carolina, causing serious pollution and killing countless fish.

What's worse, factory manure is more polluting than the old-fashioned kind. When on pasture, animals get a high-fiber, low-cal diet of grasses and roughage, plenty of vitamins and minerals, and traces of micronutrients like copper, zinc and selenium. When animals are confined, these elements have to be added to the feed—often in excessive amounts. And, to speed growth, factory animals get a steady diet of ground grain and other rich feedstuffs loaded with protein and energy—so loaded, in fact, that the animals can absorb only about 20 percent of all the nutrients. The rest become a very rich manure—rich not only with protein and the usual organic matter, but also with nitrogen, phosphorous, heavy metals, feed additives and other potential pollutants.

Factory super-wastes, if not carefully handled, can create havoc in the local environment. Full of organic matter, decaying animal waste uses up oxygen in fresh water. Rich in nitrates, phosphorous, and other minerals, it encourages algae to grow, which takes up even more oxygen. This can kill streams and the fish and other life in them.

The 550 million chickens in factories surrounding the Chesapeake Bay contribute heavily to nitrate pollution of the area's soils and waters. Manure from New York state dairies pollutes the watershed for cities' drinking water. And in Wisconsin, "the dairy state," cow manure carried the microorganism cryptosporidium that wound up in Milwaukee's water supply, causing death and widespread dysentery in 1993. According to a 1994 report by an Environmental Protection Agency (EPA) workgroup, animal factories are the biggest contributors to polluted rivers and streams—bigger than storm sewers, bigger than all industrial sources combined.

There's a final cost behind the much-touted "efficiency" of the factory farm: Even the best factories convert no more than 23 percent of the energy and protein fed to animals into energy and protein in the form of meat, milk, or eggs.

Contact: United Poultry Concerns, P.O. Box 59367, Potomac, MD 20859/(301)948-2406. ■

Questions

1. What percent of chicken and eggs that we consume comes from factory farming?

2. What types of areas do corporations choose in order to set up animal factories?

3. How can factory super-wastes destroy streams and the life contained in them?

Answers are at the back of the book.

38 *Germany perceives that eliminating pollution can strengthen their economy. The Gerermans believe that efforts to decrease pollution by increasing efficiency will reduce operating expenses. In turn, this will raise employment opportunities and boost income. Having the world's most rigid environmental regulations, encourages the development of "green" technologies that they can market globally. Recycling in Germany has numeric requirements. For example, 90 percent of discarded glass and metals and 80 percent of paper, board, plastics, and laminates must be recycled. Recycled plastic went from 41,000 tons in 1992 to ten times that in 1993. Germany also has an environmental-labeling program in which the government gives a brief description of the product qualities, such as "low-noise," or "100-percent recycled." This label has a figure in blue with outstretched arms encircled by the laurel wreath of the United Nations. It is aptly named the Blue Angel environmental labeling program. Germany is looking to the future through innovation. With long-term stringent environmental laws benefitting the economy and employment, Germany has become a model for the world.*

Green Revolution in the Making

Curtis Moore

Nations like Germany that mandate clean technologies are poised to profit. It's an economic lesson the U.S. has yet to grasp.

All in all, it's been a pretty tough 14 years for Americans seeking to make a buck by protecting the environment, and this factory in the heart of Germany says it all. As wide ribbons of heavy, brown kraft paper unspool from massive rolls onto a production line, a coverall-clad worker stands with his beefy left hand gripping a switch. At his elbow, a constant stream of cream-colored gypsum paste squirts between the kraft sheets, forming an endless plaster sandwich that disappears into a flat, shimmering oven where it is baked to rock hardness. With his back arched and his head cocked, the worker peers down the production line as mile after mile of what Americana call "Sheetrock" or "wallboard" thunders through the factory, bound for building sites in Germany and across Europe to form the walls and ceilings of offices and bedrooms, closets and boardrooms.

Just another factory making one of the thousands of products so common that they are scarcely worth noting, a casual observer might say—but not so. This factory represents the leading edge of a new technological revolution, one that could transform the industrial world from a cauldron of pollution to a relatively safe haven. For this wallboard is made from—are you ready?—air pollution.

This process of making wallboard, mortar, and other construction materials—and from them, homes and offices—out of the residues of air pollution is emblematic of the innovations that have sprung up as Germany, propelled by a fierce environmental ethic, has leapt to the forefront of the global environmental movement.

The German passion for environmental protection was fueled initially in the late 1970s and early '80s by reports of *Waldsterben*, or "forest death"—the widespread damage to the country's forests caused by air pollution. After that, the meltdown at Chernobyl and mounting fears of stratospheric-ozone depletion and global warming established a firm ecological consciousness, leading the *Los Angeles Times to* comment that in Germany, "environmental correctness has come to rival tidiness and punctuality as a national obsession." As ethically committed as Germany's citizens and government are to protecting the earth, they also perceive the process of eliminating pollution as an opportunity to further strengthen their nation's economy.

Already running a close race with the United States as the world's leading exporter of merchandise, Germany is convinced that its environmental regulations, easily the world's most stringent, will stimulate the development of a wide range of new "green" technologies that can be marketed globally just as demand for them is beginning to increase sharply. The Germans also believe that new efforts to curb pollution by boosting efficiency will further reduce operating expenses in their already efficient economy, providing them with a competitive edge over Japan and (especially) the United States.

The homes-from-pollution process illustrates how environmental concerns have stimulated German innovation, causing many of the country's firms not only to launch their own research programs, but to raid the workshops of less-alert competitors—including the United States, where many of these new technologies were developed. Like any number of other emerging technologies, ranging from super-efficient electrical generators to add-on pollution-control systems, the homes-from-pollution process is a product of Yankee ingenuity. It was originally installed in 1973 at the Cholla I power plant in Arizona during the first wave of air-pollution regulation in the United States, but the process was exported to Germany in 1980, where it has thrived and been perfected. This is how it works:

When coal is burned to generate electricity, prodigious amounts of pollutants pour into the air, including sulfur dioxide, which causes acid rain. Some nations, though not many, require modest controls over these emissions. If the regulations are stringent, scrubbers are usually installed to remove the sulfur dioxide by spraying the exhaust with a watery mist containing limestone. The pollution/limestone reaction produces a sludge that is usually dumped on the ground or into pits or waterways.

But in Germany, where all power plants are equipped with pollution controls, the sludge can't be dumped because the law prohibits it. Such waste must be put to some use, leaving German power plants with two options: develop a means other than scrubbers to eliminate the air pollution, or find a way to use the scrubber sludge. German industry has done both, yielding two simultaneous streams of innovation, one aimed at developing pollution-control systems superior to scrubbers, the other at devising better ways to use scrubber waste. Both streams not only help make the German economy itself more efficient, but create products that can be sold on the world market, boosting employment and income at home.

When the homes-from-pollution system was exported to Germany in 1980, it was initially marketed by Knauf-Research Cotrell (KRC), a subsidiary of its U.S. developer and Knauf Gypsum. Rapidly improved there in response to the German air-pollution and waste requirements, the technology was acquired in October 1986 by the Salzgitter Group, which now sells the system globally.

One place where the technology has been installed is New Brunswick, Canada, where the 450-megawatt, coal-fired Belledune power station went into operation in 1993. The production of market-grade gypsum was "a fundamental requirement" contained in the specifications for the Belledune plant, in the words of an executive of New Brunswick Power, because it not only solved waste-disposal problems, but was less expensive than competing systems. Thus, a North American innovation traveled to Europe and back again in the space of 20 years (though the profits are being made by Germans) and is selling globally because it is, quite simply, better than the alternatives.

Sadly, the U.S. market for the homes-from-pollution process was destroyed in the 1970s when the federal government allowed utilities to build "tall stacks" for dispersing sulfur dioxide over wide areas—thus creating a new acid-rain problem—rather than requiring them to eliminate it. Even if strict controls on power plants had remained in effect, however, lax waste-disposal regulations might have had the same ultimate impact. In Germany, though, the process has proved so effective and profitable that in 1990 Knauf Gypsum opened a British plant at Sittingborne-on-

Thames, where German air-pollution residues are made into building materials for homes and factories. Flowing in the opposite direction, of course, is profit that can be plowed back into the German economy—perhaps to acquire still more products of U.S. origin.

There are other examples of remarkable innovation stimulated by Germany's tough attitudes toward pollution:

• The Ford auto plant in Cologne complied with new requirements by modernizing its paint-spray line, cutting pollution by 70 percent and the cost of painting a car by about $60—a savings that makes German-built cars marginally more salable.

• The "4P" plastic-film manufacturing and printing plant in Forchheim, where plastic bags for frozen french fries and other foods are printed and stamped by the millions, was forced to cut pollution by 70 percent. The company installed a recycling system that reclaims up to 90 percent of the plant's solvents, saving so much money that the 4P pollution controls will not only pay for themselves, but actually start saving the company money by reducing the cost of solvents. A sister plant with a similar system already recaptures solvents—once again lowering its overhead while increasing profit.

Little wonder that Edda Müller, former chief aide to Germany's minister for the environment, declares emphatically that "what we are doing here is economic policy, not environmental policy."

She is not alone in this view of the future, nor is Germany. For example, Takefumi Fukumizu, U.S. representative of Japan's powerful Ministry of International Trade and Industry, says that industrialists in his country see "an inescapable economic necessity to improve energy efficiency and environmental technologies, which they believe would reduce costs and create a profitable world market."

With virtually no coal, oil, or natural gas, and limited mineral resources, Japan has historically been forced to do more with less than its principal industrial competitors, the United States and Germany. As a result, it makes steel, automobiles, and a wide range of other goods with greater efficiency and less pollution than any other nation. That national thrift and the technologies it has spawned are now global commodities as other nations increasingly search for cleaner, more efficient manufacturing methods and energy use. "The potential profit in such a market," explains Fukumizu, "is limitless."

In the United States, however, governments and businesses alike remain so focused on short-term profits and quarterly earnings that they overlook the true source of long-term wealth: innovation. Necessity breeds invention, and during the 1970s, when protecting the environment and saving energy were seen as essential elements of national policy, the United States brought hundreds of new products and processes to the verge of commercial reality.

These ranged from systems to generate electricity from wind and sunlight with zero pollution to little-known devices such as fuel cells that can power everything from homes to locomotives with zero or near-zero pollution and noise, while requiring minimal space. Yet these and thousands of other born-in-the-U.S.A. environmental products were abandoned during the 1980s as the Reagan and Bush administrations, the Congress, and many state officials turned their backs on environmental protection, orphaning technologies that now stand to generate billions, perhaps trillions, of dollars for their new proponents.

Solar photovoltaics, for example, were originally developed to generate electricity for space satellites, then modified for ground-based uses, making the United States the world's leading producer. But when Ronald Reagan took office, he slashed federal funding for the program from more than $150 million to zero. Then he rejected the "energy independence" policies of presidents Nixon, Ford, and Carter, substituting a "cheap oil" strategy expressly designed to increase U.S. reliance on Persian Gulf oil by driving prices down through secret negotiations with Saudi Arabia. As oil prices plummeted, they destroyed the U.S. market for solar and other forms of renewable energy, allowing the Japanese, Germans, and others to buy companies, patents, and production licenses for pennies on the dollar. Now Japan is the world's leading producer of solar cells. The United States is second, but the nation's largest factory is owned by the German conglomerate Siemens. If its production were assigned to Germany instead of the United States, America's photovoltaic sector would drop to a level on par with those of developing nations like Brazil.

A similar fate befell fuel cells, compact and virtually silent devices that chemically convert fuel to electricity. When run on hydrogen, fuel cells produce zero pollution or, if a hydrogen "carrier" such as natural gas is used, almost zero. First developed for the space program, they still meet all of the electrical needs of NASA's space shuttles. But in the 1980s, U.S. companies such as General Electric and Englehard turned their backs on fuel-cell technology. As a result, the world's first fuel-cell assembly line was Japanese, and the first zero-polluting, fuel-cell-powered bus is Canadian. Both employ technologies that were developed with hundreds of millions of U.S. tax dollars. The governments of both Canada and Japan helped their nation's companies acquire and develop the fuel-cell technology.

The list goes on and on, and includes technologies ranging from high-efficiency light bulbs to new ways of burning coal, all developed in large part, with U.S. capital, but now wholly or partially in the hands of others. In the United States, the cheap-oil strategy remains in place, energy taxes have been rejected by Congress, and environmental laws continue to fall further and further behind those of Germany, Japan, Sweden, the Netherlands, and other industrialized nations. Once the world's environmental leader, the United States is now a laggard, its political landscape hostile to those seeking to pioneer in what many regard as a new industrial revolution greening the global economy.

Germany, meanwhile, has been restructuring the technological basis of its economy to make it sustainable over the long run, leading to a profusion of new environmental products and processes spurred by the world's most aggressive protection programs. Consider, for example, the following:

• It is retrofitting all power plants. While politicians in North America were arguing about whether acid rain was real, Germany listened to its scientists and adopted rules requiring every power plant within its borders to slash the air pollutants that cause acid rain by 90 percent. By 1989 the German retrofit was complete. Today, seven years before the U.S. control program will take full effect in 2002, Germans are selling Americans and the rest of the world anti-pollution technology and know-how.

• It is aggressively phasing out chemicals that destroy the ozone layer and cause global warming. In 1989 Germany mandated a ban by 1995—five years before the rest of the world—on chlorofluorocarbon (CFC) gases, the primary culprits in the destruction of the ozone layer that protects Earth from solar radiation. It had also committed to reducing emissions of carbon dioxide, the principal cause of global warming, by 25 percent by the year 2010. These are the swiftest and toughest phase-downs in the world, and they required German industries to respond quickly, creating new products and processes that can be marketed globally as other nations begin to follow suit.

• It is revolutionizing the trash business. Aiming not only to reduce the volume of trash swelling landfills and clogging incinerators, Germany has also fostered a new industry by adopting a "take back" program that requires everything from cameras to yogurt cartons—and the scrubber sludge from which Knauf Gypsum makes wallboard—to be collected by manufacturers and recycled. The program was so fabulously successful that the volume of trash quickly outstripped the nation's recycling capacity, thus creating even further pressure for industry to minimize packaging and other waste. Although the program was originally meant to include cars as well, German manufacturers staved off formal government action by agreeing to mount voluntary take-back programs, thus starting a global movement among car makers to develop vehicles that can be recycled. Some are already rolling off the assembly line with bar-coded parts and instructions for dismantling an auto in 20 minutes.

Like most of Germany's environmental laws, the take-back rule imposes explicit, numeric requirements: as of this year, 90 percent of all discarded glass and metals must be recycled, as well as 80 percent of all paper, board, plastics, and laminates. Incineration, even if used to generate power, has been ruled out as a solid-waste-disposal method because the burning of materials pollutes the air, especially with highly toxic dioxins and furans.

Because the take-back law sweeps virtually every form of waste into its ambit, its results were almost immediate: 400 German companies randomly surveyed less than 18 months after the law took effect on December 1, 1991, said they had completely abandoned use of polyvinyl packaging, plastic foams, and 117 other types of packaging. All but one of 146 companies had stopped using "blister" packs, which are both tough to recycle and yield dioxins when burned. One of every four companies was using packaging made at least in part from recycled materials. Companies were running full-page newspaper advertisements touting the recyclability of their products. And with good reason, as almost two-thirds of Germany's consumers shop for environmentally friendly products. Indeed, public insistence on recycling has become so widespread in Germany that the amount of recycled plastic rocketed from 41,000 tons in 1992 to ten times that amount a year later. Now, the nation has become a favorite testing ground for products to be marketed as environmentally safe. When in 1991, for example, Procter & Gamble introduced Vidal Sassoon "Airspray" hair spray in the United States, it did so only after testing in Germany. Similarly, two years earlier, P&G launched its "Enviro-pak" containers for laundry detergents in the United States only after testing in Germany.

Companies such as P&G have no choice but to develop such products if they wish to do business in Germany. One reason for this is the government's Blue Angel environmental labeling program. Introduced in 1977, Blue Angel is a symbol owned by Germany's environment ministry, which describes it as "a market-oriented instrument of government" that informs and motivates environmentally conscious thinking and acting among manufacturers and consumers. The ministry licenses the label's use for about 3,500 products selected on a case-by-case basis by the independent, nine-member Environmental Label Jury. The label depicts a blue figure with outstretched arms encircled by the laurel wreath of the United Nations. Inscribed in the border for each product is a brief explanation of the product's qualities, such as "low-polluting," "low-noise," or "100-percent recycled." Although there are imitators in other countries—the Green Cross of Canada, for example—Germany's Blue Angel remains far and away the most famous and successful environmental-labeling program.

In the United States, efforts to establish such a government-sponsored labeling system have foundered on industry opposition. A private effort, Green Seal, is struggling to establish itself, but is hampered by high testing costs and a lack of publicity. In Germany, however, the Blue Angel offers its bearers the prospect of winning an edge over competing brands and products in the environmentally conscious German marketplace. The prospect of its award—and attendant profits—has made it possible for a wide variety of environmentally friendly products ranging from low-pollution paints to mercury-free batteries to establish themselves. Public recognition of and enthusiasm for the Blue Angel program has boosted the market share of many products. For instance, before water-soluble lacquers were awarded a Blue Angel in 1981, these products commanded a meager one-percent market share. Today, 40 percent of Germany's do-it-yourself-wood finishers and 20 percent of its professionals buy the less-toxic coatings. Similarly, biodegradable chainsaw lubricants are in high demand by the foresters who manage virtually every acre of Germany's woodlands. First introduced in 1987, the formula eliminates up to 7,000 tons a year of highly toxic oil otherwise absorbed by forest floors and nearby streams. After receiving a Blue Angel, these oils achieved a dominating position in the market.

Having deployed these weapons in their battle against pollution of air, water, and soil, some German officials believe they have all but exhausted the reductions that can be achieved through conventional cleanup means such as wastewater-treatment plants and scrubbers. Nevertheless, pressured by voters to cut pollution further, the government is imposing a wide range of increasingly tough requirements designed to reduce pollution by further increasing efficiency both in factories and on the highways.

Regulations now being implemented, for example, will force drivers out of gas-guzzling cars and onto energy-efficient public transit. Inner cities are being systematically closed to auto traffic, while highway, bridge, and other tolls are being raised, and long-term "green passes" for public transportation are sold in all of Germany's major cities. A Berlin green pass costs about $40 a month, and is valid for an unlimited number of rides. A comparable pass in Washington, D.C., costs roughly twice as much. Germany also intends to increase bicycle ridership by providing specially marked lanes on sidewalks and at intersections.

Systematically shifting people and goods in this fashion not only reduces pollution, but boosts the overall efficiency of Germany's economy by cutting transportation costs. Commuting by train, for example, slashes both fuel consumption and air pollution by up to 75 percent compared with cars, and 90 percent compared with planes. Traffic congestion and the pollution it generates, as cars creep through crowded roadways or idle at stoplights, is also cut, because trains occupy only a quarter of the road space required by buses and 1/13th that needed by cars. Because trains run on electricity generated by Germany's domestic coal, oil imports required to fuel diesel buses or gasoline cars are likewise reduced.

The nation's self-imposed target of reducing carbon dioxide emissions from the former West Germany by 25 percent and from the former East Germany by 30 percent—both by the year 2010—requires the economy to become even more efficient. One way the country intends to achieve this is to put energy that is now being wasted to some useful purpose.

In most power plants and factories, only about one-third of the energy in coal, oil, or gas is actually used. The rest escapes as waste heat. The German government has prepared regulations that will require large- and medium-size industries and utilities to market this waste energy. It can be used to heat homes and factories (or, by running CFC-free "absorption chillers," to cool them), operate paper mills and chemical plants, and even generate a few more kilowatts with super-efficient technology. Officials estimate that use of this waste heat will boost efficiency to roughly 90 percent and that air pollution—already at the world's lowest levels—will be chopped by at least half.

Because of the immense cost of bringing the former East Germany into compliance with the environmental requirements of the former West Germany, the waste-heat law has been delayed while officials turn to more pressing needs. Work is already under way, for example, on shutting down 80 percent of the former East Germany's power plants while retrofitting the remaining facilities that generate 10,000 megawatts—that's slightly more than Thailand's entire electricity consumption—with state-of-the-art pollution controls.

The cumulative effect of all these programs is to place Germany in a commanding position as nations beset with environmental problems search for ways to reduce pollution quickly and inexpensively. Thailand, for example, decided to install scrubbers on its coal-fired power plants after a single episode of air pollution in Mae Mo District sent more than 4,000 of its citizens to doctors and hospitals. Smog-bound Mexico City has been forced to implement emissions controls on cars and factories. Taiwan is even going so far as to require catalytic converters for motorcycles. Such mandates will almost inevitably benefit Germany because, as Harvard Business School economist Michael Porter explains, "Germany has had perhaps the world's tightest regulations in stationary air-pollution control, and German companies appear to hold a wide lead in patenting—and exporting—air pollution and other environmental technologies."

In the United States, however, where environmental standards were relaxed by a succession of Reagan/Bush appointees, often in the name of competitiveness, "as much as 70 percent of the air pollution control equipment sold . . . is produced by foreign companies," according to Porter, whose 855-page study of industrial economies, *The Competitive Advantage of Nations,* examines the impact of environmental regulations on competitiveness.

Germany's actions continue to contrast sharply with those of the United States, even under President Clinton, whom most environmentalists supported as the green answer to George Bush. Germany's emissions limits on power plants and incinerators are 4 to 300 times more stringent than those of the United States. German companies that generate electricity from wind, solar, or other renewable forms of power are reimbursed at twice to three times U.S. levels. German recycling is mandatory, while American programs are usually voluntary where they exist at all.

Still, support for Germany's environmental initiatives is by no means unanimous. Wolfgang Hilger, for example, the chairman of Hoechst, Germany's largest chemical company, complained bitterly in 1991 that the government had lost all sense of proportion. He claimed that regulations had jeopardized 250 jobs at his company, and threatened it with a $100-million loss. But Hilger represents a minority view. Most German citizens and businesses remain convinced both that environmental protection is essential and that the technological innovation stimulated by stringent environmental requirements will, over the long term, strengthen their national productivity and competitiveness.

Tragically, U.S. political leaders continue to embrace the outmoded and false view that the environment can be protected only at the expense of the economy, when the truth is precisely the opposite. Meanwhile, products of American genius continue to depart for Japan, Germany, and other nations, only to be sold back to U.S. industry sometime in the future. So far, the homes-from-pollution process hasn't traveled full circle back to its place of invention in the United States. But don't be surprised if sometime soon you see a piece of wallboard being nailed into a new of office or a remodeled home only to find it boldly emblazoned: "Made in Germany." ■

Questions

1. What is the Blue Angel environmental labeling program?

2. How will waste heat be marketed and what will be its effect on efficiency and air pollution?

3. How much more stringent is Germany's emissions limits on power plants and incinerators than the Unites States?

Answers are at the back of the book.

Section Four

Social Solutions

39

China is employing economic incentives to improve environmental compliance. Some incentives include: pollution levies, discharge permits, environmental achievement rewards for officials and managers, price reform, security deposits, environmental taxes and tariffs, environmental labels, and fines. But China's pollution problems remain significant. Approximately 75 percent of China's energy needs depend upon coal consumption. This contributes enormously to air pollution. Pollutant levels in most of the large cities exceed World Health Organization standards. The 1979 Environmental Protection Law (EPL) set guidelines to emphasize pollution prevention. Also included is a "polluter pays" principle that forces an enterprise to pay a fee if they discharge pollutants exceeding state standards. In 1989 a new EPL law was introduced. It addressed weaknesses in the 1979 laws, and laid the groundwork for new pollution control policies. The development of environmental laws, regulations, and infrastructure has greatly slowed environmental degradation in China. With economic reforms decreasing the central government's control, the promotion of economic development over environmental protection is becoming evident.

China Strives to Make the Polluter Pay

H. Keith Florig, Walter O. Spofford, Jr., Xiaoying Ma, Zhong Ma

Are China's market-based incentives for improved environmental compliance working?

Following the birth of the People's Republic of China in 1949, an emphasis on the development of state-owned heavy industry in urban areas left a legacy of large point sources of air and water pollution. Economic reforms introduced in 1978 have sparked rapid growth in industrial output accompanied by increasing pollution discharges into urban airsheds and water courses. Residential waste has intensified the industrial pollution problem. Consumption increases caused by growth in household incomes are propelling concomitant growth in per capita solid waste generation. As millions of unemployed farmers migrate to urban areas, pollution problems are exacerbated by untreated sewage and the burning of coal for domestic heating and cooking. A rapidly expanding transportation sector threatens to become a significant source of air pollution in China's larger cities.

Since the enactment of its first trial environmental legislation in 1979, China has moved aggressively to develop ambient and emission-effluent standards, establish an infrastructure for ambient environmental monitoring, and experiment with new programs for pollution reduction. Although these programs depend in part on central planning and moral suasion, the Chinese are harnessing economic incentives as environmental protection instruments.[1] As China moves toward a full market economy, these incentives become more attractive. Economic instruments that are in use or being tested include pollution levies, discharge permits, environmental achievement rewards for officials and managers, price reform, security deposits, environmental taxes and tariffs, environmental labels, and fines.[2]

Although China has made notable progress in environmental protection, her pollution problems remain serious. China depends on coal for about 75% of primary energy needs, including residential heating and cooking and industrial process heating. This use contributes heavily to air pollution in urban areas, particularly in winter. Air pollutant levels in most large cities exceed World Health Organization guidelines. High levels of airborne particulates are thought to be largely responsible for a death rate from chronic obstructive pulmonary disease five times that of Western industrialized countries. In many urban areas, surface waters are polluted by sewage and industrial wastewater. In 1993 only 55% of all industrial wastewater discharges met discharge standards.[3]

Evolving Environmental Regulations

After three decades of central planning and isolation from the Western world, China in the late 1970s moved quickly to address the environmental legacy of the past and stave off the environmental effects of rapid economic growth sparked by economic reforms. China's first modern environmental legislation was passed for trial implementation in 1979. The 1979 Environmental Protection Law (EPL) described goals and principles of environmental protection and authorized a system of environmental regulation, monitoring, and enforcement. The law emphasized pollution prevention by prohibiting the siting of facilities with noxious emissions in residential areas; exhorting enterprises to control pollution or face closure, relocation, merger, or mandatory process or product changes; requiring environmental impact assessments for new or upgraded industrial facilities; requiring new or expanded facilities to include pollution prevention and control in design, construction, and operation (dubbed the "Three Simultaneous Steps" policy); requiring new facilities to meet discharge standards; and promoting development of an indigenous environmental technology industry. The law also embraced the "polluter pays" principle by collecting fees from enterprises that discharge pollutants in amounts exceeding state standards. China's State Council issued rules clarifying the discharge fee system in 1982.

In 1989 a new EPL was passed that superseded the 1979 trial law. It reaffirmed existing policies on siting restrictions, environmental impact assessments, the Three Simultaneous Steps, and discharge fees. The 1989 law also addressed some shortcomings of the 1979 law and paved the way for new policies on pollution control. A system was established to reward factory managers and government officials for meeting environmental goals and to punish those failing to meet the goals within a specified time. Greater attention was devoted to gaining economies of scale by the centralized treatment of industrial wastes. Incentives to recycle wastes were instituted, and a system was developed to quantitatively rank urban environmental quality and services, thus harnessing the power of publicity to encourage environmental improvements.

To implement the 1979 EPL, environmental protection units at the state, provincial, city, and county levels were formed. The 1989 EPL required individual enterprises to set up environmental units to collect and report effluent and emissions data. China's environmental bureaucracy grew rapidly through the 1980s and currently employs more than 200,000 people nationwide, including about 60,000 in industrial environmental protection offices and 80,000 in provincial, city, and county environmental protection bureaus. In 1984 China's State Council formed the Environmental Protection Commission to oversee China's institutional environmental protection structure. Composed of all relevant ministry and agency heads, it sets broad policy directives and resolves interagency disputes. In 1988 the Environmental Protection Commission elevated the Environment Office in the Ministry of Urban and Rural Construction and Environmental Protection to full agency status, forming the National Environmental Protection Agency (NEPA) to take responsibility for environmental policy.[4]

China's first environmental quality standards were developed in the early 1970s. During the 1980s more comprehensive standards were developed for ambient air and water quality, and new standards were introduced for industrial pollutant emissions, vehicle emissions, and environmental monitoring.[5] The standards are stringent enough to require significant pollution reductions without being unrealistic given China's economic base.[6] In general, Chinese discharge standards are somewhat weaker than those of Western industrialized countries.[7]

China's environmental policies are being developed and tested in a rapidly changing economic system. Since the beginning of economic reform in 1978, China has liberalized or freed prices on many commodities. Many state-owned enterprises may sell a portion of their products in markets outside the standard quota system and exercise autonomy in choosing suppliers and customers. Taxes and accounting rules for state-owned enterprises now encourage more cost reduction. Collective and private enterprises, operating in largely free markets, are growing in number and now comprise more than half the economy. Despite these changes, substantial remnants of the planned economy remain, particularly in heavy industry, energy and transportation.

Pollution Levy System

Article 18 of China's 1979 EPL specifies that "in cases where the discharge of pollutants exceeds the limit set by the state, a fee shall be charged according to the quantities and concentration of the pollutants released." Although local environmental protection bureaus (EPBs) began collecting fees in 1979, formal procedures interpreting the 1979 law were not issued until 1982. These measures required that a pollution levy be paid by any enterprise that discharged pollutants above relevant standards. A state fee schedule came with the regulation, but provincial and local governments could charge higher rates with approval from the central government. For industrial wastewater discharges, the fee for any given pollutant was based on the multiple by which the pollutant concentration exceeded the standard. For air emissions, fees were based on multiples by which standards were exceeded, but air emissions standards included a mix of concentration-based and mass-based limits. "Overstandard" fees were assessed only on the pollutant most in violation in a waste stream.

As established in 1982, the discharge fee system provided incentive to reduce discharges only to levels specified in standards. No incentive existed to further reduce discharges, even if the marginal cost of control was small. To provide incentive for further pollutant discharge reductions, a volume-based industrial wastewater discharge fee was introduced in 1993.[8] This fee is charged on the total quantity of wastewater discharged. (.05 yuan per ton; 8.5 yuan = $1). However, a factory is not required to pay both an overstandard pollution fee and a wastewater discharge fee. The overstandard fee supersedes the wastewater discharge fee if the effluent standards are violated. In 1993 collections of this "within standard" fee amounted to about 10% of the collections of the overstandard fee.[9]

The 1982 and subsequent procedures defining the pollution levy system specify four additional categories of penalty. Following a three-year grace period, effluent fees increase 5% per year for enterprises that fail to meet effluent and emission standards. Double fees are assessed both for facilities built after passage of the 1979 EPL that exceed standards and for old facilities that fail to operate their treatment equipment. A fine of 0.1% per day is specified for delays of more than 20 days in paying discharge fees. Penalties for false effluent and emission reports or for interfering with EPB inspections are mandated. These four penalties are referred to as the "four small pieces" component of the pollution levy system.

Fees and fines collected support EPB operations and subsidize pollution control projects for enterprises that have paid into the system. Local EPBs may use 20% of the fees and 100% of the fines to fund their operations. Up to 80% of the fees are allocated for pollution control loans to enterprises that have paid into the pollution levy system.

Thus, as established in 1982, the pollution levy system was intended to provide both "stick" and "carrot" pollution control incentives. However, the system turned out to be more of an EPB funding source than an incentive for reducing industrial emissions

Design Limitations

A number of observers have noted that the pollution control incentive provided by the levy system is low because fees are small relative to pollution control costs. Fees are not indexed for inflation, and, for state-owned enterprises, they can be included under costs and later compensated through price increases or tax deductions.[6,7,10,11] Thus, many enterprises choose to pay the fees rather than incur pollution control costs, which often require significant capital investment. Because the fees and fines can be lower than operating and maintenance costs, enterprises that install pollution control equipment have little incentive to operate it.[10]

Nationwide, collections of pollution fees have grown steadily over the past decade but recently have fallen behind inflation in most provinces. Most of the growth resulted from increases in the number of enterprises assessed rather than fee rate increases.[9,10] In 1993 about 2.7 billion yuan were collected in discharge fees and penalties from 254,000 enterprises. This represents an average annual fee per paying enterprise of only 10,500 yuan ($1200-$1700 depending on the exchange rates). The average annual gross output value for a state-owned industrial enterprise is about 15 million yuan. For firms that pay discharge fees, the fees typically make up less than 0.1% of the firm's total output value. By comparison, U.S. industry environmental compliance costs are estimated to be a little more than 1% of total manufacturing costs.

Since the pollution levy system was established in 1979, about 60% of the total amount collected has gone to pollution abatement grants and loans for existing (old) enterprises.[8] Nationwide, grants and loans from pollution discharge fees contribute about 8% of China's total capital investment in pollution control and about 20% of China's pollution reduction investment in existing factories.[12] Thus, although the discharge fee system provides a significant fraction of China's pollution control spending, the amount are small compared to the 12 billion yuan per year NEPA estimates is needed to meet industrial effluent and emissions standards by the year 2000.

Industrial wastewater effluent standards currently are specified in terms of allowable concentrations rather than mass flow rates. In the past, some enterprises could meet discharge standards and avoid fees by diluting their waste stream with fresh water. Therefore some local governments imposed fines for diluting waste streams.[5]

Although the discharge fee system provides significant operating revenues for local EPBs, it provides little incentive for enterprises to invest in pollution prevention or control. Because the policy is applied uniformly, it does not account for regional differences or the assimilative capacity of local environments. To address these two concerns, a discharge permit system was introduced to control pollution from large enterprises. Water pollution permitting was implemented on a trial basis in 17 cities in 1987 and now is operating in 391 cities.[9,13] Trial air pollution permitting was begun in 16 cities in 1991 and has expanded to 57 cities.[9] The discharge licenses specify both the maximum pollutant concentrations and a factory's maximum annual wastewater discharge volume. Criteria for setting limits vary from city to city. Some call for apportioning total allowable loads within a region to achieve ambient environmental quality standards; others are based on the emissions status quo or on the capabilities of available and affordable technology.[10,13] Fines are levied for failure to meet permit conditions.

To effectively negotiate a permit, EPB personnel need technical knowledge of industrial processes, but few of them have had the opportunity to gain such information. Therefore, the EPBs are susceptible to industry-biased claims.

In many regions, the pollution levy and discharge permit systems operate simultaneously. Enterprises thus can face conflicting or discontinuous incentives from a concentration-based fine applied to behavior out of compliance with discharge standards and a mass-based fine for exceeding permit limits. Conflicting incentives also have arisen between the levy and the tax systems. In earlier years, some enterprises shielded profits from taxes by paying into the levy system and then recovering 80% of their payments under a rebate program.

Local Resistance

In the early 1980s some enterprises resisted fee collection by local environmental protection units.[14] The collectors often were turned away when they appealed to higher local officials. Because of the threat to the economic welfare of vital local enterprises, local officials often preferred to waive the fees, particularly for unprofitable enterprises. This problem still exists in many rural areas but has improved substantially in urban areas.

Enforcement of the pollution levy system is weak for China's 8 million township and village industrial enterprises (TVIEs), which are significant sources of rural pollution. Because most TVIEs are small, local EPB revenues from fee collection are less than those from larger state-owned enterprises. Therefore, because they have limited personnel, EPBs concentrate on the largest polluters first. The TVIE sector is growing rapidly and contributes about half of China's industrial output.

Low collections per unit output can reflect either high compliance rates, such as in Beijing, or weak enforcement, such as in Liaoning where many large and bankrupt heavy industries are excused regularly from making payments.

In December 1994 NEPA began a two-year study of the pollution levy system. The objective is to correct deficiencies and propose changes to improve effectiveness and efficiency consistent with a market economy and with ongoing economic and institutional reform. NEPA's Department of Supervision and Management is conducting the study with a World Bank technical assistance loan. The technical portion of the study is being conducted by the Chinese Research Academy of Environment Sciences in Beijing with the assistance of foreign consultants.

Intended to be both comprehensive and specific, the study will address four main areas: designing mass-flow levy formulas for pollutants with fee schedules based on the marginal costs of pollution control; designing a pollution levy fund to include institutional arrangements, technical assessment of loans, and priorities for the use of the fund; designing an information management system for calculating fees and maintaining billing and receipt records; and practical issues of implementation. The latter includes emissions and effluent monitoring, calculating fees, fee collection, and fund management. The goal of the study is to develop a pollution levy system for China that reduces emissions and effluents, achieves environmental goals with the least cost, and imposes minimal administrative burdens on local EPBs and regulated enterprises.

Other Incentives

The 1989 EPL specified that government at all levels should be responsible for environmental quality in their jurisdiction. In 1990 this was formalized as a contract system in which officials from mayors to enterprise managers agree to work toward environmental goals. Depending on the organization, goals can include objectives for environmental quality, pollution control, facility construction, and environmental administration. Although there is no penalty for failing to meet contract goals, rewards for doing so can include grants, bonuses, or special status that offers tax breaks and control of foreign exchange.

In the autonomous regions Inner Mongolia and Guangxi Zhuang and in Fujian Province, a compensation fee is being imposed on sales of products made outside the province by highly polluting industries.[15] The levy, approved by the State Council, is applied to commercial coal, steel, crude oil, and electricity. Revenues will be used for the region's pollution prevention and control and recycling projects.

Recognizing that the growing demand for environmental products would otherwise be filled by foreign firms, China is encouraging the growth of an indigenous environmental products industry. Now more than 4000 enterprises in China make pollution control, monitoring, and recycling products.[9] These enterprises have fixed assets of about 3 billion yuan and an output of roughly 6 billion yuan per year. The Chinese government encouraged the growth of such firms in several ways. In Ningbo, a port city in Zhejiang Province, China established its first environmental products market. Elsewhere, special industrial parks for environmental protection industries are being established to take advantage of synergies and economies of scale.

The inefficiencies of materials use associated with China's centrally planned economy prompted the government to advocate "comprehensive utilization of materials" in the 1970s. Incentives such as tax breaks on recycled materials sales, subsidies for recycled materials sold at a loss, reduced prices for recycled materials, and low-interest loans for recycling projects were introduced.[11] This is one of several programs managed by a new NEPA division concerned with clean production.

Monitoring and Enforcement Problems

China's environmental policy is comprehensive, complex, rapidly developing, and contains a mix of command and control, moral suasion, and economic incentive. The development of environmental laws, regulations, and infrastructure has greatly improved, but monitoring and enforcement still are major problems. Compliance has been best for large new facilities because abatement equipment was incorporated in the early design stages. But many medium- and small-scale enterprises, particularly rural ones, feel little or no pressure to comply, and larger enterprises operating at a loss can maneuver for exemptions.

With economic reforms diminishing the central government's control, a strong bias favoring economic development over environmental protection, and no support from an independent environmental movement, environmental protection in China faces an uphill battle.[11,16] Until the Chinese legal system is strengthened, it may be necessary to employ short-term means such as taxes on inputs that do not rely heavily on monitoring and enforcement.

A number of macroeconomic incentives drive environmental protection in China. Price reform for raw materials and profit retention for state enterprises induced manufacturers to be more efficient, resulting in less waste per unit of output. Price reform still has a long way to go, however. For instance, water still is priced below its scarcity value in most areas.

The most important accomplishment of the discharge fee system has been to provide operating revenue for local EPBs, whose collection activities contributed significantly to enhanced public and official awareness of environmental problems.[11,14] But this arrangement can distort incentives within the collection bureaus, which might focus on the largest industrial facilities where payoffs are greatest or tolerate continuous violations to maintain an income stream. However, linking fee collection to EPB revenue does provide incentive for the agency to do its job.

As in most industrialized countries, command and control is the principal instrument of Chinese environmental policy. The 1979 and 1989 EPLs both require all enterprises to meet discharge standards established by NEPA. But this "legal" requirement is enforced weakly through low-level fines targeted at the largest and most egregious polluters. The discharge permit system will attempt to address this problem by collecting enough in permit fees to finance enforcement. But this probably will not extend to the smallest rural enterprises.

Although local EPBs are receptive to citizens' environmental complaints, the Chinese government is reluctant to allow formation of independent activist environmental groups. If allowed to flourish, such organizations could ease the government's enforcement burden by bringing additional pressure to bear on noxious facilities.

The Environment and Resources Protection Committee of China's National People's Congress expects to approve 14 new or revised environmental laws in the next few years.[17] These should address gaps in existing legislation such as controlling disposal of solid and hazardous waste, bring China into compliance with its international commitments, and improve existing pollution control measures. ■

Acknowledgments

This paper was supported in part by a grant from the United Nations Environment Programme through the Organization for Economic Cooperation and Development. Views expressed do not necessarily reflect those of UNEP, OECD, or any of their member countries. The authors thank Xia Kunbao and participants at the OECD/UNEP Paris Workshop on the Application of Economic Instruments for Environmental Management, May 1994, for helpful comments.

References

1. Ross, L. Environmental Policy in China; Indiana University Press: Bloomington, IN, 1988.
2. Zhang, K. M. "Apply Economic Measures to Strengthen Environmental Protection Work, Promote Sustained, Stable, and Coordinated Development of the National Economy"; Zhongguo Huanjing Bao [China Environment News], July 23, 1991; English translation in JPRS-TEN-92-005, U.S. Joint Publications Research Service, March 3, 1992.
3. "Zhongguo Tongji Nianjian [China Statistical Yearbook]"; State Statistical Bureau, distributed by China Statistical Information and Consultancy Service Center: Beijing [various years].
4. "Introductions to the Environmental Protection Organizations in China"; National Environmental Protection Agency: Beijing, 1992.
5. "Huanjing Baohu Zhengce Fagui Biaozhun Shiyong Shouce [Handbook of Environmental Protection Laws, Regulations, and Standards]"; Beijing Municipal Environmental Protection Bureau; Jilin People's Publishing House: Changchun, 1987.
6. "China Environmental Strategy Paper"; World Bank: Washington, DC, 1992.

7. Krupnick, A. Incentive Policies for Industrial Pollution Control in China; presented at the annual meeting of the American Economics Association: New Orleans, LA; January 2–5, 1992.
8. Ma, X. Y. Effectiveness of Environmental Policies for the Control of Industrial Water Pollution in China; working paper, Civil Engineering Department, Stanford University; March 1994
9. Zhongguo Huanjing Nianjian [China Environment Yearbook]" National Environmental Protection Agency; China Environment Yearbook Publishing House: Beijing [various years].
10. Sinkule, B. Implementation of Industrial Water Pollution Control Policies in the Pearl River Delta Region of China; Ph.D. dissertation, Department of Civil Engineering, Stanford University, August 1993.
11. Vermeer, E.B. China Information 1990, 5(1), 1–32.
12. Qu, G. P. Basic Analysis and Assessment of China's Environmental Protection Investments and Policies; published in three parts in Huanjing Baohu [Environmental Protection], Nos. 3-5, March 25, April 25, and May 25, 1991. English translation in JPRS-TEN-91-014, July 9, 1991, and JPRS-TEN-91-016, August 22, 1991, U.S. Joint Publications Research Service.
13. Rozelle, S. et al. Natural Resources Modeling 1993, 7, 353-78.
14. Jahiel, A.R. The Deng Reforms and Local Environmental Protection: Implementation of the Discharge Fee System, 1979–1991; presented at the 46th Annual Meeting of the Association for Asian Studies: Boston, MA; March 1994.
15. Xia, K.B. "Fines: Forcing Polluters to Listen"; China Environment News (English ed.), January 1993, 42,6.
16. Ross, L. Policy Studies Journal 1992, 20(4), 628–42.
17. Ross, L. China Business Review 1994, 21, 30–33. ■

Questions

1. Why is the enforcement of the pollution levy system weak for China's eight million township and village industrial enterprises?

2. What is the objective of the two-year study of the pollution levy system that NEPA began in 1994?

3. What is the goal of this study?

Answers are at the back of the book.

40

Many scientists think that people are victims of "environmental hypochondria." Scientists assert that people are irrational and inconsistent because they fear smaller risks and dismiss the larger ones. Comparative risk assessment (CRA) is a set of quantitative techniques that are used to estimate numerous environmental threats. These are ranked from highest to lowest based on the annual probability for causing human fatality. This allows a quantitative, objective way to measure the actual risk of environmental threats to humans. Many current political proposals are based on the idea that environmental laws should incorporate more objective risk assessments. But we must be careful not to assume that objectively measured risks are the only ones that matter. This is because all risks cannot be measured accurately. In measuring risks, we also omit ethical considerations.

Comparative Risk Assessment and the Naturalistic Fallacy

Kristin S. Shrader-Frechette

Have we humans become risk-averse wimps? Many scientists think so. Nuclear physicist Alvin Weinberg, for example, claims that contemporary lay-people are often victims of 'environmental hypochondria,' similar to the hysteria that drove 15th- and 16th-century witch hunts. Weinberg notes that witch hunts subsided only after the Inquisitor of Spain convened a group of savants who proclaimed there was no proof that 'witches' caused misfortunes. He concludes that environmental hypochondria likewise will disappear only after contemporary risk assessors dismiss most of the health problems allegedly caused by environmental threats such as pollution.[1]

Arguing along similar lines, numerous scientists have claimed that if members of the public understood the extremely low probabilities associated with environmentally induced health hazards, they would not fear threats from sources such as pesticides, chlorinated drinking water and global warming. They reason that—because laypeople are much more likely to die in a traffic accident than in a nuclear core melt—they are irrational and inconsistent if they accept automobile travel but oppose commercial atomic energy. Fully rational people, they maintain, do not fear small risks while they dismiss much larger ones. Instead, they say that people who evaluate risks consistently and rationally do so on the basis of their respective probabilities of death. Rational people rank risks according to their likelihood of fatality, say many risk assessors, and they exhibit risk aversion that is proportional to the respective probabilities (see, for example, Ref. 2). In other words, a number of experts claim that rational and consistent risk behavior is based on comparative risk assessment.

Comparative risk assessment (CRA) is a set of quantitative techniques that scientists and engineers use to evaluate various environmental threats, from owning handguns or occupying a home with out a smoke alarm, to eating meats treated with nitrates or living near a chemical plant. CRA ranks risks, highest to lowest, on the basis of their average annual probability of inducing (human) fatality. The rationale for such ranking, and for CRA-based public policymaking, is that it encourages more rational and consistent risk decisions; it helps prevent lawmakers from implementing 'irrational'

Kristin Shrader-Frechette, Comparative Risk Assessment and the Naturalistic Fallacy, *Trends in Ecology and Evolution*, January 1995. Reprinted with permission.

policies of regulating supposed small risks while ignoring larger ones; it assists in reducing overall risk; it forces society to set environmental priorities according to science rather than politics; it saves industry from excessive spending to reduce very small risks; and it enables governments to use risk-abatement expenditures wisely and efficiently. Jumping on the bandwagon, a variety of industrial, scientific and governmental groups have endorsed CRA-based policymaking. There are at least three different 'risk' bills, for example, before the 1994–1995 US Congress, and all of them mandate both performing CRA as part of quantitative risk assessment and setting standards and regulations on the basis of CRA.

Despite the merits of rational policymaking regarding environmental risk, CRA probably raises more questions than it addresses. One problem is that CRA is easier said than done. For the approximately 50000 human-made environmental contaminants, such as benzene and vinyl chloride, no more than a handful have gone through epidemiological testing. The newness of such hazards, and the expense of testing, mean that there are reliable frequency records for only a few risks. In the absence of empirical data, experts routinely use subjective probabilities in their risk assessments. On the few occasions when assessors have been able to check their risk predictions, they typically err by four to six orders of magnitude.[3]

Another difficulty with CRA ranking is that death is not the only important 'end point' whose probability rational people might want to evaluate. Other end points and/or effects include a variety of health, ecological and welfare risks—from eggshell thinning, mutagenic injury and loss of ecosystem functions—to inequities in risk distribution, lack of citizen consent to certain threats, and inadequate compensation (or denial of due process rights to recover damages) after exposure to hazards. Indeed, rational and consistent people arguably might prefer to take a voluntarily chosen risk (like driving an automobile) rather than an involuntarily imposed one (like living near a nuclear power plant), even though the probability of dying in a traffic accident is greater. In other words, rational people might ask not only 'how safe is safe enough?' but 'how safe is voluntary enough?', 'how safe is equitable enough?', 'how safe is fair enough?', 'how safe is compensated enough?'.[4]

In its simplistic 'body-count' approach to environmental hazards, CRA ignores important questions, such as who calculates the relevant risk probabilities, who bears the risks, who benefits from the risks, whether the risk is avoidable, how the risk figures are averaged, who would pay for risk reduction, and who has the right to give or withhold consent to the risk. If one assumes a multi-attribute view of risk and denies that probability of fatality is the only relevant variable for comparing environmental risks, then one could just as well rank societal threats in terms of the respective threats they pose to the gene pool, to future generations, to political autonomy, to justice, or to some other value. All these considerations suggest that, like beauty, risk may be—in part—in the eye of the beholder.

By assuming that a wide variety of social and environmental values can be reduced merely to probability of fatality, proponents of CRA err in assuming that quantitative information—even when it is uncertain—trumps qualitative factors. This bureaucratic enthusiasm for number crunching in CRA illustrates Gresham's Law: monetary drives out nonmonetary information, and quantitative drives out nonquantitative information.[4] Assessors who follow Gresham's Law err in reducing questions of ethics (is this risk acceptable?) to questions of science (what is the probability of this risk?). British moral philosopher G.E. Moore warned, earlier in this century, that anyone who attempts to reduce an ethical question to a scientific one commits 'the naturalistic fallacy'.[5] Obviously we cannot learn how things ought to be, ethically, by looking only at how they are, scientifically. And if not, then although science and CRA are essential tools for risk comparisons, they are not the only tools. Indeed, they err in telling us that we ought to accept preventable environmental risks just because we cannot eliminate all risks.

References

1. Weinberg, A. (1988) in *Phenotypic Variations in Population* (Woodhead, A., Bender, M. and Leonard, R., eds.), pp. 121-128, Elsevier
2. Whipple, C. (1989) in *The Risk Assessment of Environmental and Human Health Hazards*

(Paustenbach, D., ed.), pp. 1116-1123, *Wiley*
3. Cooke, R. (1992) *Experts in Uncertainty: Subjective Probability* and *Expert Opinion*, Oxford University Press
4. Shrader-Frechette, K. (1991) *Risk and Rationality*, University of California Press
5. Moore, C. (1959) *Principia Ethica*, Cambridge University Press ■

Questions

1. What is comparative risk assessment?

2. When will environmental hypochondria disappear according to Alvin Weinberg?

3. What mistake does science and comparative risk assessors make?

Answers are at the back of the book.

41

Some environmentalists assert that debt promotes the depletion of resources and the increase of pollution in developing countries. The "exports promotion hypothesis" correlates debt with environmental degradation. This hypothesis contends that resources are used for export rather than for domestic consumption. This causes environmental degradation. Some environmentalists thus suggest that lender nations like the United States write loans off because the loan is irretrievable and relieving debt payment will encourage sustainable development. They also suggest that the developing nations establish individual ownership of resources, terminate environmentally damaging subsidies, institute market-based pollution-control mechanisms, and make direct, necessary payments to preserve environmental assets of global consequence.

Debt and the Environment

David Pearce, Neil Adger, David Maddison And Dominic Moran

Loans cause great human hardship, but their connection to ecological troubles is hard to prove.

Some environmentalists claim that loans to developing nations have led to a spiral of debt and environmental degradation. According to their argument, the domestic policy changes that countries make to generate cash for loan payments—often under duress from the International Monetary Fund or the World Bank—hasten the depletion of natural resources, increase pollution and harm the poor, who may be uprooted in ways that cause further environmental damage. Many critics also contend that lenders should write off the loans because the money, in any case, is effectively irretrievable and because relieving countries of repayment obligations will encourage "sustainable development."

Although economic theory does not automatically render repayment incompatible with full employment, steady prices, economic growth or an equitable distribution of income, in reality these goals have suffered. As a result, most debtor nations continue to rely on outside funds, even though additional loans only make their predicament sharper. Whether the environment has also been harmed directly is less clear. There is scant empirical evidence to suggest that the connection between debt and the environment is significant. Indeed, in some cases, the fiscal discipline imposed by debt may rein in environmentally harmful spending.

The debt crisis has its origins in the oil-price shock of 1973, when energy prices roughly doubled in a matter of months. Commercial banks, flush with deposits from oil producers, were eager to lend money to developing countries, especially as they took it as an article of faith that nations always repay their debts. The borrowers, meanwhile, were glad to see money plentifully available at low interest rates. In 1979, when oil prices doubled again, industrial nations raised interest rates to slow their economies and thus reduce inflation. This action spurred a global recession that stifled demand for the raw materials developing nations were producing.

As interest rates rose, debtor nations faced higher payments on their outstanding loans but had less income with which to pay. Many found themselves unable to meet their current obligations, much less get new loans. Repayment became an overriding policy objective, affecting both government and private spending, because only wide-ranging changes in developing economies could generate the needed hard currency. A large fraction of many countries' earnings continues to be earmarked for the repayment of debt.

Exports at Any Cost

The most commonly held view linking debt with environmental degradation is known as the exports promotion hypothesis. To earn foreign exchange with which to repay international debts, a country must divert resources away from production of domestic goods to sectors generating commodities for export. According to this theory, production of goods for export causes more environmental degradation than does production of goods for domestic consumption, and so debt repayment harms the environment. There is no a priori reason to expect such a difference, but some environmentalists contend that it still does occur. They point to the possibility, for example, that countries will raze their forests for tropical timbers or to open up land for cash crops.

Nevertheless, statistical analysis of data from many developing countries suggests that national income and commodity prices have just as much influence on levels of exports as debt does. Raymond Gullison of Princeton University and Elizabeth C. Losos, now at the Smithsonian Tropical Research Institute, examined the effect of debt on timber exports from Bolivia, Brazil, Chile, Costa Rica, Paraguay, Peru, Colombia, Ecuador and Mexico but found only minimal correlations overall. Furthermore, in Paraguay, the only country on which debt apparently did have a significant effect, increased debt was associated with reduced production.

More recently James R. Kahn of the University of Tennessee and Judith A. MacDonald of Lehigh University found more concrete evidence of a correlation between debt and deforestation, although they also found that country-specific factors played a strong role. They estimate that reducing a country's debt by $1 billion might cut annual deforestation by between 51 and 930 square kilometers. Brazil currency clears more than 25,000 square kilometers a year and Indonesia more than 6,000.

Deforestation is only one aspect of environmental degradation in developing countries. Unfortunately, the impact of indebtedness on other environmental indicators such as pollution, biodiversity or depletion of other resources has not been tested.

Reductions in Domestic Spending

Evidence for or against other mechanisms by which debt repayment might damage the environment is largely anecdotal and speculative. Some observers have placed blame on reductions in domestic spending by nations that shifted money toward financing their debt. In sub-Saharan Africa, outlays for health, education and other public services decreased by more than 40 percent during the 1980s, in parallel with a sharp rise in money spent on repayment of debt. Yet the effects on the environment remain unclear.

Cuts in government spending may fall on measures designed specifically to enhance the environment, such as schemes to improve water quality and sanitation. On the other hand, some cuts may cancel large capital projects, including the construction of dams and roads, that have often been criticized for causing environmental devastation far in excess of any financial returns they may bring. Elimination of road-building programs in the Brazilian Amazon, for instance, may have helped curtail deforestation. From a theoretical point of view, then, reductions in government spending can act to either improve or degrade the environment.

The same uncertainty emerges when one looks specifically at the effect of debt on spending for environmental protection. Some economists assert that environmental quality should increase with national income. (Their claim, called the Kuznets curve effect, grows out of the observation by Simon Kuznets, who won the Nobel Prize in Economics in 1971, that richer nations have more equitable income distributions.) Conversely, a nation faced with a huge debt is likely to divert money from the environment to more pressing problems.

But does the relaxing of environmental standards inevitably increase pollution? Perhaps not. Many debtor nations have probably always spent little on protecting the environment, and so decreases in spending could have a minimal impact. Moreover, a country repaying debt might not be able to afford high standards for pollution control, but it also might be unable to afford goods whose production damages the environment.

It is conceivable, however, that absent or unenforced environmental regulations or their lax enforcement might also make for the establishment of "pollution havens"—a situation that many claim has occurred in the Maquiladora export-processing zone of northern Mexico. Supposedly, U.S. companies have been attracted to this area because of the lower environmental standards. Yet a 1992 study by Gene M. Grossman and Alan B. Krueger of Princeton found little statistical support for this claim. They argue that low wages and easy access to U.S. markets spurred investment. Thus, the net effect that reduced domestic spending has on pollution and resource degradation is not clear.

In addition to causing ecologically unsound export drives or budget cuts, excessive debt can potentially exacerbate local practices that already put the environment at risk. Diversion of money to debt service has triggered massive unemployment in many countries, sometimes prompting poor people to migrate in search either of work or land on which to grow food to live. Marginal lands and fisheries whose ownership was indeterminate have often attracted migrants, until the topsoil or fish have been depleted.

Overexploitation would have come about even without these evils, however; the inefficiency results from the lack of established rules for access to land, water and other resources. Many indebted countries seem to be characterized by the hallmarks of shared resources, a high level of migratory subsistence farming, overgrazing and declining yields. Ill-defined or nonexistent ownership has often precipitated the exhaustion of land. The enforcement of property rights would in many cases have prevented overuse by giving people incentives for proper husbandry.

Structural Adjustment

Nations that face increasing poverty are often forced to try to secure additional loans. As a condition of attaining this money, heavily indebted countries have often had to make "structural adjustments" to their economies: eliminating subsidies, removing tariffs and privatizing government-owned enterprises. These reforms aim to help them grow out of indebtedness by removing glaring economic inefficiencies.

Some observers contend that these "conditionality" programs let governments push ahead with policy changes that were previously impossible, by putting multilateral institutions in the position of political lightning rods. Others think conditionality makes all objectives, including environmental ones, subordinate to debt repayment. Systematic cuts in such vital services as education, health and food (which would typically be higher on a government's list of priorities than the environment) give some weight to the latter view. If it is correct, one would expect to find environmental repercussions in countries that have received significant structural adjustment loans.

The evidence is equivocal. Case studies of conditionality programs of the International Monetary Fund in Mexico, Ivory Coast and Thailand, sponsored by the World Wildlife Fund International, suggest that structural adjustment programs have on balance benefited the environment. In Thailand the removal of indirect irrigation subsidies has helped reduce waterlogging and salinization. Yet in Malawi, which has undertaken four International Monetary Fund restructuring packages and six World Bank structural adjustment loans since 1979, the Overseas Development Institute found many more negative results than positive ones. In the Philippines, a study by the World Resources Institute found that structural adjustments undertaken in the 1980s encouraged overexploitation of natural resources and resulted in increased emissions of pollutants, concentrated pollution and congestion in urban areas.

Whether the consequences for the environment are good or bad seems to depend a great deal on the particular provisions of the loan agreement and on the individual circumstances of a country or even a region within a country. In Ivory Coast, controls on food prices and subsidies for fertilizers and pesticides have been minimized. These changes could cause the abandonment of some environmentally harmful farming practices, but lower yields may also bring about cultivation of remaining forestland. In Malawi, currency devaluation and agricultural reforms have contributed to an increase in tobacco farming and created incentives for planting such crops as cotton and hybrid maize, which tend to be grown in a manner that promotes erosion. Malawi's case cannot be taken entirely at face value, however: large-scale migration of refugees from Mozambique has placed an unprecedented strain on the country's resources, unrelated to structural reform.

In other parts of the world, structural adjustment appears to have helped the environment. Agricultural subsidies, which adjustment programs curtail, have played a significant role in deforestation and the destruction of soil in areas subject to erosion. Until the late 1980s, for example, Brazil gave tax credits for cutting down forests, and it subsidized loans for crops and livestock development. These government incentives typically covered more than two thirds of the cost of cattle ranches, which reportedly accounted for 72 percent of all deforestation in the Brazilian Amazon up to 1980. Toward the end of the 1980s, after the government abandoned many of the clearance subsidies, deforestation slowed. Other evidence from Brazilian Amazonia also supports the idea that public expenditure on infrastructure, such as on roads, is linked to migration onto marginal land that results in deforestation, erosion and other deleterious effects.

Reduction of subsidies for energy use could also have a salutary effect on the environments of developing nations. Anwar M. Shah and Bjorn K. Larsen of the World Bank have calculated that total world energy subsidies amounted to $230 billion in 1990. Eliminating them could cut emissions of carbon dioxide (the main greenhouse gas) by 9.5 percent and improve prospects for economic growth by freeing the money for other uses. Indeed, the developing world has many resources that could be used more efficiently, thereby enabling nations to repay foreign debts without reducing domestic consumption.

Structural adjustment programs can potentially prove beneficial in another way as well. Recent structural loans have been made on the condition that governments clarify land-ownership questions. Such arrangements should reduce somewhat the environmental degradation brought about by shared use of land that seems to belong to anybody and to nobody in particular.

Debt Forgiveness

At best, then, evidence on the extent to which environmental degradation in the developing world can be attributed to debt repayment remains inconclusive. Nevertheless, given the fact that debt repayment is largely to blame for the drastic reductions in per capita spending on social programs in these countries, cannot a strong case be made in favor of debt forgiveness? In secondary markets, debt of developing countries often changes hands at a fraction of its contractual value. This steep discount is an indication that the commercial banks expect that the loans will never be repaid in full.

Banks do not simply write off the uncollectible debt, for two simple reasons. First is the problem of moral hazard: forgiveness may encourage countries to get into more debt in the expectation that it, too, would be forgiven. Such profligacy would obviously jeopardize other assets held by the banks. The second reason is uncertainty. There is a slight possibility that unexpected favorable developments will eventually enable developing countries to repay more of their debts, and so it is not in the interest of any bank to deprive itself of the opportunity of benefiting from any windfall to the borrower.

Conversely, a reduction of part of the contractual debt that the debtor is not expected to repay anyway is of little value. It neither reduces current cash requirements nor makes it easier to get new loans. When Bolivia spent $34 million to buy back $308 million in bonds in 1988, the price of the remaining bonds rose from six to 11 cents on the dollar. As a result, the real value of outstanding debt declined from $40.2 million ($670 million at six cents on the dollar) to $39.8 million (362 million at 11 cents), less than $400,000.

The same problem afflicts other debt-reduction mechanisms, such as debt-for-nature swaps. Until 1992, 17 countries had participated in such swaps: donors spent $16 million to retire nearly $100 million of debts in developing countries in return for the establishment of national parks and other environmental improvements. Although the swaps do preserve some environmentally vulnerable regions, the large nominal reduction barely touches nations' real burdens. Indeed, they may even have increased expected repayments.

No Easy Path

It seems, then, that excessive debt inevitably causes radical restructuring of a nation's economy. Because economic policies play a crucial part in determining how natural resources are used, the environment is bound to be affected. Yet it is very difficult to predict whether any particular change will cause harm or prevent it.

Such links as exist between indebtedness and damage to the environment stem largely from structural adjustment programs and their requirement that money reallocated from government subsidies and other spending repayment. Even then, most environmental degradation in the developing world probably has causes other than the servicing of debt. Structural adjustment is more justly criticized on humanitarian grounds than on environmental ones. Instead of alleviating unemployment and equitably redistributing income, price reforms—particularly elimination of subsidies for food and fuel—have fallen most heavily on the poor.

Given that the connection between debt repayment and environmental degradation is tenuous at best, attempting to improve the environment by debt forgiveness would probably be futile. The most effective way to confront pollution, deforestation and similar problems in debtor nations is to establish individual ownership of resources that are currently open to all, to end environmentally damaging subsidies, to institute market-based pollution-control mechanisms (in which those who produce toxic substances pay for their effects) and to make direct payments where necessary to preserve environmental assets of global significance. Undoubtedly, the flow of funds from the South to the North causes poverty, malnourishment, ill health and lack of educational opportunity. These consequences make a compelling case for debt relief. But such aid is not a panacea for environmental degradation. ■

Questions

1. When did the debt crisis begin?

2. What is the exports promotion hypothesis?

3. What is the most constructive way to deal with resource depletion and pollution in debtor nations?

Answers are at the back of the book.

42

The federal Farm Bill is coming up for reauthorization. One of its provisions is the Conservation Reserve Program (CRP). Through this program, the federal government pays farmers to replace their crops with grasses. This turns farm acreage into wildlife habitats and is the largest and most successful provision of the Farm Bill. CRP has annually prevented an estimated 700 million tons of soil from being eroded. Grassland bird species that have been in decline are now more abundant on CRP land than on surrounding cropland. Its defeat could prompt farmers to return their fields to agricultural use and to dependence on costly commercial pesticides, fertilizers and antibiotics.

Growing Crops, and Wildlife Too

Greg Breining

[Ed. Note: Congress was still debating the Farm Bill amendments when this book went to press.]

Though far from perfect, the Farm Bill has made the nation's farmers into some of its most important conservationists. Is Congress about to change all that?

A dark cloud hangs over the wildlife habitat on Mel Rodenburg's North Dakota farm, though you wouldn't know it on a bright spring day as sharp-tailed grouse and meadowlarks flush from roadsides. Migrating sandhill cranes, grounded by wind, forage in grassland. Northern harriers tilt like kites on the gusts. In dips and swales, mallards, pintails, gadwalls and shovelers swim in temporary wetlands barely waist-deep to a duck. "You just can't beat this habitat for waterfowl," says Skip Baron, Central Division Staff Director for the National Wildlife Federation in Bismarck, North Dakota. "It is literally a duck factory."

Ten years ago, this land was covered only with wheat, barley and corn stubble, with barely a bird in sight. Then the federal government paid Rodenburg to replace some of his crops with grasses. Observes Rodenburg, "Wildlife here has really built up in the last few years." The same has been true on other farms too, on a nearly unimaginable scale. On more land than exists in the National Wildlife Refuge system, wildlife has found dramatically improved farm habitat over the past decade.

The cloud over Rodenburg's land and all the nation's farm habitat takes the form of a question: When the federal Farm Bill comes up for reauthorization in 1995, will Congress renew and strengthen enforcement of landmark legislation designed to protect wetlands and other critical areas? Among the provisions are the Conservation Reserve Program (CRP) that now protects the habitat on Rodenburg's farm, as well as Swampbuster and Sodbuster regulations. From the perspective of wildlife advocates like Kevin Lines, farmland wildlife program leader for the Minnesota Department of Natural Resources, the fate of those provisions—particularly that of CRP—is "the most important wildlife conservation issue of the century."

It is, however, only a small part of a larger question. Many of the farmers, lawmakers and others who cope with the Farm Bill—shorthand term for all federal farm legislation—are asking whether the complicated tangle of regulations can be improved as a whole. Just about everyone involved thinks it can be, and suggestions of reform are pouring in from all directions. They range from a group of Iowa farmers that essentially wants an insurance system for farmers who hit hard times, to The Campaign for Sustainable Agriculture, a network of grassroots and national groups planning a sustainable food-and-farm system.

Not all the possible fixes take wildlife into account. For one thing, Congress could respond to budget pressures by reducing spending on the Farm Bill's conservation provisions. The cost of CRP alone adds up to $20 billion over the 10-year life of the program, which, as critics like to point out, makes it one of the most expensive conservation programs in the federal budget. However, CRP payments are offset by savings on amounts farmers can receive under other programs in the current system, Taking that into account, the Food and Agricultural Policy Research Institute at the University of Missouri has calculated that continuation of CRP would effectively cost far less (about $220 million a year) than its budget alone indicates.

And supporters of CRP say the program is worth its price. In one study, federal researchers concluded that the value of CRP's environmental benefits over its life ranges from $6 billion to $13 billion. Commented *The Detroit Free Press* last winter in an editorial, "It would be a small natural disaster to kill the program,"

One fix-it strategy for farm legislation is an alternative-agriculture or "green" Farm Bill, Variations of such a bill, which would shift the very goal of farm policy from production control to stewardship, are championed by groups ranging from NWF, to state natural resource agencies, to the nonprofit Soil and Water Conservation Society. A green Farm Bill would encourage farmers to adopt practices designed to cut costs and reduce dependence on commercial pesticides, fertilizers and antibiotics.

Making stewardship a priority would be a huge shift. Since the 1920s, the Farm Bill basically has been a price-support system aimed at boosting farm income by driving up crop prices. The legislation now costs taxpayers as much as $25 billion a year in subsidies and $10 billion a year in higher food prices.

The bill's most recent version, passed in 1990, influences nearly every aspect of agriculture. Its backbone is the Acreage Reduction Program, the latest name for a longtime practice of setting aside cropland to reduce production. The program pays farmers roughly $10 billion a year *not* to plant certain major crops. For the crops they are allowed to grow, farmers in the program receive "deficiency payments" that make up the difference between an average market price and a target price set by Congress. The farmers then sell their crops on the open market.

Setting aside that land often creates what Al Berner, group leader at the Minnesota Department of Natural Resources Farmland Wildlife Populations and Research Station, calls a "death trap" for nesting birds. The government requires that millions of acres be planted in cover such as oats to prevent erosion. The crops attract nesting birds such as ducks. But farmers, required either to destroy the cover crops or pay for an inspection to prove they aren't furtively harvesting them, often disc or cut the crops by early summer—destroying nests and crushing young and sometimes females.

According to a 1988 study of pheasant-nesting habitat in 11 Midwest states, more than three-quarters of all set-aside fields were seriously disturbed before the end of nesting season. During any given year during which the program idles land in Minnesota, pheasant numbers are an estimated 30 to 50 percent lower than if there were no set-asides.

The Acreage Reduction Program also discourages crop rotation, hastening soil depletion on millions of acres. Overall, federal efforts to control production and to set aside acres on a yearly basis have worsened erosion, polluted surface and groundwater and destroyed wildlife habitat. Says Berner, "From an environmental standpoint, it has been a total disaster."

Enter the 1985 Farm Bill, with its new conservation provisions, which came about through an unusual alliance of conservationists wanting to protect habitat and farmers needing financial help. For the first time, Congress made conservation requirements a condition of eligibility for most federal farm programs. To qualify, farmers must abide by conservation requirements when converting highly erodible lands to cropland (known as the Sodbuster provision) and not convert wetlands to crop production (Swampbuster) to remain eligible for two dozen major farm programs. The Conservation Compliance regulation requires farmers with highly erodible cropland to develop and follow conservation plans.

The biggest and most successful provision has been CRP, which has paid hundreds of thousands of farmers like Rodenburg to remove highly erodible and environmentally sensitive land from production and plant it with grass or trees. Since 1985, the federal government has purchased more than 375,000 CRP contracts on more than 36 million acres, 8 percent of all U.S. cropland—an area equaling the size of Iowa. The contracts are concentrated in the Midwest and Great Plains. Five states (Iowa, Texas, Kansas, North Dakota and Missouri) account for more than 40 percent of total CRP payments.

Before CRP, North Dakota farmer Rodenburg set aside a portion of his wheat, corn and barley "bases" (the set acreages he planned in theory for each crop). The more the government wanted to cut production, the more Rodenburg could set aside—which meant that it would be largely without cover. When CRP began, Rodenburg opted to enroll first his corn base and then his barley base, keeping only wheat for the short, dry growing season. "CRP was sure income," he says. "And if we hadn't put the land into CRP, we would have had to buy additional machinery. It was a good financial decision." That was, of course, the aim of farmers when they first supported CRP.

As for the goal of conservation, CRP has prevented the erosion of an estimated 700 million tons of soil a year nationwide, according to the U.S. Department of Agriculture. Grassland bird species that have been in long-term decline—such as the clay-colored sparrow, bobolink, dickcissel and Baird's sparrow—are much more abundant on CRP land than in surrounding cropland, according to research by the National Biological Survey. Federal research has also found increased populations of several species, including grasshopper sparrows and eastern meadowlarks, as CRP grasslands have spread throughout the Great Plains.

North Dakota and South Dakota game officials estimate the states are producing I million more ducks each year because of CRP. In Iowa, Minnesota and Wisconsin, ring-necked pheasants increased tenfold or more on CRP land. In California, CRP lands are providing habitat for reintroduced tule elk and pronghorns. Several endangered species, such as the San Joaquin kit fox and the blunt-nose leopard lizard, also benefit from CRP habitat.

Despite all these success stories, even the staunchest supporters of the Farm Bill's conservation provisions see room for improvement. Though erosion has slowed, it is still a problem. And enforcement of Sodbuster and Swampbuster has been weak, in part because committees that rule on cases involving drainage of wetlands are made up of local farmers. Says NWF's Baron, "There's a lot of pressure put on those individuals to bend the rules." A nearly endless appeals process also allows swampbusted farmers to shop for favorable rulings. Conservation Compliance has fared no better. According to a 1992 survey by the Soil and Water Conservation Society, little more than 50 percent of farmers followed conservation plans, and monitoring and enforcement were "uneven."

As for CRP, a 1989 report by the U.S. General Accounting Office found that to meet acreage goals, about 25 percent of the land enrolled by the Department of Agriculture was unsuitable. Also, many CRP acres are planted in exotic grasses, when native prairie plants would make better habitat. In the Southeast, many CRP lands have been planted in pine. In Georgia, for example, 89 percent of CRP land is forest, mostly monocultures of pine of little use to wildlife.

The General Accounting Office report also found that the Department of Agriculture often paid two to three times the local rental rate to enroll land in CRP. Although the government has since begun driving a harder bargain, critics point out that CRP pays more to retire land for 10 years than the land would cost as an outright purchase. Defenders point out that taking CRP land out of production is saving the commodities price-support program as much as $19 billion.

Congress soon will begin shaping the 1995 Farm Bill. Unless CRP is renewed, its contracts are due to start running out in 1995, and all will finish by 2003. "Without CRP, you're going to see a dramatic decrease in wildlife populations," says NWF's Baron. Farmer Rodenburg, like others, would probably plow his grasslands. "I'm sure a high percentage would go into crop production," he says.

If the tight budget doesn't lead Congress to reduce the Farm Bill's conservation programs, perhaps it will instead force a cutback on price supports—along with a push for stewardship. Baron, for one, is hopeful. "There will be some kind of stewardship provisions," he says. "I believe the public and the agricultural community want that." Ultimately, America's duck factory—and other farmland wildlife habitat—will depend on them. ■

Questions

1. What would a green farm bill encourage?

2. Even though CRP is one of the most expensive conservation programs how are these payments offset?

3. What is the goal of the green Farm Bill?

Answers are at the back of the book.

43

For the environmental technology industry to work, it needs to tie its products and services to profitability. Often regulated industries will buy environmental technologies if they increase production, save money and clean the environment. The U.S. environmental technology market in products and services is worth roughly $135 billion per year. Traditional environmental industries domintated 1994 revenues. However, companies that handle hazardous waste have been hurt by competition and overcapacity. This is because many factories now produce less waste. Companies that are involved in producing pollution equipment had total revenues of approximately $60 billion last year and are expected to grow worldwide.

Selling Blue Skies, Clean Water

Jeff Johnson

Will deregulation, pollution prevention, and international competition hold promise or problems for a stagnant environmental compliance industry?

A decade ago, the environmental compliance business—a newborn industry spawned by legislation such as Superfund, the Resource Conservation and Recovery Act (RCRA), and the Clean Air Act—was said to face a brilliant future. RCRA's ban on dumping hazardous waste on land held great promise for the incineration industry, Superfund was expected to send revenues for remediation businesses sky high, and Clean Air Act requirements were thought to provide solid revenues to many environmental technology industries.

But today, the overall market is predicted to grow at a humble 3 or 4% annually. What happened? Were projections and expectations unrealistic? Was the promise undercut by an easing of regulations? Did the regulated industry just find a cheaper way to meet compliance targets?

The dynamics that drive the environmental compliance business are complex and in flux, making it difficult but important to ferret out where the industry is headed and what it needs to prosper. Adding to the confusion is today's antiregulatory climate and the regulated industries' desire to soften if not roll back the laws and regulations that have created this industry. However, experts see a way out. Call it pollution prevention or common sense, they say the environmental technology industry has to tie its products and services to production efficiency to beat out the competition. Increasingly, regulated industries will buy environmental technologies only when they increase production and save money as well as clean the environment. Environmental technology market analysts predict a bright future for companies that combine environmental and production improvements rather than sell end-of-pipe compliance hardware.

Overall the U.S. environmental technology market in products and services is worth about $135 billion a year, according to Environmental Business International, Inc. (EBI), an environmental research, consulting, and publishing firm in San Diego. About 74% of that total is services and 26% is equipment.

Looking at estimated 1994 revenues, traditional environmental industries dominate: solid waste management ($30.6 billion, 23%); water utilities ($24.3 billion, 18%) and water equipment and chemicals ($13.8 billion, 10%); resource recovery ($16 billion, 12%); hazardous waste management ($8.6 billion,

6%); waste management equipment ($11 billion, 8%); and remediation and cleanup services ($8.9 billion, 6%).

Suffering a down market, says Grant Ferrier, head of EBI, are companies that manage hazardous waste and have been hurt by stiff competition and treatment overcapacity; the remediation industry, reeling from a stall in Superfund reauthorization and cuts in federal cleanup funding; and the air pollution control market, dragged down by delayed regulations and weak enforcement. Doing well are companies in the water sector, which he calls the "sleeping giant" of the U.S. environmental industry. Its total revenues were about $60 billion last year, nearly half those of the total environmental industry, and the sector is expected to grow in the United States and abroad.

Ferrier points to other bright spots in addition to designers and sellers of wastewater treatment systems and supplies, including solid waste handlers, the recycling industry, and engineering firms that provide companies with technical advice and engineering services.

The total environmental market will grow, Ferrier says, but not very fast. He notes that three of eight mutual funds that dealt in environmental stocks dropped those portfolios last year because of poor stock performance. But consolidation and mergers in this historically fragmented industry are taking place, and that may help buoy up survivors. Four years ago, for instance, 1400 environmental testing labs competed for about $1.5 billion in annual revenues, and today 1150 chase $1.6 billion.

Ferrier and other analysts predict a strong future for pollution prevention equipment (1994 sales of $800 million) and engineering services (1994 sales of $15 billion)—a 15% annual growth for pollution prevention and at least a 5% yearly increase for consulting and engineering services over the next five years. More important, however, is the potential these sectors hold for combining compliance with production efficiency and beginning to move past a regulation-driven marketplace.

The Role of Regulations

"You must remember, there has been no natural market for environmental technologies, like there is for food or clothing," says Donald L. Connors, head of the International Environmental Business and Technology Institute, a Boston environmental technology research and promotion firm, and a founder of one of the first environmental business trade associations, the Environmental Business Council of Massachusetts.

"The demand for clean air and water is created by regulations, and a change in regulations has a major impact on this business. Weakening environmental laws and regulations is a negative for the million or so people who make a living in the environmental industry," Connors adds.

Although a regulatory floor of some sort is necessary to drive the environmental market, where to fix the bottom is a subject of much debate in the new Congress, the regulated community, and the environmental movement.

However, even when the floor was set high, such as under the 1990 Clean Air Act Amendments (CAAA), compliance dollars fell far short of expectations of environmental technology marketers, who too often hung their hopes on cost estimates from regulated industries.

When the CAAA were debated in Congress in the late 1980s, regulated industries and their trade associations said compliance costs would top $100 billion a year, more than what was then spent on all environmental compliance. EPA estimates were much lower, about $25 billion a year. Actual spending is now half of EPA's estimate, according to Joan B. Berkowitz, an analyst with Farkas Berkowitz & Co., who says spending is about $13.5 billion a year and only in part a result of the CAAA.

Although cost estimates from regulated industries were considered inflated, they looked pretty good to environmental compliance businesses. However, environmental industry analysts now wonder where those big compliance dollars fled.

Analysts blame EPA for the down market for air pollution controls and single out the Clinton administration's slow release of air regulations and weak enforcement resulting from the reorganization of EPA's enforcement office in 1993. Now that enforcement is back on track, the Agency has been forced to delay issuance of new regulations to avoid stirring up the antiregulatory Congress.

The mood in Congress is sure to hurt Clean Air Act-related businesses, says Jeffrey C. Smith, executive director of the Institute of Clean Air Companies. "A corporate executive would be foolish to buy new air pollution control equipment if the Act may be frozen, rewritten, or a new cost-benefit requirement tacked on," he says.

Berkowitz predicts that regulatory confusion will continue. For instance, compliance dates faced by the majority of industries covered by the CAAA's maximum achievable control technology standards remain "uncertain," she says. Even when the rules are promulgated, industries will have many routes of compliance, Berkowitz believes, such as process changes and emissions trading as well as end-of-pipe controls. "There are hundreds of technologies out there seeking market share, assuming there is a market."

Beyond Compliance

To survive in this uncertain climate, where regulations cannot be relied on to drive the market, the challenge for environmental compliance businesses is to find ways to move beyond compliance. Or in the words of Ferrier, firms must "sell value" and find the "economic validation" for environmental investments. "Environmental companies must find economic drivers, not regulatory ones, for environmental technologies," he says.

"Most [pollution] generators view environmental expenditures as a cost and something that takes away from the bottom line, not as an investment," Ferrier says. "The trick for environmental service and technology providers is to change that mindset in customers to see expenditures as an investment in a shift to long-term efficiency and a phase-out of certain chemicals."

Ferrier points to water treatment as one sector in which economic drivers are likely to take precedent over regulations. As clean water becomes more expensive and discharge fees to sewage treatment facilities increase, he believes industrial polluters will be more likely to turn to firms that sell water treatment or water recycling systems as a way out of dependence on treatment facilities and their fees. As a result, he predicts, this business will grow.

Another example in which economic concerns are beginning to outweigh regulations is in the consulting field. Engineering firms increasingly are providing industry with long-range guidance that couples compliance with production efficiency. One business that has moved into this area and is profiting is TRC Companies, Inc., headquartered in Windsor, CT, with 900 employees and S100 million in revenues. TRC specializes in air pollution control, pollution prevention, and hazardous waste engineering, says Vincent A. Rocco, TRC chair and chief executive officer. Despite the weak national market for air pollution control equipment, Rocco says, TRC increased revenues from air pollution control engineering by 30% last year. It is the fastest growing part of TRC's business.

He describes TRC's work as a process in which company engineers examine a client's production methods and then design manufacturing improvements that lower production costs and cut pollution. In particular, the firm develops a production management plan that combines production efficiency and compliance with the CAAA's operating permit provisions. The permitting provisions, Title V, require companies to determine their air emissions and lay out a comprehensive operating plan to control them.

According to Rocco, the permit provisions provide his clients with an "opportunity to define a broad range of products and select the most cost-effective control options and ultimately to be in the position where a client doesn't have to worry about the regulatory policeman for the five-year life of the permit."

TRC is preparing operating permits for 300 factories and has coupled the requirements to a review of a company's total operations, both production and environmental strategies.

What TRC is selling is basic pollution prevention by examining a company's manufacturing process and seeing how it can be made better. This analysis, Rocco says, requires a manufacturer to closely consider what it is making, what its chemical mass balance is, and what it is going to make over the next five years. Then these components must be woven into a permit.

"Companies have to anticipate process changes and write them into the permit," Rocco says. "They must plan for flexibility, see product reformulation, controls, pollution prevention as options, and include strategic planning as part of environmental controls." He calls the CAAA's air toxics provisions simply "good housekeeping."

"Probably the biggest challenge to industry is to stop and think about where it wants to be in five years in terms of productivity," he says. "That's future shock. Nine out of 10 plants don't even have final as-built plant drawings. If it wasn't for Title V, they might not know how they make a product, how the plant is piped, or where they have leaks."

Ferrier sees a solid future for engineering consulting firms that specialize in pollution prevention and other adjustments to the manufacturing process. But he warns that many of the companies that generate pollution may have already captured much of the "low-hanging fruit"—the first 20-30% of easy emissions reduction. "The next 20% will be much tougher and more expensive," Ferrier says. "We are talking about redesigning, rebuilding a whole factory, but this is the future—a focus on process, on energy use, higher productivity, efficiency, and not only environmental compliance."

Regulatory Rollback?

Tough regulations and strict enforcement have been the primary drivers for the environmental market, but times are changing. In recent years, many companies have significantly reduced toxic emissions in response to voluntary programs promoted by EPA as well as community and stockholder pressure that springs from the annual publication of the Toxics Release Inventory of emissions to the environment. Connors notes that many companies have incorporated an environmental ethic into how they plan production, and they are not solely driven by environmental regulations. "But," he adds, "lots of companies are not working under that model."

Despite regulations and voluntary cuts in emissions, TRI figures show U.S. companies still put 2.8 billion lbs. of pollutants into the environment each year, which signals plenty of business for environmental industries. However, as Ferrier points out, each incremental reduction will get more difficult and costly. How aggressively manufacturers pursue pollution reduction programs in the new era of deregulation may determine the environmental industry's future.

Looking at the CMA permit provisions, which have been a boon to companies like TRC, manufacturers charge that the requirements are too bureaucratic and place manufacturers in a regulatory straitjacket when they make normal product changes over the five years a permit is in force. EPA has never finalized the regulation, but states have moved ahead with their own rules based on Agency drafts and the law. Industry permit applications are due to the states by the end of this year under the Act.

Several corporations—Occidental Petroleum, Procter & Gamble Co., Intel Corp., and others—recently formed the "air implementation reform coalition" and announced their intention to change CAAA Title V. They said they doubt a federal hammer is needed and argue that most states have adequate permit programs in place without the Acts new requirements. Title V, according to the coalition, creates debilitating delay that could cripple industry's ability to react quickly to marketplace changes."

They have a strong ally in newly elected Rep. David McIntosh (R-IN), who has vowed to block the regulation, if not change the law. McIntosh first gained a reputation as a deregulator when he headed the Bush-era Council on Competitiveness and opposed the permitting provisions of the CAAA. Now he heads the House Government Reform and Oversight Subcommittee on National Economic Growth, Natural Resources, and Regulatory Affairs.

Resistance to regulations runs deep, and a bumpy road may lie ahead in the form of legislative proposals before the new Congress, according to Alan Miller, director of the University of Maryland's Center for Change and co-author of *Green Gold*, a study of international competition for environmental technologies.

In his role as researcher, Miller interviewed many officials in regulated industries, and he says, "If you turn off the tape recorder, you will find deep pent-up anger from a lot of frustrated people—even from some of the most environmentally progressive companies in America. They see themselves as victims of regulatory rigidity, and if offered a chance to get rid of all those forms and requirements in one fell swoop, they would jump at it. This is not necessarily a clear strategy, but more of a feeding frenzy. I don't think companies have thought through that they may be giving a competitive advantage to their less progressive competitors, or that they might wind up with 50 different sets of regulations."

Even officials at companies that have been held up as models of environmental compliance say they have trouble with Title V. One of them is Merck & Co., Inc., the large pharmaceutical company, which, its officers say, has voluntarily phased out air emissions of carcinogens and has set a goal of a 90% reduction of emissions of chemicals listed on the Toxics Release Inventory by year's end.

Merck Vice-President for Environmental and Safety Policy Dorothy P. Bowers says putting together a Title V plan is a "very, very major operation." She would like to see the title changed, even if it takes rewriting the Act. Meanwhile Bowers says Merck is preparing five Title V applications for its companies operating in states where the plans are required.

Generally, Merck supports regulatory standards, Bowers says, emphasizing how the company has examined its manufacturing activities to see whether process changes, reduction in toxic chemical use, or pollution controls can help cut pollution. The problem, she says, "is the bureaucracy and process of getting us to meet standards."

Although she says "very few responsible companies support a rollback of standards," she adds that it might be worthwhile to revisit "at least some Clean Air Act regulations. The United States seems to lead the world in environmental bureaucracy."

Help in the International Market

A way to ease the impact of today's backlash against environmental regulations in the United States may be to turn to the international market. And U.S. companies increasingly are looking to foreign shores to fill marketing voids at home. To exploit this $300 billion market, U.S. companies have begun to get a helping hand from the federal government and even states.

Trudy Coxe is not your average environmental technology booster, but last year the Massachusetts secretary of environmental affairs found herself in India trying to help sell the Bombay port authority on a satellite-based ship-tracking system to avoid environmental disasters from ship collisions and oil spills.

"There we were, saying 'Rah, Rah, Raytheon. There's no company that can do the job better than Raytheon,'" Coxe recalls, referring to the Massachusetts-based high-tech company. Having the state's top environmental official on a trade mission with the governor and some 30 business representatives may seem unusual, but for Coxe it has become almost commonplace. She has traveled with the state delegation to Brazil, Argentina, Chile, and Sardinia.

Governor William Weld, she says, has singled out sectors of the Massachusetts economy that have the potential to merit international promotion: finance, health care, biotechnology, computers, and environmental services. The promotions seem to be working. Following a visit by Coxe and other state and business people, she notes, Raytheon sold the Brazilian government a satellite system to monitor the rain forest. The system is similar to that proposed in Bombay and to one the company developed following the *Exxon Valdez* oil spill.

Over the past few years, the Clinton administration Department of Commerce has begun to help U.S. environmental technologies get a foothold in the international market, which is expected to double in size by the turn of the century (*ES&T*, January 1995). But Coxe says Massachusetts is the first state to actively promote the industry.

Market analysts predict great growth for the international sector. Connors says many nations are only now developing environmental policies and they lack an infrastructure and trained government and industry staff to comply with pollution reductions. "Many companies there are like we were 20 years ago," he says.

Ferrier also predicts more emphasis on the international marketplace for U.S. environmental technologies, in part because of a slowdown in the U.S. market. He notes, for instance, that businesses that sell environmental instruments and testing equipment are doing half their sales outside the United States, as are providers of solar and wind energy systems. U.S. water treatment systems companies are doing one-third of their business outside U.S. borders and the percentage is expected to grow, he adds.

Coxe believes support for environmental technologies will benefit the overall Massachusetts economy. She wants to roll environmental technologies firmly into the state's economy and encourage development of an "atmosphere" in which environmental technology businesses can thrive with universities and research facilities and in which jobs are the result. As an example, Coxe points to the state's decision to join California's clean vehicle demonstration program, which will lead to production of a small number of electric cars. That decision gave a boost to Massachusetts' Solectria Corp., the developer of an electric powertrain.

"Solectria may give us a chance to bring back the thousands of auto industry jobs lost when General Motors shut down a big plant in the state a decade ago," Coxe notes. "And this car would be green."

But a successful environmental business is based on a firm regulatory floor, Coxe acknowledges. She worries that a congressional attack on environmental regulations will hurt states like Massachusetts that have tried to move ahead in the environmental technology business arena.

"Rules set a standard for people to strive for and set a standard for new technologies," Coxe says. Because many of Massachusetts' environmental laws are stronger than federal ones, Coxe says, Washington events may have less impact on Massachusetts than on states where standards are not as high.

"In the longer term, though, I don't want to see the states in economic warfare with one another, and I think that is where we might end up if federal laws get weakened. The justification may be to give states and businesses more flexibility, but an overall dismantling would pit one state against another to lower standards. If that happens, I know there will be a constituency of people who will be very vocal and persuasive in seeking weaker regulations that could be very detrimental to us in the long term."

The Next Century

What will the environmental technology market look like in the next century? Will it grow? Will it disappear and be absorbed into the nation's overall capital spending accounts? If analysts are right and pollution prevention proves to be the industry's future, will that mean more revenues?

Analysts describe a tug of war over waste, with one sector of the compliance industry wanting to handle it and another sector wanting to eliminate it. In the long run, some analysts predict pollution prevention will kill off the environmental technology industry.

"In general, producing less waste is bad for the environmental technology industry," says Jonathan Naimon, corporate program manager of Investors Responsibility Research Center Inc., a Washington, DC, firm that assesses companies' corporate environmental performance. "Process modification, better valves, and so forth lead to less waste to handle or treat, with the result that pollution prevention will harm a lot of things that [Vice President] Al Gore thinks are going to get much bigger. Ultimately, these technologies, even advanced ones, will run out of work."

Naimon says companies he tracks are producing fewer toxics per unit of revenue than they ever have before. "Pollution controls are not dying but they are slowing dawn, and the seeds of pollution prevention signal the demise of the environmental technology industry as we know it. In the future, we will have better seals and flanges and process engineers to cut waste and raise productivity."

His views are borne out by an annual pollution abatement spending survey of 17,000 companies conducted by the Commerce Department Bureau of Economic Analysis. The figures show a shift in corporate spending over the past 10 years from end-of-pipe pollution controls to what the bureau calls "integrated technologies" that combine environmental with manufacturing process changes. In 1983, integrated technology spending made up about 15% of compliance spending and controls made up the rest. The most recent figures, from 1992, show that integrated technologies reached 36%. Commerce Department officials also note that at some point the environmental compliance spending figures will have less meaning because they will be thoroughly blended into other technological production spending.

"In the long run, in the macro sense, effective

pollution prevention will eliminate the environmental technology market in the United States," Ferrier says. Today, about 90% of these revenues are focused on waste, pollution, and cleanup, he says, but if pollution is prevented and Superfund sites eliminated, there will be a peak to the curve.

"The services and equipment sectors are growing in the short term, but by the turn of the century we will see their revenues start to decline, and in the future they will be engulfed by general manufacturing, driven by production and material resource efficiency systems." ■

Questions

1. What is the biggest challenge to industry?

2. What have the primary drivers for the environmental market?

3. Generally, what is bad for the environmental technology industry?

Answers are at the back of the book.

44

Environmentalism and big business working together may seem paradoxical. Profit margins often encourage companies to contaminate the land, sea, and air. However, opponents of the green movement have found that support of the environment is sound business—either by creating new markets or by protecting existing markets against competitors. Government regulation is behind the growth of the clean-up industry. America, Japan, and Germany have the largest share of the world environmental market, and all have strong environmental laws. In Germany, 2,500 companies earn more than half their revenues from green technology. Insurance firms and even large oil firms are rethinking their position. Even though oil demand would be hurt, the clean-up industry would help natural-gas business(es). DuPont, the world's largest producer of CFCs, backed a ban on their use and now leads the market for CFC substitutes created by the ban. These are a few examples of how the green movement and big business have come together, not only to make money but to help save the environment.

How to Make Lots of Money, and Save the Planet Too

The Economist

In principle, you might expect "greens" and businessfolk to be at one another's throats. A blind pursuit of profit, say environmentalists, encourages companies to foul up the land, sea and air. Likewise, few things annoy the average capitalist more than rampant tree-huggers and their ludicrous owl-protecting, business-destroying rules. Across America, businessmen are cheering the efforts of Republicans in Congress to make a bonfire of green regulations.

Or so it seems. Yet a strange love affair is growing between some firms and some parts of the green movement. In places such as Washington and Brussels a fast-growing army of business lobbyists is working for tougher laws. Many firms have discovered that green laws can be good for profits—either by creating new markets or by protecting old ones against competitors.

Whenever a green law forces a company to change its machinery, clean up some manufacturing process, decontaminate a site or even just "consider" the environmental impact of something it is doing, it adds to the clean-up industry. Defining this industry is difficult (does it, for example, include clean fuels, such as solar power, as well as technologies which reduce emissions from dirty fuels?), but one report from the Organisation for Economic Co-operation and Development put its value at $200 billion in 1990.

The OECD thinks it might grow to $300 billion by the end of the decade, and some experts are even more bullish, seeing a rising demand for clean-up services from fast-growing countries such as China, Taiwan and South Korea, and from the former Soviet Union as it undoes the pollution inflicted by communism.

The driving force behind this industry's growth is government regulation. In America, its godfather was California's Jerry Brown, who as governor pushed through clean-air rules that led indirectly to Los Angeles's "Smog Valley," where many clean-up firms started. America, Japan and Germany—the three countries with the largest share of the world environmental market—all have particularly stringent environmental laws. "It is an industry uniquely dependent on government policy," says Adrian Wilkes, director of the Environmental Industries

Commission (EIC), a new British lobby group launched last month. The EIC argues for tougher environmental standards, more rigorous enforcement, and investment subsidies. Its impressive list of supporters includes 25 green campaigners and parliamentarians. But its money comes from the clean-up firms. British firms that manufacture pollution-control equipment have been complaining that the National Rivers Authority makes it too easy to discharge pollutants into rivers, and that air-quality standards are too weak.

In America the Environmental Technology Council, which represents firms dealing with hazardous and industrial waste, has pushed for tougher regulation since 1982. In 1992, together with several big green groups such as the Sierra Club, it sued the Environmental Protection Agency (EPA) for allowing firms to dilute waste rather than treat it. It won the case, thus boosting business for its members. Last year, in another argument with the EPA, it again joined forces with mainstream green groups and won tougher regulation on the burning of hazardous waste in cement kilns.

In Germany, the 2,500 companies that earn more than half their revenues from green technology are beginning to organise themselves. The Environment Industry Association, founded in January, already has 50 members; by next year, says Helmut Kaiser, a consultant who founded the group, it will have 1,000. Environmental businessmen have been heartened by the recent success of the Green Party in regional elections. The party has long argued that green industry can create jobs. Indeed, Mr. Kaiser's group will argue that new technologies, such as the recycling of industrial waste water, save more money than they cost. And it will also lobby for more regulations—with plenty of advance notice so that polluters can ready their chequebooks.

Mr. Kaiser complains that a government decision to back away from a requirement that electronic equipment be recycled cost the green industry "millions." Even so, German waste-management companies profited hugely from a stringent law in 1991 that forced companies to recycle the packaging in which their goods were sold. Many of the firms are now lobbying East European countries such as Slovakia and Czech Republic to adopt a similar law.

Even on global environmental issues some businesses are beginning to lobby for tougher agreements. The main opponents of international targets to reduce greenhouse-gas emissions are coal producers and oil-producing countries. Yet other businesses are siding with the greens. The Business Council for a Sustainable Energy Future, a group of American clean-energy firms formed in 1992 has been calling for international targets on greenhouse gases, which would boost demand for clean energy. Earlier this year it launched a European offshoot.

Insurance firms, worried by a spate of natural disasters, have begun to campaign on climate change. Even big oil firms are thinking twice about their stance. Tough targets would hurt demand for oil, but could help their natural-gas businesses. Significantly, the Montreal Protocol on curbing the use of ozone-eating CFCs was secured with support of big business. In 1988 Du Pont, the world's largest producer of CFCs, backed a total ban on their use. Du Pont, alongside Britain's ICI, now leads the large market for CFC substitutes created by the ban.

Protect me, I'm Green.

Environmental regulation can also raise barriers to entry in established markets. This is most stark when green rules protect domestic producers from imports. Last year the European Union complained unsuccessfully about American standards on car fuel-efficiency. Ostensibly aimed at conserving energy, these happened to protect American car makers from imports of large, upmarket European cars. Another dispute involves Germany's 1991 packaging ordinance, which forces brewers to use refillable bottles. Apart from its green merits, the rule also protects Germany's small brewers which, unlike foreign competitors, already have local distribution in place.

Green laws can split domestic industries too. American greens are urging the EPA to toughen limits on chlorine emitted by the paper-making industry. Though some big paper companies are opposing tougher standards, others, who have already invested in chlorine-free technologies, are siding with the greens.

Even firms traditionally opposed to environmental regulation are becoming more pragmatic. In recent years in America, for example, alliances of mainstream companies (including oil and chemical firms), environmental companies, and government regulators have sprung up to promote better forms of regulation. In particular, they want laws which allow polluters to choose the most economic way of reducing emissions—rather than specifying a particular green technology or product to be used.

Traditional "polluters" also want to see the same laws enforced on their competitors. In America, points out Daniel Esty, a former senior official at the EPA, the pressure for federal environmental regulations in the 1960s and 1970s came not just from green groups but from firms anxious that differing state rules were putting some of them at a competitive disadvantage. Now the same complaint is made on a global scale: many firms in countries where green rules are stringent say they will lose out unless poorer countries follow suit.

In other words, even greenery's most vigorous opponents now direct a lot of their energy towards trying to influence how laws are written rather than whether they are written at all. For the mainstream green movement, this is splendid; environmentalists now have rich allies in smart suits. Whether the emergence of the green business lobby is good news for environmental policy-making, however, is another question. Governments should forever be wary of lobbyists, even those in suits. ■

Questions

1. How can green laws be good for profits?

2. What is the driving force behind the clean-up industry's growth?

3. Who are the main opponents of international targets to reduce greenhouse-gas emissions?

Answers are at the back of the book.

45

An international effort must be made to stabilize human interactions with the Earth. A global partnership will be needed to emphasize several solutions. These solutions include a newrelationship between the industrial North and the developing South, a division of responsibility between different levels of government, and an active participation of citizens. So far, the world community has responded to the environmental challenges through more than 170 ecological treaties. This affects roughly one-quarter of the Earth's land area. If these treaties are obeyed, humans may prevent global ecological collapse and social disintegration. The success of the Montreal Protocol is an encouraging example that we may succeed.

Forging a New Global Partnership to Save the Earth

Hilary F. French

An international effort must be made to stabilize the planet before environmental deterioration reaches a point that it becomes irreversible.

In June 1992, more than 100 heads of state and 20,000 non-governmental representatives gathered in Rio de Janeiro for the United Nations Conference on Environment and Development (UNCED). It resulted in the adoption of Agenda 21, an ambitious 500-page blueprint for sustainable development. In addition, Rio produced treaties on climate and biological diversity, both of which could lead to domestic policy changes in all nations. Significantly, the conference pointed to the need for a global partnership if sustainable development was to be achieved.

Since Rio, a steady stream of international meetings have been held on the many issues that were on its agenda. For instance, the September 1994, International Conference on Population and Development in Cairo put the spotlight of world attention on the inexorable pace of population growth and the need to respond to it through broad-based efforts to expand access to family planning, improve women's health and literacy, and ensure child survival.

The pace of real change has not kept up with the increasingly loaded schedule of international gatherings, though. The initial burst of international momentum generated by UNCED is flagging, and the global partnership it called for is foundering due to a failure of political will. While a small, committed group of individuals in international organizations, national and local governments, and citizens' groups continues trying to keep the flame of Rio alive, business as usual largely is the order of the day in the factories, farms, villages, and cities that form the backbone of the world economy.

As a result, the relentless pace of global ecological decline shows no signs of letting up. Carbon dioxide concentrations are mounting in the atmosphere, species loss continues to accelerate, fisheries are collapsing, land degradation frustrates efforts to feed hungry people, and the Earth's forest cover keeps shrinking. Many of the development and economic issues that underpin environmental destruction are worsening. Income inequality is rising, Third World debt is mounting, human numbers continue growing at daunting rates, and the amount of poor people in the world is increasing.

The global partnership that is needed to reverse these trends will have several distinct features. It will involve a new form of relationship between the industrialized North and the developing South. Another feature will be a division of responsibility among different levels of governance worldwide. Problems are solved best at the most decentralized level of governance that is consistent with efficient performance of the task. As they transcend boundaries, decision-making can be passed upward as necessary—from the community to the state, national, regional, and, in some rare instances, global level. A third requirement is the active participation of citizens in village, municipal, and national political life, as well as at the United Nations.

Above all, the new partnership calls for an unprecedented degree of international cooperation and coordination. The complex web of ecological, economic, communication, and other connections binding the world together means that no government can build a secure future for its citizens by acting alone.

Protecting the Global Environment

One of the primary ways the world community has responded to the environmental challenge is through the negotiation of treaties and other types of international accords. Nations have agreed on more than 170 ecological treaties—more than two-thirds of them since the 1972 UN Conference on the Human Environment. In 1994, the climate and biological diversity conventions as well as the long-languishing Law of the Sea treaty received enough ratifications to enter into force. In addition, governments signed a new accord on desertification and land degradation.

These agreements have led to some measurable gains. Air pollution in Europe has been reduced dramatically as a result of the 1979 treaty on transboundary air pollution. Global chlorofluorocarbon (CFC) emissions have dropped 60% from their peak in 1988 following the 1987 treaty on ozone depletion and its subsequent amendments. The killing of elephants has plummeted in Africa because of the 1990 ban on commercial trade in ivory under the Convention on International Trade in Endangered Species of Wild Flora and Fauna. Mining exploration and development have been forbidden in Antarctica for 50 years under a 1991 accord.

The hallmark of international environment governance to date is the Montreal Protocol on the Depletion of the Ozone Layer. First agreed to in September, 1987, and strengthened significantly twice since then, it stipulates that the production of CFCs in industrial countries must be phased out altogether by 1996. It also restricts the use of several other ozone-depleting chemicals, including haloes, carbon tetrachlorides, methyl chloroform, and hydro-chlorofluorocarbons. Developing countries have a 10-year grace period in which to meet the terms of the original protocol and its amendments.

While this is a momentous international achievement, the world will have paid a heavy price for earlier inaction. Dangerous levels of ultraviolet radiation will be reaching the Earth for decades to come, stunting agricultural productivity and damaging ecological and human health.

The lessons learned in the ozone treaty are being put to a severe test as the international community begins to confront a more daunting atmospheric challenge—the need to head off climate change. Less than two years after it was signed in Rio, the Framework Convention on Climate Change became international law in March 1994, when the 50th country (Portugal) ratified it. The speed with which the treaty was ratified was in part a reflection of the fact that it contains few real commitments.

The pact's deliberately ambiguous language urges, but does not require, industrial nations to stabilize emissions of carbon—the primary contributor to global warming—at 1990 levels by the year 2000. Developing nations face no numerical goals whatsoever, though all signatories must conduct inventories of their emissions, submit detailed reports of actions taken to implement the convention, and take climate change into account in all their social, economic, and environmental policies. No specific policy measures are required, however.

As of late 1994, most industrial countries had established national greenhouse gas targets and climate plans, but they vary widely in effectiveness. Among the most ambitious and comprehensive are those of Denmark, the Netherlands, and Switzerland, none of which have powerful oil or coal industries to contend with. Through the use of efficiency standards, renewable energy programs, and limited carbon taxes, these plans are likely to limit emissions significantly in those nations.

According to independent evaluations by various nongovernmental organizations (NGOs), most of the climate plans issued so far will fall short of stabilizing national emissions and the other goals they have set for themselves. For example, Germany and the U.S., two of the largest emitters, have issued climate plans that fail to tackle politically difficult policies—the reduction of coal subsidies in Germany and the increase of gasoline taxes in the U.S. Neither country is likely to meet its stated goals. Reports from Japan suggest that it, too, is unlikely to achieve its stabilization target. In another failure of will, long-standing efforts by the European Union to impose a hybrid carbon/energy tax have failed so far, despite strong support from the European Community.

Even if the goal of holding emissions to 1990 levels in 2000 is met, this falls far short of stabilizing atmospheric concentrations of greenhouse gases, which will require bringing carbon emissions 60-80% below the current levels. As a result, several European countries and the U.S. have voiced cautious support for strengthening the treaty to promote stronger actions, though they have not said exactly how.

As with protecting the atmosphere, preserving biological diversity is something all nations have a stake in and no one country effectively can do alone. One of the most important achievements of the 1993 Convention on Biological Diversity was its recognition that biological resources are the sovereign property of nation-states. When countries can profit from something, they have an incentive to preserve it.

Genetic diversity is worth a lot. The protection that genetic variability affords crops from pests, diseases, and climatic and soil variations is worth $1,000,000,000 to U.S. agriculture. Over all, the economic benefits from wild species to pharmaceuticals, agriculture, forestry, fisheries, and the chemical industry adds up to more than $87,000,000,000 annually—over four percent of the U.S. gross domestic product. Though international pharmaceutical companies have been extracting genes from countries without paying for years, the convention says that gene-rich nations have a right to charge for access to this valuable resource and encourages them to pass legislation to set the terms.

One widely publicized model of this is a 1991 agreement between Merck, the world's largest pharmaceutical company, and Costa Rica's National Institute of Biodiversity (INBIO). Merck agreed to pay the institute $1,000,000 for conservation programs in exchange for access to the country's plants, microbes, and insects. If a discovery makes its way into a commercial product, Merck has agreed to give INBIO a share of the royalties. Discussing how to replicate such agreements likely will be a high priority for countries that have signed the convention.

Besides providing a forum for future negotiations, the convention calls for a number of actions by governments to preserve biological wealth. Possible steps in the future include discussions of a protocol on biotechnology, as well as deliberations on international standards for biodiversity prospecting agreements.

The oceans are another natural resource whose protection requires international collaboration. Not only did the Law of the Sea receive sufficient ratifications to enter into force in 1994, agreement also was reached on modifications to the original agreement that are expected to mean that the U.S. and other industrial countries will join in. The rebirth of this treaty comes just in time for the world's oceans and estuaries, which are suffering from overfishing, oil spills, land-based sources of pollution, and other ills.

The Law of the Sea contains an extensive array of environmental provisions. For instance, though countries are granted sovereignty over waters within 200 miles of their shores (called Exclusive Economic Zones, or EEZs), they also accept an obligation to protect ecological health there. The treaty contains pathbreaking compulsory dispute resolution provisions, under which nations are bound to accept the verdict of an international tribunal.

Just as the Law of the Sea is coming into force, however, its rules are being overtaken by events in one important area—overfishing. In particular, the original treaty failed to resolve the issue of fish stocks that straddle the boundaries of EEZs and species that migrate long distances. The UN has convened a series of meetings to discuss possible international action to deal with a situation that has seen seafood catch per person fall eight percent since 1989.

Curbing Land Degradation

The latest addition to the international repertoire of environmental treaties is a convention intended to curb land degradation, adopted in June 1994. According to the UN Environment Program, the livelihoods of at least 900,000,000 people in about 100 countries are threatened by desertification, which affects about one-quarter of the Earth's land area. The degradation—caused by overgrazing, overcropping, poor irrigation practices, and deforestation, and often exacerbated by climatic variations—poses a serious threat to efforts to raise agricultural productivity worldwide.

The desertification treaty supplies a framework for local projects, encourages national action programs, promotes regional and international cooperation on the transfer of needed technologies, and provides for information exchange and research and training.

Protecting the environment and combating poverty are recognized to be interlinked priorities. The Cairo conference looked at the complex interconnections among population growth, deteriorating social conditions, sexual inequity, environmental degradation, and a range of other issues. A sustainable future can not be secured without an aggressive effort to fight poverty and meet basic social needs.

Trends during the last several decades suggest a mixed record on improving human welfare. Even though impressive progress has been made in boosting immunization rates, reducing infant mortality, and increasing life expectancy, one in three children remains malnourished, more than 1,000,000,000 people lack safe water to drink, and about 1,000,000,000 adults cannot read or write. The share of the world's population living in poverty has declined steadily, but the actual numbers continue to rise to more than 1,000,000,000 individuals. Rather than shrinking, the gap between the rich and the poor is growing. In 1960, the richest 20% of the world earned 30 times as much income as the poorest 20%; by 1991, the difference had risen to 61 times as much.

A crucial first step toward turning these statistics around was taken in Cairo, when more than 150 countries approved a World Population Plan of Action aimed at keeping human numbers somewhere below 9,800,000,000 in 2050. It covers a broad range of issues, including the empowerment of women, the role of the family, reproductive rights and health, and migration. The plan calls for expenditures on population programs to more than triple by 2000—from $5,000,000,000 to $17,000,000,000. Of the total, $10,000,000,000 is intended for family planning programs; $5,000,000,000 for reproductive health; $1,300,000,000 for prevention of sexually transmitted diseases; and $500,000,000 for research and data collection. The action plan also calls for accelerating existing UN initiatives aimed at expanding women's literacy and improving their health—though it fails to provide spending targets for doing so.

Vatican opposition to proposed language on abortion rights captured headlines during the conference, but the real news was the consensus forged between the industrial and developing worlds and among representatives of population, women's, and human rights groups during two years of preparation for the meeting. Key elements of this include a recognition that slowing population growth and making progress on a range of social fronts are inextricably linked challenges. It follows from the new consensus that reaching population stabilization goals will require a far different approach than in the past and that family planning programs alone will be insufficient to do so. Equally important are investments in changing the conditions that generate demand for large families—such as illiteracy and a low status of women. In addition, there was widespread agreement that family planning efforts must be noncoercive and integrated broadly with reproductive health programs.

At the Cairo conference, 10 diverse developing nations representing Muslim, Buddhist, and Christian religious traditions joined together to share their experiences with others. Each has achieved considerable success in recent years in bringing fertility rates down. In Indonesia, for instance, the birth rate dropped from 5.6 births per woman in 1971 to three in 1991. In Colombia, it declined from 7.1 to 2.9 over 30 years.

As for poverty, unemployment, and social integration, efforts to combat these problems have decreased in recent years, as recession-ridden nations have found it harder and harder to appropriate funds. Few countries have reached the international target of devoting .07% of their gross national product to development assistance, and the amounts that are spent often are not targeted well. Because donor nations have tended to skew their disbursements toward their own security interests, the 10 countries that are home to two-thirds of the world's poorest people get just 32% of total aid expenditures. The richest 40% of the developing world receives twice as much aid per person as the poorest 40%.

Under the proposed 20:20 Compact on Human Development, developing countries would agree to devote 20% of their domestic resources to human priorities and donors would target 20% of their aid funds for such purposes. If this initiative succeeds, it will be making a major contribution to a more sustainable world.

Additional priorities include progress toward alleviating debt burdens and addressing unfavorable terms of trade for developing countries. Though the financial crisis has been eased for some of the largest debtors, such as Brazil, it remains very much alive in many of the poorest nations. The total external debt of developing countries has grown sevenfold during the past two decades, from $247,000,000,000 in 1970 to more than $1.7 trillion in 1993.

Though the ratio of debt-service payments to foreign-exchange earnings has been declining globally in recent years, it still is on the rise in sub-Saharan Africa, which spends some 25% of export receipts on debt repayments. For many countries, this number is far higher.

To generate the hard currency required to pay back loans, the International Monetary Fund (IMF) and others have urged debtor nations to undertake export-promoting reforms, such as devaluing exchange rates, and fiscal reforms to reduce public-sector deficits. The strategy has been only partially successful. A handful of countries in East Asia and Latin America have boosted exports dramatically, but others with the poorest 20% of humanity have not, accounting for just one percent of world trade.

Trade barriers to developing-country products continue to be a major impediment to boosting exports. Restrictions on textiles and clothing alone are estimated to cost the Third World $50,000,000,000 in lost foreign exchange annually. Though recent negotiations under the General Agreement on Tariffs and Trade (GATT) made modest inroads into the problems, developing countries and the former Eastern bloc are expected to account for a mere 14-32% of the projected global income gains from the revised GATT by 2002. Africa is projected to lose $2,600,000,000 a year as a result of the agreement, as rising world agricultural prices due to the mandated subsidy cuts will boost its food import bill.

Where the push to expand exports has been successful, the benefits often have been unequally distributed. In Latin America, for instance, economic growth has picked up in recent years, but the share of the population living in poverty is projected to hover near 40% through the end of the decade. For some, the strategy is a net loss. Subsistence farmers—frequently women—often are displaced from their land so it can be devoted to growing crops to please the palates of consumers in distant lands. Indigenous peoples are forced from their homelands as forests are felled for foreign exchange revenue.

Grassroots Opposition to Selling Resources

The uprising in the Mexican state of Chiapas in early 1994 was a wake-up call to some of the failures of this development model. In terms of resources, Chiapas is rich, producing 100,000 barrels of oil and 500,000,000,000 cubic meters of natural gas daily; supplying more than half of the country's hydropower with its dams; and accounting for one-third of the nation's coffee production and a sizable share of cattle, timber, honey, corn, and other products. However, the benefit from selling these resources is not flowing to many of the people who live there. According to Mexican grassroots activist Gustavo Esteva, "Rather than demanding the expansion of the economy, either state-led or market-led, the [Zapatista rebels] seek to expel it from their domain. They are pleading for protection of the 'commons' they have carved out for themselves.... The [Zapatistas] have dared to announce for the world that development as a social experiment has failed miserably in Chiapas." World leaders would do well to heed his warning that the existing economic orthodoxy needs some fundamental rethinking.

Achieving sustainable development requires protecting the rights of local people to control their own resources—whether it be forests, fish, or minerals. Yet, nations and individuals also are discovering that, if today's transnational challenges are to be mastered, a wider role for international institutions is inevitable.

To respond to this need, considerable reforms are necessary in the United Nations to prepare it for the world of the future. The UN Charter, for example, was written for a different era. Neither "environment" nor "population" even appear in the document. Moreover, though the need for more effective international institutions is clear, people the world over justifiably are worried by the prospect of control of resources being centralized in institutions that are remote from democratic accountability.

As the 50th anniversary of the UN approaches, many ideas are being floated for changes to prepare the world body for the future. Some proposals concern the need to expand membership on the Security Council to make it more broadly representative of today's world. Others focus on the economic and social side of the organization's operations. The UN Development Program (UNDP), for instance, is advocating a Development Security Council—a body of about 22 members to promote the cause of "sustainable human security" at the highest levels.

While these proposals are being debated, another idea merits consideration—the creation of a full-fledged environmental agency. The UN Environment Program (UNEP) has contributed a great deal considering a limited budget which until recently was smaller than that of some private U.S. environmental groups. UNEP does not enjoy the stature within the UN system of a specialized agency, meaning it has few operational programs of its own. Though charged with coordinating the UN response to environmental issues, it has little ability to influence the programs of other agencies with much larger budgets. The time has come either to upgrade UNEP to specialized agency status or create a new environmental agency.

In considering what the functions of such an organization might be, Dan Esty of Yale University suggests that a Global Environmental Organization (GEO) might develop basic environmental principles analogous to widely recognized trade principles advanced by GATT, such as most-favored-nation status and nondiscrimination. High on such a list would be full-cost pricing, the idea that environmental costs should be internalized in the prices of products, rather than passed on to taxpayers. Other proposals include the precautionary principle—that decisions to take preventative action sometimes cannot await conclusive scientific proof—and a right to public participation. Governments already have endorsed these ideas and others in the Rio Declaration, but have not given an organization the task of seeing that they are respected.

In addition, a GEO could play a critical role by serving as an information clearinghouse—as UNEP's Global Environmental Monitoring System already does on a small scale. It also might serve as the implementing agency for some UNDP-financed projects. A GEO could be a partner in recycling or land reclamation. It also might elaborate some common minimum international environmental production standards.

Finally, the time has come for governments to create some form of dedicated funding mechanism to finance investments for the transition to a sustainable society—including environmental expenditures, social initiatives, and peacekeeping costs. Among the possibilities are a levy on carbon emissions, international air travel, or flows of money across national borders. To discourage destabilizing currency speculation, Nobel-laureate economist James Tobin has suggested that a .5% tax be placed on foreign-exchange transactions. This would raise more than $1.5 trillion annually. Even a smaller levy would raise far more funds than are available today. A tax of .003% of daily currency transactions would raise $8,400,000,000.

Even in the best of circumstances, the slow pace of international diplomacy and the rate at which environmental and social problems are growing worse are difficult to reconcile. The best hope for improving the process of global governance lies with people. Just as national policymaking cannot be considered in isolation from public pressure, global policymaking increasingly must consider an organized and influential international citizenry.

The most familiar role for nongovernmental organizations and grassroots groups is within national borders. Around the world, there is an encouraging growth in such activities. In addition to this critical work, citizens' groups are beginning to make their influence felt in international forums. In Rio, the 20,000 concerned citizens and activists who attended from around the globe outnumbered official representatives by at least two to one. More than 4,000 NGOs participated in the Cairo conference, where they widely were credited with helping to shape the terms of the debate. Some of the organizations at these meetings—such as Friends of the Earth, Greenpeace, the International Planned Parenthood Federation, and the World Wide Fund for Nature—represent global constituencies rather than parochial national interests. Taken together, all this activity adds up the creation of a bona fide global environmental movement.

Working through international coalitions such as the Climate Action Network and the Women's Environment and Development Organization, these groups are a powerful force. Daily newsletters produced by citizens' groups, including *Eco* and the *Earth Negotiations Bulletin*, have become mainstays of the international negotiating process. Widely read by official delegates and NGOs during international meetings, they reveal key failures in negotiations and prevent the obscure language of diplomacy from shielding governments from accountability for their actions.

The participation of the international scientific community also is critical. International panels of scientists convened to study both ozone depletion and climate change played instrumental roles in forging the consensus needed to push these political processes forward. The treaties on these two problems created scientific advisory groups that meet regularly and offer advice on whether the agreements need to be updated in light of new scientific information.

The interests of the business community sometimes can be harnessed to positive effect. The Business Council for Sustainable Development, 50 chief executives from the world's largest corporations, were active in the lead-up to the Earth Summit. Though the council opposed language that would have advocated developing standards to regulate multinational corporations, it argued persuasively in its report, *Changing Course*, that sound environmental policies and business practices go hand in hand. The U.S.-based Business Council for Sustainable Energy—a coalition of energy efficiency, renewable energy, and natural gas companies that favor taking action to avert global warming—has begun to participate in international climate negotiations, counterbalancing the lobbying efforts of oil and coal companies.

Formidable Obstacles

Despite their impressive contributions, citizens' groups working at the global level face formidable obstacles. International law traditionally has functioned as a compact among nations, with no provisions for public participation comparable to those that are taken for granted at the national level in democracies around the world. There is nothing yet resembling an elected parliament in the United Nations or any of its agencies. Though the UN has begun to experiment with occasional public hearings on topics of special concern, these continue to be rare events. No formal provisions are made for public review and comment on international treaties or is there a mechanism for bringing citizen suits at the World Court. International negotiations often are closed to public participation, and access to documents of critical interest to the public generally is restricted.

The UN Economic and Social Council is reviewing the rules for the participation of citizens' groups in the UN system at large. Some of those involved in the debate advocate making it easier for groups to be involved, taking the Rio experience as their guide. Others resist this view, worrying about the system being overwhelmed by sheer numbers or about whom the citizens' groups are accountable to. The outcome of these deliberations remains to be seen, but it seems likely that the UNCED process has set a new standard for participation that the UN system will have difficulty backing away from.

When it comes to openness and accountability, GATT has been subject to particularly strong criticism for its secretive procedures. When a national law is challenged as a trade barrier under GATT, the case is heard behind closed doors by a panel of professors and bureaucrats steeped in the intricacies of world trade law, but not in the needs of the planet. Legal briefs and other critical information generally are unavailable to the public, and there is no opportunity for citizens' groups to testify or make submissions. Governments are discussing rules on public participation for the Trade and Environment Committee of GATT's successor, the World Trade Organization. Preliminary reports suggest that the fight for public access will be a long and hard-fought battle.

Despite a checkered history regarding openness, the World Bank has instituted two new policies that others would do well to emulate. Under an information policy, more of its documents will be available publicly and an information center has been established to disseminate them. The second change—the creation of an independent inspection panel—will provide an impartial forum where board members or private citizens can raise complaints about projects that violate the financial organization's policies, rules, and procedures. Though both initiatives were watered down in the negotiating process, they nonetheless represent sizable chinks in the World Bank's armor. It will be up to the concerned public to test the limits of these new policies and to press for them to be strengthened—and replicated elsewhere.

Besides access to information, the public must become a fuller partner in the development process itself. All too often, "development" has served the purposes of a country's elite, but not its poorest members. A growing body of evidence suggests that, for a project to succeed, the planning process must include the people it is supposed to benefit. In other words, aid should be demand-driven, rather than imposed from above. Several bilateral aid agencies have developed new ways of fostering widespread participation in the development planning process, and the World Bank has come up with a new strategy along these lines. The challenge, as always, will be moving from words to action.

Despite public support for far-reaching changes, the international response to the interlinked threat of ecological collapse and social disintegration remains seriously inadequate. Fifty years ago, with large parts of Europe and Asia in shambles in the wake of World War II, the world community pulled together with an impressive period of institution-building that set the tone for the next half-century. The time has come for a similar burst on innovation to forge the new global partnership that will enable the world to confront the daunting challenges that await it in the next millennium.

If the changes called for in this article are made and the power of public commitment to sustainable development is unleashed, the planet can head off global ecological collapse and the social disintegration that would be sure to accompany it. However, if complacency reigns and international forums generate lots of talks and paper, but little action, the future does not look bright. The choice is ours to make. ■

Questions

1. What does the Montreal Protocol stipulate and restrict?

2. What is degradation caused by?

3. Under the 20:20 Compact on Human Development, what would developing countries have to agree on and how would this affect the world?

Answers are at the back of the book.

46

Ecologists and economists usually have different thoughts concerning environmental issues. Economists think that technological advancements can be counted on to solve environmental problems. Ecologists are less inclined to count on technology to cure or circumvent problems. But recently there has been increasing cooperation between ecologists and economists. Members of these two professions maintain that although they may have different backgrounds, there is no proof that they hold significantly different values. Both groups have failed to sufficiently consider the input of the other. Better communication between both is needed. There are often conceptual differences between population, resource, and environmental models due to the lack of understanding of each other's fields. Both ecologists and economists can productively work together. Ecologists can guide society in the direction it needs to go in order to avert environmental catastrophe, and economists can devise strategies that would influence that direction.

Ecologists and Economists Can Find Common Ground

Carl Folke

In recent years there has been an encouraging trend of increased interdisciplinary collaboration between ecologists and economists. But, as a general rule, the writings on environmental matters by ecologists and by economists are different in tenor and message. The economists on the whole appear to be more optimistic when regarding the condition of the human environment; in particular, the economists tend to believe that technological innovations can be relied upon to solve environmental problems, while ecologists are less inclined to trust technology to cure or bypass problems.

Do the different attitudes of ecologists and economists reflect different value systems, that is, do members of these professions hold different world views? In short, are ecologists nature lovers, while economists are materialists? And if so, is there self-selection of these types into the two professions? In a discussion in Askö, Sweden, prominent members of the two professions[1] concluded that although members of the two groups may on average come from different backgrounds, there is no evidence that they hold substantially different values. Ecologists have been known to enjoy high consumption levels, while economists have been known to love nature walks. More likely, it was felt, the professional differences were generated elsewhere.

Neither discipline has sufficiently considered the inputs of the other discipline, according to the discussion participants. Models in each field are constructed for specific ends. Not all economic models need include environmental variables, any more than there need be an economic element in all ecological models. However, the economists felt that too many economic models ignore the environmental resource base of material production (e.g., deterioration of mangrove ecosystems into shrimp aquaculture) and the consequences of that production for critical environmental systems employed as sinks (e.g., the atmosphere as a sink for carbon dioxide emitted in the process of burning fossil fuels in rich countries).

Those models continue to postulate unlimited growth in population, unlimited growth of the physical economy by means of capital accumulation and substitution, improved organization, and technological progress. The nature of these models, the

Carl Folke, Ecologists and Economists Can Find Common Ground. *BioScience* 45:283–284.

economists suggested, may well adversely affect their profession's perception of the natural world. They noted too that, as a profession, economists tend to overly stress the ability of markets to allocate resources efficiently. For example, in view of the lack of well-defined property rights in most environmental resources and sinks, failure to consider externalities (which exist when prices do not reflect true social costs) is pervasive. These externalities are not given the prominence they require and in textbooks are still regarded as aberrations.

Another long-standing weakness of the economist's modeling of the environment is that, on the whole, it ignores possible threshold effects, a central concern of ecologists. The economists explained that if threshold effects are significant, the price mechanisms on which economists rely cannot perform well.

The ecologists in turn feel that, when searching for solutions to environmental problems, members of their discipline all too often fail to take advantage of the knowledge of economists and other social scientists, in particular regarding the importance of markets for allocating environmental resources. Frequently ecologists do not bring critical scrutiny to bear on environmentalists who may support central command and control measures when a market mechanism may be more efficient, and vice versa. Ecologists often do not appreciate the underlying economic causes or other driving forces of environmental problems or the indirect effects of remedial measures proposed.

Better communication is needed between ecologists and environmentalists, as well as between ecologists and economists. The ecologists at the Askö meeting also felt that on issues such as the relationship between complexity and stability, they have not sufficiently informed environmentalists of the current state of their science. As a result, environmentalists may use as slogans some notions that are outdated. Ecologists have also tried to move quickly from an understanding of small-scale case studies to predictions about large-scale systems. They may therefore appear more certain than is justified about what they know about the behavior of large systems. In view of the limited ecological understanding, the ecologists at the meeting recommended a precautionary approach to treatment of large systems out of a fear of the consequences of their possible collapse.

The ecologists and economists agreed that there are often substantial conceptual differences between their population, resource, and environmental models due to ignorance of each other's fields. It is bad science when economists build models that are oblivious of ecological knowledge (such as the limited substitution possibilities among resources), and bad science can lead to bad policy and faulty management. By the same token, if ecological models ignore the ways in which economic institutions operate, they too can have unsatisfactory consequences.

The ecologists expressed concern that most economists continue to view as an unalloyed good the growth of the gross national product in rich nations with high levels of wasteful consumption. They emphasized that it was necessary for the material economies of poor nations to grow but that this growth should be balanced by decreasing throughput in rich nations—something they claimed could be achieved with an improvement in the quality of life. They argued that increasing the scale of the global human enterprise (that is, the product of population size and per capita consumption) is a recipe for environmental disaster. Therefore, they are alarmed that economic analyses of the global economy often do not capture the critical relationship of the scale of the human economy to the scale of the ecosystems that support it.

The economists agreed that the gross national product is not an ideal measure of human welfare and that it is all too often misinterpreted in the press and by politicians. They noted, however, that economists have in recent years put considerable effort into devising improved measures. The economists shared the ecologists' concern on the importance of global-scale issues, because they agree that the world's natural capital is increasingly becoming scarce. The two groups also agreed on the need for a careful reconsideration of where and how economic growth and shrinkage should be pursued. An overall conclusion was that the ecologists know more than enough to alert a risk-adverse society to directions in which it should move in order to avoid serious environmental catastrophes and that economists' expertise would be critical in designing mechanisms that would encourage that movement.

[1]The group comprised ecologists R. Costanza, Maryland International Institute for Ecological Economics, P. R. Ehrlich, Stanford University; C. Folke, Beijer International Institute of Ecological Economics and Stockholm University; C. S. Holling, University of Florida; A.-M. Jansson and B.-O. Jansson, Stockholm University; and J. Roughgarden, Stanford University; and economists P. Dasgupta, University of Cambridge, G. M. Heal, Columbia University; K.-G Mäler, Beijer International Institute of Ecological Economics and Stockholm School of Economics; C. Perrings, University of York, and D. A. Starrett, Stanford University. The meeting, sponsored by Beijer International Institute of Ecological Economics (which is part of the Royal Swedish Academy of Sciences), was held at the Askö Laboratory of the Stockholm Centre for Marine Research, Askö, Sweden, 5–7 September 1993. ■

Questions

1. How do ecologists and economists differ in their solutions for solving environmental problems?
2. How do economists feel about many of their economic models?
3. Why are there conceptual differences between the ecologists and economists models regarding population, resource, and the environment?

Answers are at the back of the book.

47

Globally, biological diversity continues to be lost at a rapidly increasing rate. Efforts to establish protected areas have consequently grown. Approximately 8,000 protected areas are now in existence worldwide. This accounts for roughly four percent of the Earth's land surface. Even though there has been a large degree of success, the long-term results have been insufficient. There are limitations to the protected-area strategy: insularity, scope, feasance, justice and presumption, and conflict. Protected-area strategy has tended to ignore human needs. Efforts must be taken to address this problem. This will require people to become both the beneficiaries and the custodians of conservation efforts. A key issue affecting conservation of biological diversity is the tensions between national and local interests. These interests include the rights and responsibilities over authority and accountability, and over their costs and benefits. These interests must be addressed politically at the local and national levels.

The Man with the Spear

James Murtaugh

A little more than a year ago, an unusual collection of people converged at a rural Virginia conference center called Airlie. They were a small group—only about 60 individuals—representing a broad range of experiences and contexts. They included tribal leaders, village organizers, regional activists, ministerial officials, field-tested biologists and anthropologists, a few eminent academic thinkers plus representatives from both public and private donor organizations. They came from 20 countries and 6 continents.

With all their differences, they shared two concerns: the conservation of biological diversity and the alleviation of human poverty in rural landscapes. They shared also the dawning belief that these two concerns not only can be, but must be, addressed as a single challenge. Driven by that belief, the 60 people had assembled to ponder its implications.

On the first afternoon of the gathering at Airlie, a distinguished professor from Zimbabwe, Marshall Murphree, told an emblematic story. Murphree's son, an ecologist, had been escorting a delegation of European Community officials on a tour across Omay Communal Land, a tribal reservation that borders Matusadona National Park. The EC delegation was much concerned over this region's wildlife. Driving along a dusty road, while some members of the delegation held forth on the subject of the African elephant and its conservation, they passed an Omay man, walking alone with a spear on his shoulder. Ultimately it will be *that* man, and not us, who decides the fate of the elephants, said Murphree's son. The delegates came alert. "*What* man, *what* man?" They hadn't even seen him.

The common purpose, as Professor Murphree understood, must be to bring the man with the spear back into view.

Conservation Is in Crisis

The conventional approach isn't working.

Throughout most regions of the globe, plant and animal species continue to go extinct, ecosystems continue to be destroyed or degraded, biological diversity continues to be lost, despite elaborate efforts to the contrary.

Why is conservation in crisis? One crucial reason is strategic. The chief strategy of conservationists for more than a century has been exclusionary. The idea has been to establish *protected* areas, encompassing great natural beauty or fecundity and (more recently) high floral and faunal diversity, and then to safeguard these areas by carefully limiting human use. The tangible products of this strategy have been national parks, wildlife refuges, designated "wilderness" areas in the American sense and other types of statutorily defined reserves, tracing back to Yellowstone, the world's first national park, established in 1872.

The methods of this strategy have been proscription and enforcement—laws and penalties, in some cases fences and gates. The efforts have been great. Roughly 8,000 protected areas now exist worldwide, constituting about 4 percent of the planet's land surface. Although the protected-area approach has yielded certain important successes, overall and for the long-term the results have been insufficient. Biological diversity can't be preserved—not enough of it, anyway, and not perpetually—by setting it aside within protected areas.

What exactly are the limitations of the protected-area strategy? First, there's the matter of insularity. Scientists now realize that small parcels of protected habitat, insularized (as most parks are) within a sea of human-modified terrain, tend to lose biological richness over time, by the same processes of extinction that affect oceanic islands. Most islands are biologically impoverished, and when parks become insularized, they suffer impoverishment, too.

Second, there's the matter of scope. The protected area strategy is too costly and too scattershot to embrace a major portion of the world's biologically rich landscapes. Eight thousand protected areas may seem a large number, but 4 percent of Earth's land surface isn't much; to be satisfied with that approach is to despair of the 96 percent that *isn't* statutorily protected.

Third, there's the matter of feasance. Many protected areas exist only on paper because governments lack the means to enforce their borders or oversee their biological riches.

Fourth, there's the matter of justice and presumption. Most wild landscapes have been anciently inhabited by indigenous peoples. In other cases, the land has been more recently occupied by needy immigrants. Who has the right to tell those people—either the indigenes or the immigrants—that they may not kill an animal or cut a plant? Arguably, a larger unit of society does have that right in some circumstances; but not always, not everywhere, and more often than not the moral questions are nightmarish. The protected-area approach, dependent on centralized power and top-down planning, has often robbed rural communities of their traditional user-rights over forests, waters, fisheries and wildlife, without offering appropriate remuneration. It has obliged poor people who are resident in contested landscapes to bear most of the costs of conservation, while larger societal interests reap most of the benefits. The result is that many rural communities now regard conservation as inherently anti-development and anti-people.

Finally, then, there's the matter of sheer conflict. As human population continues growing, pressure on protected areas will grow, too. Hungry people will take what they need. Blockading rural people against the use of their own landscape without offering them viable alternatives will always, to the blockaded, seem perverse and intolerable. And will always, consequently, be futile.

The real trouble with the protected-area strategy is that it tends to omit humanity from the realm of nature and from the enterprise of nature conservation. Humanity can't be omitted. *Homo sapiens* is an ecological reality, an ineluctable part of the larger landscapes outside of protected areas, where most of Earth's biological diversity abides. Realism, not to mention justice, therefore demands that efforts to conserve biological diversity must be efforts to address human needs, too.

But it's a hard truth to implement, and an easy one to ignore. It requires that local people become both the *beneficiaries* and the *custodians* of conservation efforts.

During the first session of the Airlie workshop, Perez Olindo, a senior conservationist from Kenya, suggested the conceptual key to this new approach. "I lived in a community that had the use of certain resources. Not the ownership but the use," he explained. "That community was then invaded by foreign laws, foreign values, that changed the meanings of words. *Use. Collect. Hunt.*" And there was one utterly alien word that had arrived with the invaders, he added: "*Poach.*" The missing word among this colonial lexicon, as Olindo well knew, was *tenure*.

Tenure is a complicated and variable concept, implying arrangements more subtle than mere ownership. Tenure doesn't define relationships between people and resources so much as it defines relationships between people and other people. It specifies who may use, who may inhabit, who may harvest, who may inherit, who may collect, who may hunt, under what circumstances and to what extent; it also specifies, implicitly, who may not. A tenure system, in any given situation, is the traditionally accepted (and, in some cases but not all, legally codified) understanding of user-rights, interests and limits. So tenure is central to the issue of who can and should conserve *what* resources for *whom*.

Land itself is the underpinning resource and the most elemental focus of tenure concerns. But the list continues with water, forests, fisheries, minerals, wildlife and other categories of resource, not the least precious of which are the genetic resources inherent in biological diversity. Who holds the user-rights, and therefore the rights of commercial exploitation, over germplasm and phytochemicals that might derive from tropical forests, or over the knowledge of indigenous people who inhabit those forests? Who holds the user-rights over the wet breath of the Amazon forest, exhaling oxygen and moisture back into the sky above central Brazil? These are intricately consequential questions. Perez Olindo stated why: "Use is a necessary ingredient of the protection of biodiversity. If you outlaw use, it will be a recipe for the most rapid depletion, degradation, extinction." And the systematic understanding that legitimizes and limits use is what's called tenure.

All rural communities have their systems of tenure. Whether ancient or recently evolved, those systems derive from direct experience at using, maintaining and apportioning particular resources. In many cases, especially among indigenous peoples, they entail communal, as distinct from individual, forms of proprietorship. Also, traditional tenure systems are often framed in unique language and complex cultural practices, and, like any aspect of culture, they're constantly evolving. Furthermore, they aren't readily compatible with the concept of ownership as codified in developed countries and as imposed on developing countries during colonial periods. Traditional tenure systems don't generally confer exclusive ownership of resources within neatly mapped boundaries. The only boundary applicable to many traditional tenure systems may be political rather than geographic: the boundary of participation within the system.

But national governments often refuse to recognize such traditional tenure systems. As governments impose their own codified systems of resource ownership, traditional rights and culturally enforced limits are disallowed, and resources are thrown open to exploitation without the constraints of long-term self-interest and place-specific knowledge that characterize traditional systems.

Without secure tenure, rural communities have no standing in the decision-making process that determines use or protection of resources. Equally baneful, those communities have no incentive to manage their resources sustainably. The people who live in the forest, along the banks of the river—the people whose daily lives and traditions are inseparable from the biological systems at issue—need and deserve assurance that conservation efforts will promote their long-term as well as their short-term interests. For that assurance, tenure is the crucial first element. Without secure tenure, the members of rural communities *can only* afford to consider their own short-term interests. They are compelled to exploit resources for maximum immediate gain, regardless of future consequences for themselves, for the resource base or for biological diversity. Cindy Gilday, an aboriginal-rights activist from Canada with her own roots among the Déné people of Canada's Northwest Territories, spoke passionately about one consequence of cross-jurisdictional tenure disputes. Why should the Déné, she asked, maintain their own traditional strictures on salmon-harvesting along the Mackenzie River, conserving that resource for their posterity, if a new pulp mill has already begun poisoning the waters upstream?

Margaret Taylor, then ambassador from Papua New Guinea to the U.S., sounded another cautionary note that echoed through subsequent discussions at Airlie. Taylor is a member of a traditional clan-group in the highlands of Papua New Guinea. She understands intimately the interests of highland villagers. "The basic desire," she said, "is to improve their material wealth." And what does material wealth mean in their terms? It means "a bit of cash to buy trade goods, a school, a health clinic, a road, an airstrip"—in short, access to the larger society and to the larger economy. "We are prepared to give up our forests and our way of life for economic development," she recognized. Her point was this: Traditional tenure systems may unravel under stress from outside market forces and rising local expectations.

At Airlie, the small group that considered the tenure issue returned to the plenum with a number of comments and recommendations. It's essential, they advised, that governments "recognize existing, community-based tenure systems." The nationalize-and-regulate approach often misfires. Governments tend to claim authority whether or not they have the means to exercise it effectively. Failure to exercise authority allows open access to resources. Forests and fisheries are over-exploited. Protected areas suffer ecological erosion. And rural communities, with their traditional tenure systems overridden, "seldom are compensated directly for their loss of access."

Not that rural communities can't make bad choices as easily as national governments, and with widespread consequences. Deforestation in a highland drainage of the Guaraqueçaba River in Brazil's Atlantic Coastal forest could deprive farmers downstream of perennial water supplies, or destroy fisheries in the Bay of Pinheiros with silt.

The growing tension between national and local interests—over rights and responsibilities, over authority and accountability, over costs and benefits—has become a determinant issue affecting conservation of biological diversity. It is as clear in the management of public lands in the American West as it is in the development of resources in the Amazon Basin. The process of balancing those interests is not scientific, although it must be informed by science. It is not legal, although it must be legally framed. It is political, and it must be addressed politically at the local and at the national level. And it is the northern industrial countries that must set the example for the rest of the world in the conduct of their own political affairs.

Just as cordoning off natural areas from human use no longer adequately serves the conservation of biological diversity, isolating the practice of conservation from full engagement in the political process begs the central questions: What is the relationship between people and nature? How has it gone wrong? How can it be made right? How do we bring the man with the spear back into view? ■

This article was adapted from "The View from Airlie: Community Based Conservation in Perspective," a sampling of discussions on community-based conservation sponsored by the Liz Claiborne and Art Ortenberg Foundation.

Questions

1. Why do many rural communities regard conservation as inherently anti-development and anti-people?

2. What is the real trouble with protected-area strategy?

3. What does tenure define and specify?

Answers are at the back of the book.

48

Everyone is guilty of pollution in one form or another. We are a throwaway culture. However, the reuse industry is solid. It is evolving into a new and innovative industry from the traditional fix-it businesses. In this time of extreme waste, reuse has become very fashionable for Americans and the trend is definitely towards the preservation of resources.

The Real Conservatives

Jim Motavalli

Whatever became of fixing things? A noble tradition is dying, but reuse is still alive and well.

Everybody on Earth is guilty of fouling the environment to some extent—to be alive, after all, is to be a polluter—but some of us are much worse offenders than others. And what's true for individuals is true for countries. The African nation of Togo, for instance, has 3,500 cars, one for every 200 people. The U.S. has 137 million cars, 35 percent of the world's total, or one for every *two* residents, including people too young to drive.

We here in the U.S. lead the world in fossil fuel pollution, acid rain generation, production of industrial waste and energy consumption. The U.S. creates 19 percent of the world's garbage—compared to 4.4 percent for Japan, 1.1 percent for Australia, and 2.9 percent for West Germany. The U.S. is among the elite 20 percent of the world's population that takes in 82.7 percent of global income. The fortunate few also consume 10 times as much energy and one-and-a-half times more food than people in the developing world.

There were, at last count, over 250 million Americans, and we've created the biggest throwaway culture the world has ever known, with close to 200 million tons of municipal solid waste generated every year—three-and-a-half pounds per person, per day. An incredible 30 percent of the garbage rapidly filling up our 5,800 landfills is packaging. We're chucking out 10 to 20 billion disposable diapers, two billion razors, 1.7 billion pens and 45 billion pounds of plastic every year. Indeed, our plastic waste disposal problem has gotten so serious that we're now exporting 200 million pounds of it every year—mostly to Asia, where it merely gets landfilled there instead of here.

If there's a bright spot in all this gloom, it's that we still retain a solid—though evolving—"reuse" industry. David Goldbeck, in his new book, *Choose to Reuse* (Ceres Press), describes reuse as "making a worn-out product new again, as in retreading a tire." There's also what he calls secondary reuse, in which "the tire is used for something else, like helping to form an artificial reef, or ground up and used in surfacing roads."

Unfortunately, the traditional reuse business is being put on life support just as environmentalists are recognizing its significance. It's unlikely that Greenpeace is going to start a campaign to save the jobs of cobblers and Maytag repairmen, but environmental groups are beginning to encourage the growth of small and innovative reuse industries. One such, The Tutwiler Quilting Project, was launched in the Mississippi Delta in 1988 as a way for women to make money for themselves and their families. Using largely donated textile scraps, about 40 local quilters earn their livelihood sewing blankets, wall hangings, handbags, potholders, table runners and placemats. They're making money and conserving resources at the same time.

And the reuse message is getting through to people. According to Michael Lewis, a senior research engineer at the Institute for Local Self Reliance in Washington, "A new type of reuse organization is emerging to take the place of traditional fix-it shops. We had reuse in the past and now we're getting back to it." Lewis cites extensive returnable bottle programs, as well as nonprofit groups around the country that acquire unwanted office goods or appliances and pass them on to low-income organizations. The Institute is preparing a book on just this kind of reuse.

Reusers don't necessarily identify themselves that way. Joe of Joe's Fix-It Shop often just does what he does, taking quiet pride in returning a worn or broken widget to service. Their crafts very widely, but they share a commitment to reducing the waste stream and smokestack pollution, and saving on landfill space and raw materials. If these people constitute a movement, it's a threatened one. Many of those *E* interviewed say their work is increasingly embattled by cheap imported goods—which make fixing something broken more expensive than buying it new—and disposable designs that don't even allow some new products to be taken apart, let alone repaired.

Back into Service

It's becoming a throwaway world," says Dan McMillion, who runs Dan's Volvo Service in Tampa, Florida. "I see big billboards advertising contact lenses that you use for just a day or two and then throw away." McMillion's parents raised him better. He likes to take broken-down Volvos and make them ready for service again. "I've probably done ground-up restorations on 20 or 30 cars," he says. "If we restore them, it keeps them out of the ground for a while." The problem, McMillion says, is that the sturdy 122S and 544 Volvos he favors "aren't worth tremendous amounts restored. You end up with more time into it than the car is worth."

In much the same way, Tom Migliaccio of Beacon Electronics in Westport, Connecticut, admits that the ancient Philco and RCA tube radios he fixes amid the nostalgic clutter of his repair shop are a labor of love. "I don't make any money on fixing them," he says. "But I like the idea that they'll be around after I am."

Migliaccio has been fixing radios and TVs in Westport since he got out of the Navy in 1946. "TV was a novelty then. They said it would last a few years and then fade. But look what it's become. People come in here, see the old stuff from the 20s, 30s and 40s and it brings back their childhood," he says. Finding parts for old electronic equipment takes some ingenuity. Migliaccio says his replacement tubes come from Russia and China. He also sells such obsolete items as phonograph needles and rooftop TV antennas.

Migliaccio says today's electronics are too complicated for the storefront fix-it man. "I used to have 18 competitors in the phone book, but now there are very few. The new stuff, it's like today's cars—you need computers to fix them. In our day, we'd take a screwdriver and bailing wire and fix our cars ourselves."

Morris Campbell is another take-charge guy who opened his store—Campbell's Clock Shop in Grand Rapids, Michigan—right after World War II. "I've been fixing clocks for 50 years," he says. As he talks, Campbell is surrounded by the gentle revolution of 30 or 40 glass-domed anniversary clocks, most of them German-made. He estimates he is only one of 30 or 40 people in the country who know how to fix this popular postwar accessory.

"It's becoming a lost art, it really is," Campbell says. "Most shops just clean 'em up, put in a little oil and hope for the best. I tried to get my son interested in the business, but he said there wasn't enough money in it and got involved with computers instead." Modern digital clocks, says Campbell, "are really junk," and with their electronic circuitry beyond the country clockmaker's art. Campbell, who's 79, is going to retire and let other people worry about the future of clock repair.

"I hate to see repair places go by the wayside, because they serve a valid purpose," says Allen Blakey, a spokesman for the Environmental Industry Association in Washington who shakes his head at Campbell's story. "But sometimes the economics of the times are stronger than our sense of what we ought to be doing."

Joseph Ancona is another victim of "the economics of the times," but he still needs to feed his family. Ancona runs Economy Shoe Repair in Norwalk, Connecticut, a business his father started in 1929. Americans don't get their shoes fixed anymore, he says, they just replace them. "That's my problem," he said. "They make these injection-molded shoes in China or wherever and because of the molding they can't be repaired. Even if they could be fixed, it would cost more than a new pair. Now we sell workboots, make keys, anything to get a day's pay. The business is just not as good as it used to be—in the last two years we've seen a big drop." Ancona cites a stark example. The new (and fixable) Chinese-made workboots he sells are $35. The cost of resoling those same boots? $35.

Ancona may be gloomy about his prospects, but he's still proud of the work he does. "What we do saves on landfill space," he says. "When you rebuild a shoe, only a very little bit of material is thrown away. And the shoe is often better than it was new."

Invisible Repairs

Nancy Molleur, who runs AA Reweaving in Albuquerque, New Mexico, is a practitioner of an even more endangered craft. Clothes reweaving, she says, "is a declining skill." Molleur learned her art from 75-year-old Nina Davis, who started the company in 1954.

"It's a very old craft that was used originally during the industrial age to repair fabrics that had flaws in them," Molleur says. "Now we use it mainly to repair garments that have holes in them, or are fraying. Other people reweave Navajo rugs out here. It's a difficult and somewhat secretive skill, and quite wonderful. People are amazed when they see what I can do." Molleur's clients bring in $2,500 suits that ripped the first time out, or priceless heirlooms. But her problem is similar to other fix-its: New products are more difficult to work on. "With polyester, for instance, the threads are not very tangible," Molleur says. "The clothes tend to fray rather than unravel, and it's hard to reweave them back into a garment." Blends of synthetics and natural fibers are better but, with mass-produced imported clothes getting cheaper and cheaper, reweaving appears to have a cloudy future.

If people aren't fixing damaged clothing anymore, the Asian factories are working overtime to keep up with increased demand—and creating a much bigger waste problem. Linda Shotwell, communications director of the Washington-based National Recycling Coalition, says that source reduction and reuse "are very important, because they reduce the need to have recycling and waste management in the first place."

Furniture reupholstering, meanwhile, remains a healthy profession, as many Americans still recover their old couches and sectionals. Ben Saiz of Ben's Upholstery ("One Call and You're Covered"), also in Albuquerque, has been on the recovery scene since 1958. "I think it's a form of recycling," he says. "I love working with solid pieces of furniture, and it's still cheaper to reupholster than to buy new." Again, though, cheap imports are clogging the market. "There's too much junk out there—from halfway around the world. That's what's polluting our planet." Saiz plans to fight back with his own line of sofas and chairs. "You'll get a quality oak or pine upholstered sofa for $800 to $1,000—that's a good deal, isn't it?" he asks.

Donnis Samples, whose voice has the lilting music of the Appalachian Mountains, is getting ready to close the doors at the AAA Appliance Service Company in Chattanooga, Tennessee. for the last time. She and her husband have been fixing washers, dryers and refrigerators for decades, but now they'd rather go fishing. "A lot of people will just go out and buy a new appliance when it breaks down now," Samples says. "We always tell them their unit can be fixed unless there's a bad compressor or something like that. That kind of work can cost $300 or $400 and you might as well buy new."

Jerry Powell, editor of *North America's Recycling Journal* in Portland, Oregon, says shoppers should not be using price as their only motivator. "Consumers don't do enough forward thinking when buying things," he says. "They think about durability and quality only when buying clothes and cars. People care more about what's cheaper or more efficient." Powell says that federal and state governments must lead by example, as in replacing paper milk cartons in schools with reusable lexan plastic designs.

The American Rental Association certainly thinks it's setting a good example. It blankets its member stores with big Earth posters emblazoned with green emblems and the message, "Rental is Good For All of Us."

At the Archdale Rental Center in Charlotte, North Carolina, Manager Kellie Brown says her customers don't have to buy machines or equipment that they'd really only use once a year. "We're talking about things like drain snakes," she said. "Or very specialized equipment like lawn aerators, which you use only in the spring. People will probably only re-tile their bathrooms once, so do they need to buy ceramic tile cutters?" She also cites big construction equipment like jackhammers that the crews don't have the time or expertise to maintain. "I would say that renting helps save our country's resources," says Brown, echoing the association's message.

Buried Treasure

Two other businesses—pawnbroking and auctioneering—also turn one man's junk into another's treasure. Bob Peltier is the Bob of Bob's Viking Pawnbrokers, in St. Paul, Minnesota. "We recycle everything that comes in," says Peltier. "We refurnish it and it goes back into the system." Hard luck stories often come attached to gold wedding and engagement rings, and these are refined (to remove the brass and copper) and then formed into blocks of "casting grade" gold for reforming into jewelry. "I'd say 80 percent of our customers are working people who get a little short before payday," said Peltier. "We buy their stuff and help get them back on their feet. We try to fix the stuff, but if it's beyond repair, we sell it to auctioneers."

And that's how people like Hank Kessler get their hands on it. Kessler runs the Auction Barn in Cranbury Township, Pennsylvania, near Pittsburgh. "We specialize in estate work," Kessler says. "We take everything, from pots and pans to the washers and dryers. We have many different buyers, from antique dealers to appliance salesmen. Everything's sold at our Friday night auctions. Tonight we're selling the entire contents of a three-bedroom house. People sometimes write into their wills that everything should be auctioned when they die, so the relatives don't fight over who gets what." Kessler said he likes to deal with the older furniture "because it's made better than the newer stuff, which is pressed sawdust. We get young couples come in here, just starting out, with little money, and we can set them up with a solid bedroom set that's going to last at a fraction of the cost of a new one." And, auction fever being what it is, everything is sold down to the last potholder.

Reuse Reborn

But while older forms of reuse are slowly declining, new and innovative forms are springing up. The Nevada-based Ribbon Factory, for instance, reloads used "disposable" computer printer ribbons and laser cartridges and resells them at half price. Kodak remanufactures about 18,000 copiers for the North American market every year, almost double its production of new copiers. The Xerox Corporation is also going ahead with a new line of "remanufactured" copiers. Asked why American corporations build in so much waste, Xerox communications manager Daniel Minchin says, "The trend is clearly toward the preservation of resources. In the specific case of Xerox Corporation, we do fix things. The machines are returned to us in Rochester, New York, and completely rebuilt from the frame up. They have warranties that are identical to new copiers."

Remanufacturing is also available for automotive parts, office furniture, tools, vacuum cleaners and an assortment of other appliances and equipment. Laser cartridge remanufacture is one of the fastest-growing cottage industries in North America.

Waste Reduction

The most exciting inroads in refuse reduction have come from industrial "waste exchanges"—regional and national computerized matchmaking services that link businesses discarding potentially usable material with other businesses that can use it. In 1993 an estimated 15 to 25 percent of the 12 million tons of goods listed were exchanged. Even more important is the fact that, in 1993, the number of waste exchanges doubled.

Even modest commercial programs can have significant effects on the generation of garbage. The Neighborhood Cleaners Association in New York City estimates that if customers at each of the 1,100-member drycleaners used reusable garment bags, more than 6.6 million plastic bags would be taken out of the waste stream each year.

In Los Angeles an arrangement between the major movie studies and Re-Sets Entertainment Commodities brings significant cuts in the city's solid waste, particularly wood. The estimated 250,000 sheets of lauan plywood utilized by the region's entertainment industry each year are now finding a second home among nonprofit theaters and cultural groups.

Reuse is hip, and entrepreneurs all over America are coming up with novel ways of making old things dance to new tunes. With a lot of work, we might reach the happy situation of Western Europe, which is on track to recover 90 percent of its packaging waste in 10 years. While it's sad to see the sun set on traditional fix-it industries, reuse is still alive and well.

Contacts: *Choose to Reuse*, $15.95 from Ceres Press, P.O. Box 87, Department CTR, Woodstock, NY 12498; Institute for Local Self-Reliance, 2425 18th Street NW, Washington, DC 20009/(202)232-4108. ■

Questions

1. What population percent does the U.S. account for globally? And how much do we take in of the global income?

2. What types of businesses are environmental groups encouraging?

3. What are industrial "waste exchanges"?

Answers are at the back of the book.

49

Military installations comprise a huge amount of land in the United States that is potentially available as wildlife habitat. Over 25 million acres of land are managed by the Department of Defense. Much of this land is largely unoccupied for security reasons, so it has become a safe haven for many species of wildlife. In addition, military land has the advantage of restricted public access. Unlike much public land that is used by wildlife, such as national parks and national forests, military land can be readily closed to public use, thereby protecting rare and endangered species. Furthermore, the military is not just protecting habitat; biologists working for the Department of Defense have begun restoring habitat for endangered species. The Navy, for example, has a program to create pine tree habitat for the red-cockaded woodpecker of the southeastern US. It remains to be seen if such programs will continue in the face of political pressures to cut their funding.

New Defenders of Wildlife

Jeffrey P. Cohn

Practicing conservation and protecting habitat are not usually considered parts of the military's job description.

James Bailey stands on the banks of Romney Creek, a broad but short tidal stream that flows through the US Army's Aberdeen Proving Ground and into the Chesapeake Bay north of Baltimore. Here, on a sunny fall day, an eastern deciduous forest sheds its autumn leaves quietly not far from a site where Army engineers test weapons, munitions, and other military equipment.

A short distance upstream, some 30–40 roosting bald eagles eye the water for signs of fish. Downstream, great mounds of dried guano on Pooles Island in the Chesapeake Bay attest to the 600 pairs of great blue herons that nested there over the summer. At a nearby pond, some 2000 ducks and other migrating waterfowl rest on cool fall nights.

"If we weren't here, this land would be all marinas and condominiums," says Bailey, an Army wildlife biologist at Aberdeen. Instead, Aberdeen's 79,000 acres hold a healthy deer herd, beaver ponds, turkey vultures and black vultures, red-tailed and marsh hawks, and the largest bald eagle population on the northern Chesapeake. "We are a de facto wildlife sanctuary," Bailey says.

When one thinks of natural lands with abundant wildlife, protected endangered species and research opportunities, images of national parks, wildlife refuges, and other preserves usually come to mind. One rarely thinks of military bases, like Aberdeen.

Yes, the US Department of Defense (DOD) manages more than 25 million acres of land nationwide. Among government agencies, that amount of area is second only to the Department of the Interior. Within that department, the Bureau of Land Management holds 270 million acres, the US Fish and Wildlife Service (USFWS) has 92 million acres of national wildlife refuges, and the National Park Service has 83 million acres of national parks.

Some DOD facilities encompass more land than most national parks or wildlife refuges. Eglin Air Force Base near Pensacola, Florida, for example, totals 464,000 acres. The Naval Air Weapons Station at China Lake, California, has 1.1 million acres, and the Army's White Sands Missile Range in New Mexico 2.2 million acres. The latter is equal in size to Yellowstone National Park in Wyoming, the largest park in the continental United States.

Jeffrey P. Cohn, New Defenders of Wildlife. *BioScience* 46:11–14.

"We've got more endangered species per acre than any other federal land management agency," says Philip Pierce, an Army natural resources officer in Arlington, Virginia. "Our bases are islands of protected species. They probably hold more species today than when we acquired them."

The reason: Most land on military bases remains largely natural. More than 90% of the 850,000 acres at the Army's Yuma Proving Grounds in southwestern Arizona is undeveloped. At Avon Park, Florida, the Air Force uses only 3000 acres of a 106,000-acre base and bombing range south of Lake Kissimmee. And, at Fort Sill in Lawton, Oklahoma, where the Army artillery trains, 86,000 of the base's 94,000 acres are wild.

Large expanses of land are needed as safety or security zones, says Lieutenant Colonel Tom Lillie, an Air Force program manager for natural and cultural resources at the Pentagon. They are also needed, the Army's Pierce adds, to train troops in environments similar to those US forces might face in future wars.

Limiting Access on Bases

Not only is land preserved on military bases, but public access is often limited. "Military installations are not national parks or zoos," says Junior Kerns, a wildlife biologist at the Yuma Proving Ground. "We can close off areas without going through public hearings."

Military personnel may also be prohibited from entering environmentally sensitive areas on bases. At Camp Pendleton, the Marine Corps' amphibious warfare training center halfway between Los Angeles and San Diego, a sign warns people to keep out of a particular beach "by order of the base commander." One-fifth of all California least terns, an endangered subspecies, nest from May to August at Pendleton along the coastal estuary where the Santa Margarita River flows into the Pacific Ocean. The sign is obeyed because "marines are used to taking orders," explains Slader Buck, natural resources manager at Pendleton.

On the other hand, natural does not mean unused, especially where ground troops operate. "All of our land may be used for training," Buck says of Pendleton's 126,000 acres, the last large reservoir of California chaparral and coastal scrub between Mexico and Los Angeles. "That's our mission. We're not a national park. It's not vacant land to us."

Nor does natural, in most cases, mean pristine. Some bases, like the Army's Fort Benning near Columbus, Georgia, were once heavily logged. Others, like the Rocky Mountain Arsenal near Denver, now a thriving national wildlife refuge, contain leftover chemical wastes and toxic pollutants. Elsewhere, troops conduct training maneuvers, drive tanks and other heavy vehicles over the ground, fire live artillery shells, and drop bombs.

Still, large, mostly natural lands with restricted human access and abundant wildlife in a variety of habitats from coast to coast provide diverse opportunities not only for wildlife conservation, but also research. "Our military bases are natural laboratories," says the Army's Kerns, president of the National Military Fish and Wildlife Association, a group for DOD wildlife biologists.

For years, the only DOD money available for wildlife conservation and research came from selling logging, farming, or grazing rights on military lands. The Sikes Act of 1960 extended that concept by authorizing DOD to collect fees for hunting and fishing on military bases. The act also required DOD to cooperate with USFWS and state wildlife agencies.

In 1991, Congress created the Legacy Resource Management Program. Until recently, the legacy program provided $50 million a year to fund projects aimed at managing wildlife and other natural, anthropological, or cultural resources on military lands. It had been viewed as a "godsend," Kerns says, because it provided money that was specifically targeted to wildlife research. (At press time, the program was facing budget cuts from Congress of 80%.)

Even with funding for research and conservation programs, DOD has sometimes had problems with managing natural resources. In 1992, three civilian Army employees at Fort Benning were charged with conspiracy to violate the Endangered Species Act, a felony. They had designated some of the base's extensive pine forests for clear-cut logging, including habitat for the endangered red-cockaded woodpecker.

Under the terms of a pretrial agreement, two of the employees were fined. One was reassigned to other duties, and the other retired. Charges against the third were dropped. Although the Fort Benning case involved civilians, it "sent a message through the military [that] this is not the way to do business," says Philip Laumeyer, a field supervisor in the USFWS's Brunswick, Georgia office.

Another case involved the commander and another officer at Fort Monroe in Hampton, Virginia. They were officially reprimanded in 1994 for ordering base workers to destroy yellow-crowned night heron nests. The birds were not harmed, but their nests were knocked down because heron droppings defiled lawns, roads, and cars. The birds are not endangered, but they are protected under the Migratory Bird Treaty Act.

Beyond civil penalties, such reprimands can affect an officer's military standing. "It could end a career, especially during times of downsizing," Kerns says. "Officers can't afford even a hint of blemish on their record."

Most military personnel now seem to understand the need to obey wildlife laws. "We hammer home the message that you need to consult with us," says Tom Campbell, a biologist at the Navy's China Lake weapons station. "Some don't like it, but they have bit the bullet. We have the commander's support, so people pay attention."

Navy Changes Practices

The attention shows in conservation and research projects on military bases from coast to coast and sometimes beyond. For starters, the Navy redesigned its tugboats and other small craft. The change came after a female manatee and her calf were killed by a ship's propellers at the Navy's submarine base at Kings Bay, Georgia. The new design adds metal shields to prevent the manatees from being drawn into ship propellers. No manatee is known to have been killed by propellers since the Navy made its changes.

The Navy's conservation efforts at Kings Bay extend into the Atlantic Ocean. Navy biologists are working with the Coast Guard and federal and state wildlife agencies to protect a right whale calving area off the Georgia and Florida coasts. Navy biologists survey the whales and monitor their movements. The information is then fed to local shippers and recreational boaters to help them avoid collisions at sea.

Habitat Conservation Is Underway

Habitat conservation and restoration has become a major focus at other military bases as well. At Barksdale Air Force Base in Shreveport, Louisiana, for example, base land managers built controls and a pump on existing canals to reflood drained wetlands along the Red River. The result: Approximately 2000 acres of recreated wetlands now attract 6000 nesting white ibises, little blue herons, and other wading birds to an area that had few if any before.

Some 2300 miles away, the Navy protects 225 acres of old-growth forest at the Jim Creek Radio Station north of Seattle. The area, a key communications link between naval shore commands and US submarines at sea, holds one of the Pacific Northwest's last stands of virgin Sitka spruce, Douglas fir, cedar, and western hemlock.

Yet other large trees are the focus of a habitat restoration project at Dare County Air Force Bombing Range in North Carolina. Forester Scott Smith has planted Atlantic white cedar trees on a 15-acre test plot in an attempt to reestablish the species. Only 10% of Atlantic white cedar habitat, which once stretched along the coast from New Jersey to Louisiana, remains. Smith hopes eventually to restore 5000 acres, including 2000 acres of white cedar, on formerly clear-cut land at the Dare Country range.

Elsewhere, near Lawton, Oklahoma, the Army has set aside 2000 acres around an artillery firing range at Fort Sill. The protected area contains the largest chunk of ungrazed tall-grass prairie in the West, says James Gallagher, an Army wildlife biologist. The preserve provides habitat for white-tailed deer, scissor-tailed flycatchers, and the largest wintering population of northern harriers in the United States.

Habitat protection is particularly difficult at Fort Sill and at Fort Knox, the Army's tank headquarters, south of Louisville, Kentucky. At both bases, heavy tanks, howitzers, rocket launchers, and other vehicles chew up the ground, dig deep ruts, and add to soil erosion. At Fort Knox, where forests rather than prairies prevail, the heavy equipment and weapons also bowl over trees and bushes.

The Army has launched active land-restoration programs at both sites. Bulldozer crews have restored natural contours on 1000 acres at Fort Knox since 1991, says Albert Freeland, the base's chief environmental manager. Ditches, some eight feet deep, have been filled, and trees and other native vegetation have been planted.

Working with Individual Species

Besides protecting and restoring habitat, DOD biologists also are concentrating on individual species. For instance, the Navy is recreating habitat for the endangered red-cockaded woodpecker at the Naval Weapons Station at Charleston, South Carolina. The small black-and-white birds, named for the males' red head tufts, usually nest in cavities in the trunks of mature pine trees. However, the birds have disappeared from large areas in the southeastern United States because pine forests have been heavily logged.

Navy biologists have affixed 16 nest boxes, modified to keep other birds away, to pine trees. The boxes replaced tree cavities lost when mature pines were destroyed during Hurricane Hugo in 1989. The biologists also have drilled holes in trees to give the woodpeckers a head start in building a nest cavity, which can take two years to complete, and they have planted pine seedlings on 385 acres to promote new growth.

The woodpecker's habitat is also protected at Fort Benning and at a dozen other military bases. One base where red-cockaded woodpeckers appear to be doing well is Fort Bragg near Fayetteville, North Carolina. Home of the Army's 82nd Airborne Division, Fort Bragg trains 40,000 troops a year. The fort also houses approximately 300 woodpecker clusters, a series of pine trees inhabited by groups of birds.

At Fort Bragg, nearly a dozen biologists study and monitor red-cockaded woodpeckers, especially during breeding season, says Army wildlife biologist Alan Schultz. To aid the woodpeckers, the biologists burn undergrowth and remove hardwood trees that make the habitat unsuitable for the pine trees the woodpeckers need for nesting. They also band the birds for population and genetic studies. Fort Bragg biologists keep tracked vehicles out of cluster sites and have closed some firing lanes to protect the birds.

To protect a butterfly species, the Navy halted construction on a pipeline at a fuel depot in San Pedro, California, 25 miles south of downtown Los Angeles. In 1994, workers discovered the Palos Verdes blue butterfly there. Thought extinct in the early 1980s when their last known habitat was bulldozed, naval biologists at San Pedro have found 270 PV blues, as the butterfly is known. Navy biologists are surveying the fuel depot for other butterflies, monitoring the population, and restoring the coastal scrub habitat that the species requires, says Dawn Lawson, a Navy botanist.

Further south, wildlife biologists at Camp Pendleton are helping conserve the least Bell's vireo. Once common in riparian forests from northern California to Baja, the least Bell's vireo was listed as endangered in 1986 when the population dropped to only 300 pairs in California. Now, the bird is making a comeback at Pendleton.

For 15 years, marine biologists and land managers have protected the Santa Margarita River and its riparian woodlands, says Pendleton's Buck. More important here, they have also trapped and removed brown-headed cowbirds, which are recent arrivals in California.

Female cowbirds lay their eggs in vireo and other birds' nests, fooling the latter into raising cowbird chicks often at the expense of their own. The least Bell's vireo numbers at Pendleton have risen from 14 in 1980 to an estimated 350 today, nearly half the species total US population.

The Marine Corps also provides money and habitat for a study of how habitat changes and fragmentation affect mountain lions in rapidly urbanizing areas adjacent to Camp Pendleton in southern California. Paul Beier, assistant professor of wildlife ecology at Northern Arizona University in Flagstaff, spent four years studying Pendleton's pumas.

Approximately 20 mountain lions wander nearly 1300 square miles of woodlands, chaparral, and coastal scrub that characterize Pendleton and the

Santa Ana Mountains to the east. Beier has found that the big cats rapidly adjust to the burning of chaparral and coastal scrub at Pendleton caused by exploding bombs and artillery shells. The mountain lions may temporarily lose the cover they need to ambush deer and other large prey, but they switch readily to raccoons, beaver, opossums, and other small animals.

Beier also learned that corridors connect separate mountain lion habitats around Pendleton. The corridors, one of which is on the base, are especially important for young mountain lions, which use them to disperse from their mother's home range. The young lions can find and use the corridors, he learned, if they contain sufficient woody cover, road underpasses, and only low-density development.

Further inland, in California's Mojave Desert, four military bases—the Navy's China Lake, the Marine Corps' Twentynine Palms, Edwards Air Force Base, and the Army's National Training Center at Fort Irwin—total 2.6 million acres of some of the nation's best desert habitat. They protect the desert tortoise, one of the three land tortoises in the United States. The animal is a threatened species. North and west of the Colorado River, desert tortoise numbers have dropped due to habitat loss, the destruction of its burrows by off-road vehicles, and a fatal upper respiratory disease spread by pet tortoises freed by well-meaning owners.

"We want to protect the tortoise and its habitat," says Roy Madden, a natural resources officer at Twentynine Palms, a 600,000-acre Marine desert training facility. "It's an important species in the desert. Military training and tortoise recovery are not mutually exclusive."

The Marine Corps has set aside 6000 acres for desert tortoises and adopted a management plan for the rest of Twentynine Palms. The designated site contains the highest tortoise density on the base. Off-road vehicles are prohibited, speed limits of 25 miles per hour have been set, and aerial bombing sites have been moved. Madden has also begun a four-year, $40,000 test project aimed at reestablishing native plants and restoring tortoise habitat.

East of Twentynine Palms, the Air Force has designated 60,000 acres at Edwards as critical desert tortoise habitat. Mark Hagen, a natural resources manager, surveys the base to identify areas requiring tortoise management and to locate burrows. He uses the information to remove any of the reptiles living near pads used to test rocket engines. The tortoises are carefully returned to their burrows following the tests.

Similarly, the Navy has established a 200,000-acre protected zone for desert tortoises at China Lake. Not only must vehicles stay on marked roads, biologist Tom Campbell says, but all work within the area is closely monitored and restricted to approved places. When necessary, Campbell also monitors weapons tests to ensure any effects on tortoise habitat are minimal and any leftover explosives or other debris are cleaned up afterwards.

Meanwhile, back at Aberdeen Proving Ground, the bald eagles are the subject of research on the importance of the base for recovering endangered species. On contract with the Army from 1983 to 1992, James Fraser, a professor of wildlife sciences at Virginia Polytechnic Institute in Blacksburg, studied eagles along the northern Chesapeake Bay. During that period, he followed some 300 eagles and radio tagged 154, including 22 at Aberdeen.

Fraser learned the Chesapeake acts as a meeting ground for three separate bald eagle populations. One comprises birds that breed or were hatched along or near the bay. A second group includes eagles that migrate to the bay each fall from northern New England and eastern Canada. The third is the birds that come north each spring, then return south in the fall.

More important, perhaps, Fraser found that bald eagles prefer the undeveloped forests and shorelines still available at Aberdeen and some other sites along the Chesapeake. They avoid places used by people, such as areas around roads, buildings, and marinas. At Aberdeen, the eagles prefer areas downrange from artillery and tank firing zones because, he says, "shooting keeps people away."

Will the Legacy Continue?

Whether such research and conservation efforts on military bases continue depends on the budget-slashing atmosphere in Washington. Republicans in Congress have criticized nonmilitary programs in DOD, and, insiders say, the legacy program remains unpopular with some military leaders because it forces them to spend time and money on nonmilitary activities.

At the same time, some military bases lack enough biologists and resource managers to run research programs. At White Sands Missile Range in New Mexico, the Army currently has only three natural resource managers for the base's 2.2 million acres. "There are so many projects that we have been unable to do," says David Anderson, White Sand's land manager.

Nevertheless, the outlook for research and conservation on military bases remains good. "We still have the habitat," says Buck. "We don't have anything you can't find elsewhere, we just have more of it. We have wild places that don't exist for many species on the outside." ■

Questions:

1. Which government Department manages the most land? Which Department manages the second-most land?

2. Where does the last large reserve of California chaparral and coastal scrub exist between Los Angeles and Mexico?

3. The death of what kind of animal caused the Navy to redesign its tugboats and other small boats? What did the redesign involve.

Answers are at the back of the book.

50

We have always had environmental problems. Originally we were concerned with the scarcity of food, shelter and other basic essentials. By the eighteenth century, western philosophers were becoming concerned with the possibility that world famine and other scarcities could jeopardize our civilization. Three theories came about in the late eighteenth to the early twentieth century. Each one of these theories has something different to contribute to the solution of environmental problems. The Malthusian doctrine of population growth stipulates that if population remains unchecked, growth will surpass the world's food supply. John Stuart Mill's steady-state economy espouses a voluntary, less consumptive existence. The third theory, the neoclassical economic notion, stresses an efficient market place in order to establish production and consumption levels attaining optimal equilibrium. Of the three theories, Mill's steady-state economy may prove to be the most promising solution.

Three Theories from Economics About the Environment

J. E. de Steiguer

For most of its time on this planet, the human race has had to deal with environmental problems in one form or another. The original concern was with the local scarcity of food, shelter, and other essentials provided at least in part by nature (Fisher and Peterson 1976). By the late eighteenth century, western philosophers were beginning to recognize the theoretical possibility of worldwide scarcities and famines that could threaten the existence of civilization (Malthus [1798] 1965).

From the mid-nineteenth century, writers expressed concern that modern industrialization and population growth was degrading the environment (Marsh [1864] 1965, Mill [1848] 1965, Pigou 1932). Environmental degradation did not, however, become a permanent public concern until biologist Rachel Carson in 1962 published *Silent Spring*. Carson's book, which focused on the irresponsible use of pesticides, generated profound public sentiment for improved environmental quality. Thus, by the mid-1960s environmental degradation had joined with the much older concern of resource scarcity to form the two interrelated, central interests of modern environmentalism.

Not surprisingly, numerous theories have evolved in an attempt to explain the causes of and solutions to environmental problems. These theories, while not always being well-developed philosophies (Norton 1991a), nevertheless represent attempts at bringing order to a confusing array of facts and hypotheses about the environment.

Three of the most influential theories contained in today's environmental literature were first formally stated in the work of English economists who lived between the late eighteenth and early twentieth centuries. These theories are: the Malthusian doctrine of population growth and resource scarcity, John Stuart Mill's theory of the steady-state economy, and the neoclassical economic notion of efficient markets as the solution to environmental and resource problems. During the period of intense environmental awareness immediately following the publication of *Silent Spring*, these theories were, in a sense, rediscovered.

In this article, I trace the history of these three schools of economic thought in relation to the environment. I begin with the origins of the theories during the Industrial Revolution (1760–1850), next

J. E. de Steiguer, Three Theories from Economics about the Environment. *BioScience* 45:552–556.

discuss their reemergence during the years immediately following *Silent Spring*, explore their present status, and finally compare their potential value in the resolution of future environmental matters. Each theory has something different to contribute, and together they present a rather comprehensive scheme for solving environmental problems.

The Industrial Revolution and the Rise of Economic Thought

Before the Industrial Revolution, Europeans had lived in a society which, for better or for worse, was characterized by stability (Fusfeld 1982). The eighteenth century, however, brought widespread change in the existing order, first in England and later on the European continent. The introduction of coal as a principal energy source followed by the invention of the steam engine and then gas lighting permitted the widespread development of industrial manufacturing (Lombroso 1931). Improved prospects for employment and wages enticed many to migrate from the country to the cities where most manufacturing facilities were located. The rapid increases in urban populations throughout western Europe were well documented (August 1975).

With this change came some improvements for people: death rates fell, and the general human population experienced the beginnings of a major expansion, which has continued to the present (Miller 1994). However, there was also an associated measure of social and economic chaos. Depressions and financial crises reduced the lower classes to a rock-bottom existence with no relief in sight, and laws passed to protect domestic production in England sent consumer prices skyrocketing (Lombroso 1931).

The magnitude of the social change and population growth was sufficient to raise concerns about the fate of civilization. A liberalized atmosphere of inquiry (Lombroso 1931) soon led scholars to wonder about how humanity in the face of change might best ensure its survival and improvement in the quality of life. A distinct academic discipline called political economy arose to address the questions of resource scarcity, allocation, and societal well-being. The objective of these new economic philosophers was as simple as it was grand: to obtain an improved understanding of the human condition (Heilbroner 1986).

Malthus and the Classical Economists

The first group of economic philosophers to emerge during the Industrial Revolution has been called the classical economists. Their method of study was based on the philosophy of natural law (Heilbroner 1986). They, thus, attempted to discern by reason a naturally endowed set of principles that order human life and community (Finnis 1980). An important member of the classical school was Thomas Robert Malthus. He expressed concern about the potential consequences of population growth. The doctrine of Malthus was set forth in his monumental treatise, *An Essay on the Principle of Population as It Affects the Future Improvement of Society* (Malthus [1798] 1965).

The common interpretation of Malthusian scarcity is: Society has the ability to increase agricultural production only at an arithmetic rate while the number of mouths to be fed increases at a geometric rate. Hence, at some point, population is likely to outstrip food supplies with calamitous results. Discussions of Malthus's ideas often suggest that a complete disappearance of resources eventually causes society to crash. Malthus, however, presented the more sophisticated concept of economic scarcity and its attendant effects on human well-being (Barnett and Morse 1963). Economic scarcity refers to the decreased availability of resources relative to the effort required to obtain them.

The example that Malthus used was agricultural production. To Malthus, land was one ingredient, like labor or tools, used in the process of growing crops. While he argued that the quantity of arable land was fixed and someday might be completely occupied by farms, Malthus also recognized that land could be made more productive through intensive cultivation. With greater effort, farmers could gradually squeeze more produce from that same fixed amount of land but—and herein lies the rub—at decreasing rates per each additional laborer. Thus each new worker sent to the field produces incrementally fewer crops or, as later economists would say, with diminishing marginal returns.

Malthus predicted diminishing marginal returns

as farmers sought ways to feed the ever-increasing masses. The economic scarcity would arise because society would have to sacrifice increasingly more to obtain less on the margin. Whether measured by the number of field hands and their hoes, or by the money required to pay for that labor and equipment, the cost of extracting agricultural produce would increase. And, as populations continued to grow and societies were pressed harder to feed their members, there would come a time when these costs would dominate the entire economy. Per capita economic growth would cease and then plummet as people struggled just to scratch a living from the earth. War, plagues, famine, and other catastrophes would periodically thin human populations, and the cycle of diminishing marginal returns to human effort would then recommence.

Mill and the Stationary State

Following the classicists came the Utopian socialist philosopher John Stuart Mill. During his life, Mill produced an impressive quantity of written work on a variety of topics dealing with ethics, law, morals, and economics (August 1975). His most important contribution with respect to economics and modern environmental thought was contained in *The Principles of Political Economy* (Mill [1848] 1965).

Like Malthus, Mill foresaw that increases in human population and wealth could not continue in perpetuity. At some time, a steady—or, as he said, "stationary"—state would eventually be reached where both population and consumption were stabilized, but perhaps at some low level of human happiness. What was needed in order to improve the human condition, wrote Mill, was a more immediate stabilization of population, reduction in aggregate consumption, and a more equitable worldwide distribution of wealth.

In order to achieve the desired steady-state, Mill advocated a voluntary, less consumptive existence. Thus, Mill did not share Malthus's pessimism regarding the ability of humanity to avoid disaster. Furthermore, the stationary economic state by no means implied to Mill a stationary state of human improvement. There would always be opportunity for elevating mental culture and for moral and social progress. In his philosophy, Mill seemed much a kindred spirit with his US literary contemporary Henry David Thoreau, who had also insisted that our lives needed "simplicity, simplicity, simplicity" (Thoreau 1854). Also, elements of Mill's philosophy anticipated wildlife biologist Aldo Leopold's call a century later for a land ethic where humanity would maintain the integrity of the environment through less resource exploitation (Leopold [1949] 1987).

Mill also introduced a concept that would seem strikingly relevant to twentieth-century economists (Barnett and Morse 1963). It was the notion that personal solitude and natural beauty could be impaired through population growth and industry. So, even though a person may not experience shortages of conventional agricultural produce and minerals, Mill saw that the process of growing crops and mining minerals could result in a paucity of quality human habitat.

The Neoclassical Economists

The end of the nineteenth century in Victorian England saw the beginnings of the neoclassical school of thought. The unofficial leader of this school was the Cambridge economist Alfred Marshall. Where the classicists had relied upon the philosophy of natural law to derive their theories, the neoclassicists employed the engineer's tools of analytical geometry and differential calculus to develop abstract models of market economies (Christensen 1991). Neoclassical economics—with its supply and demand curves, prices and quantities, and market equilibria—would eventually become the economics of the modern university classroom. All told, it was a sharp methodological departure from the classical school.

Neoclassical theory emphasized the well-functioning, or so-called efficient, market as the means of maximizing satisfaction. According to the theory, as producers and consumers meet in the market place, voluntary exchange prices are established to set production and consumption at optimal equilibrium levels. Furthermore, increasing prices spur discovery of new technologies and raw materials, and they encourage efficiency and substitution both in production and consumption. Any departure from the market system, according to the neoclassicists, results in a less than optimal allocation of resources and, likewise, lower human satisfaction.

A two-part, market-based measure of social well-being emerged from the neoclassical studies of societal welfare. The first part, called producer's surplus, was in essence a measure of profits, that is, the difference between the price received in the market for an item and the cost of producing it. The second measure, called consumer's surplus, was the difference between the value of the satisfaction received from the consumption on an item and the price the consumer must pay for it.

Producer's and consumer's surpluses provided a means for neoclassical economists to begin valuing societal welfare, but the conceptual framework was incomplete. It still lacked some means of accounting for pollution and other side effects that, although they are external to the market transaction, nevertheless still affect human welfare. The missing piece was supplied by A. C. Pigou, a former student of Marshall, in his book *The Economics of Welfare* (Pigou 1932). Pigou realized that these market-external changes in human welfare could have both beneficial and detrimental effects. These effects he called "uncompensated services and disservices" (Pigou 1932).

Pigou (1932) provided an example of an uncompensated service:

> It is true in like manner of land devoted to afforestation, since the beneficial effect on climate extends beyond the borders of the estates owned by the persons responsible for the forest. (p. 160)

And of an uncompensated disservice:

> ...smoke from factory chimneys... in large towns inflicts a heavy uncharged loss on the community, in injury to buildings and vegetables, expenses for washing clothes and cleaning rooms, expenses for the provisions of extra artificial light and in many other ways....(pp. 160-161)

Pigou had hit upon the modern notion of economic externalities—those changes in welfare due to unintended side effects, often of an environmental nature, that are not directly captured in the market transaction. Though simple as a concept, externalities provided a powerful way of incorporating environmental damage into economic assessments. Also, externalities, because of their connection to production prices and quantities, suggested remedies, such as taxes, that could be used to reduce environmental damage. Likewise, subsidies could be paid to resource owners to encourage the production of amenity goods, such as forest aesthetics, for which no cash market exists.

The Years Following *Silent Spring*

The ten or so years following *Silent Spring* were important both to the modern environmental movement and to the three economic theories. In response to Carson's book, the academic community began producing scholarly works that revived and adapted the older theories of Malthus, Mill, and the neoclassicists to modern situations.

Malthus. Certainly the most ubiquitous of the three theories in the post-1962 environmental literature was that of Malthus. However, the basic Malthusian theme of increasing resource scarcity was modified to a modern, neo-Malthusian perspective. According to the new view, in addition to the traditional population concerns, world calamity could result from environmental degradation, which also diminished the natural resource base.

Neo-Malthusian concerns were clearly evident in entomologist Paul Ehrlich's (1968) best-selling book *The Population Bomb*. Its message was apocalyptic: There were too many people on the earth and because of a decrease in the doubling time of population growth, the situation was worsening every day. Within the next nine years the world would see acute food shortages where one in every seven persons would die from nutrition-related causes. The dreadful effects of overpopulation would not be limited to starvation. The environment would deteriorate too, as people used more fertilizers and pesticides, cleared forests, increased the siltation of streams, encountered the greenhouse effect, and engaged in nuclear warfare.

Another appearance of the Malthusian theme was found in *The Limits to Growth* (Meadows et al. 1972). The study used a Massachusetts Institute of Technology–based computer model, which employed several scenarios of future world resource stocks and population. Just beyond the year 2000, according to the model, food production and industrial output suddenly would decline, triggering in approximately the year 2030 the beginnings of a 50% reduction in the world's human population. Increasing population and industrial production in grim Malthusian fashion would severely outstrip limited natural resources, thus crashing civilization.

In their attempt to save the human race, the researchers devised and fed into the model other more optimistic alternative future scenarios. One assumed a doubling, through new discoveries, of natural resource stocks. Other scenarios made energy more freely available, improved resource use, increased food yield, and instituted birth control. But the final outcome was sadly always the same. At some point during the twenty-first century, natural resource stocks would drop followed by a dramatic population decline.

A final example of the Malthusian theme can be found in the writings of economist Kenneth Boulding ([1966] 1973). In "The Economics of the Coming Spaceship Earth," Boulding explored energy entropy as a factor limiting the existence of civilization. The second law of thermodynamics states that due to entropy, the amount of energy available for work will always decrease. Without the introduction of new energy sources to our planet, the second law will see to it that existing energy dissipates into heat and thus eventually becomes useless. In theory then, Boulding thought that the second law of thermodynamics placed an absolute limit on the viability of civilization. The concern he raised regarding entropy was, in effect, the ultimate Malthusian warning. The demise of humanity was now said to be governed by the immutable laws of physics.

Mill. Mill's theory about the steady-state economy had been largely discarded by neoclassical economics. In 1973, however, economist Herman Daly published an article entitled "The Steady-State Economy," acknowledging Mill as his intellectual forerunner.

In his work, Daly lamented the problems of what he termed *growth-mania*. Among the attendant maladies were overpopulation, pollution, and human stress. His solution, like Mill's, was the steady-state economy, where humanity would strive for the maintenance of a constant population level, a constant stock of physical wealth, and a more equitable distribution of economic production among the members of society. In place of economic growth, Daly, like Mill, recommended moral growth in order to replace anxiety about material trappings with a concern for some higher and greater good. Furthermore, Daly said explicitly that the steady-state was compatible with Leopold's ([1949] 1987) land ethic. Daly's writings were responsible for the modern revival of the steady-state theory.

Neoclassical economics. Neoclassical economists made some important contributions to environmental theory during the 1950s. They recognized the common-property nature of environmental resources as the root cause of many economic externalities (Gordon 1954). Because the oceans and the atmosphere belonged to everyone, hence to no one, they were freely exploitable. Common property was then seen as a type of market failure (i.e., no defined property rights) that could reduce social well-being. Thus, economists documented in the literature the importance of common property in environmental matters. (More than a decade later, Garrett Hardin in "The Tragedy of the Commons" [Hardin 1968] presented the common property concept to a large readership of noneconomists. Such interdisciplinary transferring of knowledge appears not to have been a principal interest of the earlier academic economists.)

Despite the advancements of the 1950s, it was not until the mid-1960s that neoclassical economics developed a specialized branch of learning called environmental economics, although economic externalities had remained a part of the standard university economics curriculum. The ideas of Pigou were often cited as the foundation of environmental economics (Fisher 1981). With interest stimulated by *Silent Spring*, the study of natural resource and environmental economics accelerated rapidly at universities, government agencies, and especially at Resources for the Future, a Ford Foundation think-tank located in Washington, DC. The result was important theoretical and empirical advancements in neoclassical economics as applied to the environment.

Economics and Environmental Thought Today

It is important to understand that the Malthusian doctrine is a hypothesis (Barnett and Morse 1963). Thus, it is subject to empirical testing. Barnett and Morse (1963) tested the Malthusian hypothesis with data for the United States and found that between 1865 and 1957 most natural resources (the exception being forestry products) had not become scarcer but instead more plentiful from the economics standpoint. The lack of scarcity was evidenced by declines in both extraction costs and resource prices. The reason the Malthusian hypothesis had failed[1] was that rising resource prices had induced new resource discovery, substitution, and more efficient technologies, all of which had lowered extraction costs. Additional tests of the scarcity hypothesis are a continuing part of modern economic inquiry (Cleveland 1991).

Some 20 years ago, Daly restated Mill's thoughts on the steady-state, and the old ideas found new acceptance. The relatively new school of ecological economics, of which Daly is a founder, is undoubtedly the most rapidly growing branch of economic thought with respect to the environment. Like Mill, the ecological economists seem to support, among other things, voluntary constant rates of population and wealth accumulation and a more equitable distribution of the world's wealth. The ecological economists also have promoted the development of national income accounts that reflect resource degradation and depletion. They argue for more incorporation of biology[2] into economic studies (Costanza 1991).

Mill's theory, as rediscovered by Daly, has also greatly influenced sustainable development. This social movement, which attempts to balance economic development and environmental health, currently enjoys worldwide popularity (World Commission on Environment and Development 1987).

The importance of ecological economics notwithstanding, the neoclassical school continues to define the majority of university economics programs. Thus, modern neoclassicism must be regarded as the dominant economic paradigm. Students wishing to study environmental economics draw upon the line of scholarly thought that traces back to the contributions of Pigou and the neoclassicists (Fisher 1981). The body of academic literature generated by environmental economists includes numerous textbooks, hundreds of monographs, and annually more than 250 articles published in approximately two dozen scholarly journals (de Steiguer 1989).

A practical success of environmental economists—that is, of putting theory into action—was the provision for marketable pollution permits in the Clean Air Act of 1990. Long-favored by mainstream economists as the most efficient method of reducing pollution, marketable permits set overall acceptable levels of pollution reduction and then allow polluters to bargain amongst themselves in order to achieve these desired reductions.

Despite the academic and policy successes, however, mainstream economics has sometimes had difficulty establishing itself in major environmental science programs (NAPAP 1991). Norton (1991b) has suggested that environmentalists have not always viewed economists with favor. Society would gain by closer cooperation among economists and scientists on environmental studies.

[1]Fisher (1981) reviewed separate studies of resource scarcity conducted by economists V. K. Smith, W. D. Nordhaus, G. Brown, and B. Field in response to *Scarcity and Growth* (Barnett and Morse 1963). These studies indicated that agricultural products, forestry products, and fuels had become economically scarcer, while metals had become economically more plentiful.

[2]The idea of incorporating more biology into economic studies has been a principal theme of ecological economics. It is interesting to note however, that forest economists have routinely used biological models in the form of empirical forest growth and yield functions to determine the economically optimal timing of timber harvests.

A Comparison of Theories

Which economic theory seems best for today's world? This question may not be appropriate. Malthus had a hypothesis, Mill had a philosophy, and the neoclassicists had a quantitative model for testing hypotheses and making decisions. The more appropriate question is: What can each contribute to the solution of environmental problems?

From Malthus, we derive an unyielding sense of urgency regarding environmental matters. His hypothesis provides a haunting image of what might be should we fail to take natural resource and environmental matters seriously. Since Malthus' own time, his theory has generated reaction, and even outrage, for its implied lack of faith in humanity. Yet, Malthus continues to exert pressure on society to solve its environmental problems. Malthusian-like concerns provide the impetus for much of the modern environmental movement.

The modern neoclassicists in their role as economists have traditionally not been concerned with the philosophical and psychological factors that govern resource consumption activities. Instead, they have dealt primarily with empirical validation of that behavior in response to prices, costs, and other market-related phenomena. The ethical motivations behind an economic response have been of less concern than the response itself. Indeed, the modern mainstream economists generally conclude that the establishment of normative social goals is beyond their role as economists (Ferguson and Maurice 1974), because there is no objective way for them to establish those goals (Just et al. 1982).

Modern neoclassicists have brought practical skills to environmental matters. With rigor and mathematics, they have suggested specific methods of analysis to determine the economic importance of environmental damage, to examine the trade-offs required to control losses, and to suggest specific policy instruments for reducing damages. These mainstream economists provide essential information to elected officials who must draft and vote on environmental legislation.

Finally, the modern steady-state theorists provide important recommendations for closer working relationships between economists and biologists and for establishing better systems for measuring aggregate economic performance. However, Mill's most meaningful legacy is his expression of faith that humanity can control its destiny. Far from being simply an economic man—that pale wraith of a creature who follows his adding-machine brain wherever it leads him (Heilbroner 1986)—Mill's ideal person has a heart and a mind to make intelligent choices that might involve denial of material needs. To many people, Mill's work represents more than an optimistic ideal; it may prove the most promising solution to our environmental problems.

References Cited

August ER. 1975. John Stuart Mill. New York: Charles Scribner's Sons.

Barnett HJ, Morse C. 1963. Scarcity and growth: the economics of natural resource availability. Baltimore: The Johns Hopkins Press.

Boulding KE. [1966] 1973. The economics of the coming spaceship earth. Pages 121–132 in HE Daly, ed. Toward a steady state economy. San Francisco (CA): W. H. Freeman and Company.

Carson R. 1962. Silent spring. Boston (MA): Houghton Mifflin Co.

Christensen P. 1991. Driving forces increasing returns and ecological sustainability. Pages 75–87 in R Costanza, ed. Ecological economics: the science and management of sustainability. New York: Columbia University Press.

Cleveland CJ. 1991. Natural resource scarcity and economic growth revisited: economic and biophysical perspectives. Pages 289–317 in R Costanza, ed. Ecological economics: the science and management of sustainability. New York: Columbia University Press.

Costanza R, ed. 1991. Ecological economics: the science and management of sustainability. New York: Columbia University Press.

Daly HE. 1973. The steady-state economy: toward a political economy of biophysical equilibrium and moral growth. Pages 149–174 in HE Daly, ed. Toward a steady state economy. San Francisco (CA): W. H. Freeman and Company.

de Steiguer JE. 1989. Forestry sector environmental effects. Pages 251–262 in PV Ellefson, ed. Forest resource economics and policy research. Boulder (CO): Westview Press.

Ehrlich PR. 1968. The population bomb. New York: Sierra Club and Ballantine Books.

Ferguson CE, Maurice SC. 1974. Economic analysis. Homewood (IL): Richard D. Irwin, Inc.

Finnis J.1980. Natural law and natural rights. New York: Oxford University Press.

Fisher AC. 1981. Resource and environmental economics. Cambridge (UK): Cambridge University Press.

Fisher AC, Peterson FM. 1976. The environment in economics: a survey. Journal of Economic Literature 14: 1–33.

Fusfeld DR. 1982. The age of the economist. Glenview (IL): Scott, Foresman and Company.

Gordon HS. 1954. The economic theory of a common property resource: the fishery. Journal of Political Economy 62: 124–142.

Hardin G. 1968. The tragedy of the commons. Science 162: 1243–1248.

Heilbroner RL. 1986. The worldly philosophers. New York: Simon and Schuster.

Just RE, Hueth DL, Schmitz A. 1982. Applied welfare economics and public policy. Englewood Cliffs (NJ): Prentice-Hall, Inc.

Leopold A. [1949] 1987. The land ethic. Pages 201–226 in A sand county almanac and sketches here and there. New York: Oxford University Press.

Lombroso G. 1931. Tragedies of progress. New York: E. P. Dutton and Co.

Malthus TR. [1798] 1965. The first essay on population. New York: Augustus M. Kelley.

Marsh GP. [1864] 1965. Man and nature. Cambridge (MA): The Belknap Press of Harvard University Press.

Meadows DH, Meadows DL, Randers J, Behrens WW III. 1972. The limits to growth: a report for the Club of Rome's project on the predicament of mankind. New York: Universe Books.

Mill JS. [1848] 1965. Principles of political economy. In JM Robson and VW Bladen, eds. The collected works of John Stuart Mill. Vols. II and III. Toronto: University of Toronto Press.

Miller GT Jr. 1994. Living in the environment. 8th ed. Belmont (CA): Wadsworth Publishing Company. ■

Questions

1. What does economic scarcity refer to?

2. According to Malthus, what would be the cause of economic scarcity?

3. Describe Mill's ideal person.

Answers are at the back of the book.

Answers

Section One: Overview

1. Back to Nature

1. Disorientation in the rain forest.
2. A sense of territory and a fear of humans
3. They are nocturnal and the first generation ferrets are not territorial—they disperse for unknown destinations.

2. Chaos in the Solar System

1. Unpredictability. It characterizes the future evolution of the solar system.
2. First, recognize the difference in location between two objects. Objects can also differ in energy, orbit size, and orbit inclination.
3. If the distance between two objects (computer models were used) increases exponentially, then small changes to the system are extremely magnified over time, and the ability to predict future behavior on well-known future conditions is compromised.

3. The Andes' Deep Origins

1. The westward motion of South America "upstream" through the mantle flowing around it.
2. The leading edge of the continent is weaker and more easily deformed than both the mantle beneath the core of continent to the east and the Nazca plate and its underlying mantle to the west.
3. It is indicated in the similar shapes and great width of the Rockies and the Andes. Like the Andes, the Rockies are sharply bent.

4. The Real Message from Biosphere 2

1. It is complicated by the vast range of spatial and temporal scales over which the monetary valuations might be tabulated.
2. Massive deforestation, water and atmospheric pollution, the dumping of toxic chemicals, over exploitation of renewable and nonrenewable resources, and population growth.
3. The maximum population that can be supported without degrading the environment .

5. Mankind Must Conserve: Sustainable Materials

1. Paper, steel, aluminum, plastics, and container glass. Problems they contribute to are global warming, acid rain, and the flooding of valleys, and destruction of rivers for hydroelectric dams.
2. Develop comprehensive systems for collecting waste and transforming it into new products, which will be possible only if many consumer goods are redesigned to be re-used and recycled easily.
3. Outdated global economic framework that depresses virgin materials' prices and fails to account for the environmental costs of their extraction and processing. Prices have continued to fall even as ecological expenses of the global materials economy have risen sharply.

6. Taking Back the Nest

1. Florida.
2. Humans.
3. Restoring the habitat because it would give the songbird a breeding space, and indirectly, take the space from the cowbirds.

7. The Environmental Challenges in Sub-Saharan Africa

1. deforestation, degradation, and fragmentation.
2. 11.
3. South Africa, Nigeria, and Zaire.

8. Women, Politics, and Global Management

1. Reproductive rights can be viewed in terms of the power and resources that enable individuals and couples to make informed and safe decisions about their reproductive health.
2. High-quality reproductive health services.
3. Sexuality and reproduction.

9. Haiti on the Brink of Ecocide

1. Imported food from the U.S.
2. Political and environmental.
3. Centuries.

10. Earth is Running Out of Room

1. 1840 and 1940.
2. Oceanic fisheries and rangelands.
3. Weather and civil disorder.

11. Easter's End

1. The palm. It would have been ideal for transporting and erecting statues, constructing large canoes and it was a valuable food source.
2. Porpoises, seabirds, land birds, rats, and seal colonies.
3. Intensified chicken production and cannibalism.

Section Two: Problems of Resource Scarcity

12. USGS Oil & Gas CD Hits the Charts

1. First, the volume of assessed resources, especially undiscovered natural gas, decreased significantly from the assessment published in 1981. Secondly, certain categories of resources were not reported, and documentation of methodology and supporting data was incomplete.
2. The 1995 assessment used a geologic play-based methodology to assess 570 confirmed or hypothetical petroleum plays. 570 maps of the individual plays and all associated text; statistical charts; and graphs, including text on methodology, bibliography, and related studies needed to be published. At up to 30 pages per play, there would be more than 10,000 pages of documentation—a CD-ROM was needed.
3. Conventional, undiscovered, but technically recoverable oil and gas; additions to reserves in known fields; and oil and gas in continuous-type accumulations.

13. The Unexpected Rise of Natural Gas

1. No emissions of sulfur and negligible emissions of particulates.
2. Storage of fuel in the vehicle.
3. Russia. It is the largest producer and has the most identified reserves.

14. The Forecast for Windpower

1. Larger turbines, new blade designs, advanced materials, smarter electronics, flexible hub structures, and aerodynamic controls.
2. Because of the environmental benefits.
3. Enough for 150 to 200 homes.

15. Sea Power

1. OTEC has to move a lot of water.
2. Warm ocean water drawn from the surface flashes into 72 degree Fahrenheit steam in a vacuum chamber. The steam then drives a turbine to generate electricity; the water that doesn't evaporate is discharged. Cold deep-ocean water recondenses this steam into liquid at a heat exchanger.
3. Warm surface water boils pressurized ammonia at an evaporator. A turbine uses this ammonia vapor power to generate electricity. Cold deep-ocean water turns the ammonia vapor back to liquid at a condenser.

16. The Truly Wild Life Around Chernobyl

1. The study of how radioactive and chemical pollutants alter the life course of species.
2. This would indicate that mammals in particular are more resilient than once thought.
3. Flaws in the reactor's design combined with judgment errors by the operators. At least 10 times the amount of radiation released by the atomic bomb.

17. Bringing Back the Everglades

1. $370 million.
2. 90%.
3. The straight canal will be replaced by 103 miles of meanders.

18. Will Expectedly the Top Blow Off?: *Environmental Trends and the Need for Critical Decision Making*

1. Management plans are designed as actual experiments and the results of which can be evaluated scientifically.
2. Biological survey, ecosystem management, and adaptive management.
3. Biological diversity.

19. The World's Forests: Need for a Policy Appraisal

1. Forests protect soils, they play a major role in hydrological cycles, they exert a gyroscopic effect in atmospheric processes and other factors of global climate, they are critical to the energy budget and albedo of Earth, and they harbor a majority of species on land.
2. Lack of scientific understanding of forests' overall values and lack of economic capacity to evaluate many of their outputs.
3. The encouragement of sustainable development, enhancing their institutional status, removal of "perverse" subsidies, calculating the costs of inaction, and the promotion of forests as global commons resources.

20. Is He Worth Saving?

1. Ecosystems regulate climate, purify water, pollinate crops, maintain the gaseous composition of the atmosphere, and build and replenish soils.
2. Mandatory protection would be removed for endangered species giving the secretary of interior the power to permit a listed species to go extinct.
3. If the new law was centered on protecting landscapes that support a wealth of species. It would distribute the burden of habitat protection throughout an entire region, rather than drop it on the shoulders of a luckless few.

21. Nature, Nurture, and Property Rights

1. When the right to property is affected.
2. As an unreasonable restriction.
3. Insects, plants, and fungi.

22. Is Marine Biodiversity at Risk?

1. Because many sea creatures have larvae that can drift long distances and most are thought to have large geographic ranges.
2. Human modifications of coastal enviroments and overfishing.
3. A widespread change in the abundance of species.

23. Building a Scientifically Sound Policy for Protecting Endangered Species

1. It doubles the number of species.
2. Ecological role, taxonomic distinctiveness, and recovery potential.
3. To protect genetic diversity, and encourage people to act earlier to protect declining species.

24. Back to Stay

1. Beavers build dams to create a submerged "lodge" that protects them from predators and serves as a storage for winter food.
2. It was lowest at the end of the 19th century (late 1800s). Between 60 and 400 million beavers lived in North America.
3. As a warning about predators and to prop themselves up while eating or working.

25. Common Ground

1. It envisions a food system that nourishes the environment and local economies.
2. It concentrates on the garden itself as a source of employment and education. In addition to teaching about sustainable agriculture and environmental issues, it tutors youth in basic science and math and provides a forum to learn about economic development.
3. It builds bridges and sets agendas for environmental and environmental justice groups.

26. Will the World Starve?

1. He predicts that China's growing demand for grain imports could trigger food price shocks and cause starvation for hundreds of millions around the world.
2. By almost 90 million.
3. UN estimate—more than 700 million people. Ten to twelve million died.

27. Who Owns Rice and Beans?

1. It allows researchers to freely traverse species lines, to insert just one desired quality into a new plant rather than an uncontrolled number of traits, and to create plants that will

do just what their designers want them to do—tolerate herbicides, resists insects, or prosper in drought or heat.
2. The tend to think of patents as a matter between the company holding the patent and its competition.
3. The advent of biotechnology.

Section Three: Problesm of Environmental Degradation

28. The Alarming Language of Pollution
1. By sending false signals to the endocrine system in the body.
2. Heat produces a female and cold produces a male.
3. Dioxin.

29. What Is Polluting Our Beaches?
1. From land, usually in the form of sewage, fertilizers, and sediment.
2. Pathogens. They contaminate through raw sewage, sludge, wastewater effluent, and storm drains.
3. Hepatitis A–which is responsible for cholera, gastroenteritis and giardia.

30. Drinking Recycled Wastewater
1. Surface infiltration and direct-well injection.
2. They have built new dams, levees and canals.
3. Wastewater is usually treated with chlorine and less frequently with alternative disinfectants such as ozone, monochloramine or ultraviolet radiation.

31. Particulate Control: *The Next Air Pollution Control Growth Segment*
1. Volatile organic compounds are removed by thermal oxidation or absorption.
2. Open access transmission and premature closure of nuclear power plants.
3. 50 percent.

32. The Sound of Global Warming
1. Scientists believe that the program can help us learn more about global warming, and animal advocates are concerned that the sound waves will harm the marine mammals.
2. The average temperature can be determined by clocking the travel time of low-frequency sound from its source to a receiver.
3. Most of the heat that powers the climate is stored in the seas.

33. Dusting the Climate for Fingerprints
1. A specific pattern of global changes with temperatures increasing most dramatically in the interior of the continents.
2. Approximately 95 percent.
3. Scientists would recognize greenhouse warming through a steady accumulation of evidence.

34. Complexities of Ozone Loss Continue to Challenge Scientists
1. Chlorofluorocarbons and halons.
2. Milder northern winters.
3. Aircraft, balloons, and ground-based instruments. The results pointed to widespread chemical destruction of the Arctic ozone.

35. Ozone-Destroying Chlorine Tops Out
1. The year 2050.
2. Yes, at least 40 times more destructive, much comes from fire-protection systems.
3. Halogenated CFCs are substitutes for traditional CFCs in wealthy countries; they are weak ozone destroyers.

36. Superfund Renewal Expected to Be Rocky
1. Remedy selection
2. Flexibility in the s.election of clean–up remedies and having a reduction in the program's administrative overhead.
3. The controversy is whether or not the Superfund trust fund should be used to clean up hazardous waste sites or to lower the federal deficit.

37. Fowling the Waters
1. 90 percent.
2. Areas that have poor environmental laws, an anti-regulatory culture, low wages and a docile, anti-union labor pool.
3. Decaying animal waste uses up oxygen in fresh water. Because it is rich in nitrates, phosphorous, and other minerals, it encourages algae to grow, which takes up more oxygen.

38. Green Revolution in the Making

1. The Blue Angel is a symbol which describes it as "a market-oriented instrument of government" that informs and motivates environmentally conscious thinking and acting among manufacturers and consumers.
2. Waste heat can be used in homes and factories, operate paper mills and chemical plants, and can generate a few more kilowatts with super-efficient technology. It will boost efficiency by approximately 90 percent and air pollution will be cut in half.
3. 4 to 300 times that of the United States.

Section Four: Social Solutions

39. China Strives to Make the Polluter Pay

1. Because most of these industrial enterprises are small, local EPB revenues from fee collection are less than those from larger state-owned enterprises. Since EPB has limited personnel, they concentrate on the largest polluters first.
2. The objective is to correct deficiencies and propose changes to improve effectiveness and efficiency consistent with a market economy and with ongoing economic and institutional reform.
3. The goal is to develop a pollution levy system that reduces emissions and effluent, achieves environmental goals with the least cost, and imposes minimal administrative burdens on local EPBs and regulated enterprises.

40. Comparative Risk Assessment and the Naturalistic Fallacy

1. A set of quantitative techniques that are used to evaluate various environmental threats. These are ranked from highest to lowest based on their average annual probability of inducing human fatality.
2. After contemporary risk assessors dismiss most of the health problems allegedly caused by environmental threats.
3. They err in telling us that we ought to accept preventable environmental risks just because we cannot eliminate all risks.

41. Debt and the Environment

1. It began with the oil-price shock of 1973, when energy prices roughly doubled.
2. It links debt with environmental degradation.
3. Establish individual ownership of resources, terminate environmentally damaging subsidies, institute market-based pollution-control mechanisms, and make direct payments to preserve environmental assets.

42. Growing Crops, and Wildlife Too

1. It would encourage farmers to adopt practices designed to cut costs and reduce dependence on commercial pesticides, fertilizers, and antibiotics.
2. These are offset by the saving the farmers can receive under other programs.
3. It would shift the goal from production control to stewardship.

43. Selling Blue Skies, Clean Water

1. To stop and think where it wants to be in five years in terms of productivity.
2. Tough regulations and strict enforcement.
3. Producing less waste.

44. How to Make Lots of Money, and Save the Planet too.

1. By creating new markets or by protecting the old ones against competitors.
2. Government regulation.
3. Coal producers and oil-producing countries.

45. Forging a New Global Partnership to Save the Earth

1. It stipulates that the production of CFC's in industrial countries be phased out by 1996. It restricts the use of several other ozone-depleting chemicals.
2. Overgrazing, overcropping, poor irrigation practices, and deforestation.
3. Developing countries would agree to devote 20% of their domestic resources to human priorities and donors would target 20% of their aid funds for such purposes. It would make a major contribution to a more sustainable world.

46. Ecologists and Economists Can Find Common Ground

1. Economists tend to believe that technological innovations can be relied upon to solve problems, while ecologists are less inclined to

trust technology to cure or bypass problems.
2. They feel that many economic models ignore the environmental resource base of material production and the consequences of that production for critical environmental systems employed as sinks.
3. This is due to ignorance of each other's fields.

47. The Man with the Spear

1. The protected-area approach is dependent on centralized power and top-down planning. It has often robbed rural communities of their traditional user-rights over forests, waters, fisheries, and wildlife, without offering appropriate remuneration. It has obliged poor people to bear most of the costs of conservation, while larger societal interests reap most of the benefits.
2. It tends to omit humanity from the realm of nature and from the enterprise of nature conservation.
3. Tenure defines relationships between people and other people. It specifies who may use, who may inhabit, who may harvest, who may inherit, who may collect, who may hunt, under what circumstances and to what extent; it also specifies who may not.

48. The Real Conservatives

1. 20 percent. 82.7 percent.
2. They are encouraging the growth of small and innovative reuse industries.
3. Waste exchanges are regional and national computerized matchmaking services that link businesses discarding potentially usable material with other businesses that can use it.

49. New Defenders of Wildlife

1. Department of Interior manages the most; Department of Defense manages the second-most.
2. It exists in Camp Pendleton, a US Marine Corps training center.
3. The death of a manatee caused the Navy to add metal shields that prevent manatees from being drawn into the propeller.

50. Three Theories from Economics About the Environment

1. Economic scarcity refers to the decreased availability of resources relative to the effort required to obtain them.
2. Economic scarcity would arise because society would have to sacrifice increasingly more to obtain less on the margin.
3. Mill's ideal person has the heart and mind to make intelligent choices that might involve denial of material needs.